D0221126

Electronic Properties of Materials

Second Edition

Springer
New York
Berlin
Heidelberg
Barcelona
Budapest
Hong Kong
London
Milan
Paris
Santa Clara
Singapore
Tokyo

Rolf E. Hummel

Electronic Properties of Materials

Second Edition

With 252 Illustrations

Rolf E. Hummel, Professor
Department of Materials Science
and Engineering
University of Florida
Gainesville, FL 32611
USA

The background of the cover shows a self-timed sense amplifier magnified several thousand times from an IBM 64,000-bit chip. The inset depicts the first Brillouin zone for the face-centered-cubic structure. The background photo courtesy of IBM Thomas J. Watson Research Center, Yorktown Heights, N.Y.

Library of Congress Cataloging-in-Publication Data
Hummel, Rolf E., 1934–
Electronic properties of materials/
Rolf E. Hummel.—2nd ed.
p. cm.
Includes bibliographical references and index.
ISBN 0-387-54839-4 (U.S.)
1. Solid state physics. 2. Energy-band theory of solids. 3. Electronics—Materials. 4. Electrical engineering—Materials.
I. Title.
QC176.H86 1992
530.4'1—dc20 92-14026

Printed on acid-free paper.

Production coordinated by Brian Howe and managed by Francine McNeill; manufacturing supervised by Gail Simon
Typeset by Asco Trade Typesetting Ltd., Hong Kong.
Printed and bound by Edwards Brothers, Inc., Ann Arbor, MI.
Printed in the United States of America.

9 8 7 6 5 4 3 (Corrected third printing, 1997)

ISBN 3-540-54839-4 Springer-Verlag Berlin Heidelberg New York
ISBN 0-387-54839-4 Springer-Verlag New York Berlin Heidelberg SPIN 10575667

Preface to the Second Edition

It is quite satisfying for an author to learn that his brainchild has been favorably accepted by students as well as by professors and thus seems to serve some useful purpose. This horizontally integrated text on the electronic properties of metals, alloys, semiconductors, insulators, ceramics, and polymeric materials has been adopted by many universities in the United States as well as abroad, probably because of the relative ease with which the material can be understood. The book has now gone through several reprinting cycles (among them a few pirate prints in Asian countries). I am grateful to all readers for their acceptance and for the many encouraging comments which have been received.

I have thought very carefully about possible changes for the second edition. There is, of course, always room for improvement. Thus, some rewording, deletions, and additions have been made here and there. I withstood, however, the temptation to expand considerably the book by adding completely new subjects. Nevertheless, a few pages on recent developments needed to be inserted. Among them are, naturally, the discussion of ceramic (high-temperature) superconductors, and certain elements of the rapidly expanding field of optoelectronics. Further, I felt that the readers might be interested in learning some more practical applications which result from the physical concepts which have been treated here. Thus, the second edition describes common types of field-effect transistors (such as JFET, MOSFET, and MESFET), quantum semiconductor devices, electrical memories (such as D-RAM, S-RAM, and electrically erasable-programmable read-only memories), and logic circuits for computers. The reader will also find an expansion of the chapter on semiconductor device fabrication. The principal mechanisms behind some consumer devices, such as xerography, compact disc players, or optical computers, are also discussed.

Part III (Magnetic Properties of Materials) has been expanded to include more details on magnetic domains, as well as magnetostriction, amorphous ferromagnetics, the newest developments in permanent magnets, new magnetic recording materials, and magneto-optic memories.

Whenever appropriate, some economic facts pertaining to the manufacturing processes or sales figures have been given. Responding to occasional requests, the solutions for the numerical problems are now contained in the Appendix.

I am grateful for valuable expert advice from a number of colleagues, such as Professor Volkmar Gerold, Dr. Dieter Hagmann, Dr. H. Rüfer, Mr. David Malone, Professor Chris Batich, Professor Rolf Haase, Professor Robert Park, Professor Rajiv Singh, and Professor Ken Watson. Mrs. Angelika Hagmann and, to a lesser extent, my daughter, Sirka Hummel, have drawn the new figures. I thank them for their patience.

Gainesville, Florida

Rolf E. Hummel

Preface to the First Edition

„Die meisten Grundideen der
Wissenschaft sind an sich einfach
und lassen sich in der Regel
in einer für jedermann
verständlichen Sprache
wiedergeben.“

ALBERT EINSTEIN

The present book on electrical, optical, magnetic, and thermal properties of materials is, in many aspects, different from other introductory texts in solid state physics. First of all, this book is written for engineers, particularly materials and electrical engineers who want to gain a fundamental understanding of semiconductor devices, magnetic materials, lasers, alloys, etc. Second, it stresses concepts rather than mathematical formalism, which should make the presentation relatively easy to understand. Thus, this book provides a thorough preparation for advanced texts, monographs, or specialized journal articles. Third, this book is not an encyclopedia. The selection of topics is restricted to material which is considered to be essential and which can be covered in a 15-week semester course. For those professors who want to teach a two-semester course, supplemental topics can be found which deepen the understanding. (These sections are marked by an asterisk [*].) Fourth, the present text leaves the teaching of crystallography, X-ray diffraction, diffusion, lattice defects, etc., to those courses which specialize in these subjects. As a rule, engineering students learn this material at the beginning of their upper division curriculum. The reader is, however, reminded of some of these topics whenever the need arises. Fifth, this book is distinctly divided into five self-contained parts which may be read independently. All are based on the first part, entitled "Fundamentals of Electron Theory," because the electron theory of materials is a basic tool with which most material properties can be understood. The modern electron theory of solids is relatively involved. It is, however, not my intent to train a student to become proficient in the entire field of quantum theory. This should be left to more specialized texts. Instead, the essential quantum mechanical concepts are introduced only to the extent to which they are needed for the understanding of materials science. Sixth, plenty of practical applications are presented in the text, as well as in the problem sections, so that the students may gain an understanding of many devices that are used every day. In other words, I tried to bridge the gap between physics and engineering. Finally, I gave the treatment of the optical properties of materials about equal coverage to that of the electrical properties. This is partly due to my personal inclinations and partly because it is felt that a more detailed description of the optical properties is needed since most other texts on solid state physics devote relatively little space to this topic. It should be kept in mind that the optical properties have gained an increasing amount of attention in recent years, because of their potential application in communication devices as well as their contributions to the understanding of the electronic structure of materials.

The philosophy and substance of the present text emerged from lecture notes which I accumulated during more than twenty years of teaching. A preliminary version of Parts I and II appeared several years ago in *Journal of Educational Modules for Materials Science and Engineering* **4**, 1 (1982) and **4**, 781 (1982).

I sincerely hope that students who read and work with this book will enjoy, as much as I, the journey through the fascinating field of the physical properties of materials.

Each work benefits greatly from the interaction between author and colleagues or students. I am grateful in particular to Professor R.T. DeHoff, who read the entire manuscript and who helped with his inquisitive mind to clarify many points in the presentation. Professor Ken Watson read the part dealing with magnetism and made many helpful suggestions. Other colleagues to whom I am indebted are Professor Fred Lindholm, Professor Terry Orlando, and Dr. Siegfried Hofmann. My daughter, Sirka Hummel, contributed with her skills as an artist. Last, but not least, I am obliged to my family, to faculty, and to the chairman of the Department of Materials Science and Engineering at the University of Florida for providing the harmonious atmosphere which is of the utmost necessity for being creative.

Gainesville, Florida Rolf E. Hummel

Contents

Preface to the Second Edition v
Preface to the First Edition vii

PART I
Fundamentals of Electron Theory 1

CHAPTER 1
Introduction 3

CHAPTER 2
The Wave-Particle Duality 6
Problems 11

CHAPTER 3
The Schrödinger Equation 13
3.1. The Time-Independent Schrödinger Equation 13
*3.2. The Time-Dependent Schrödinger Equation 14
*3.3. Special Properties of Vibrational Problems 15
Problems 16

CHAPTER 4
Solution of the Schrödinger Equation for Four Specific Problems 17
4.1. Free Electrons 17
4.2. Electron in a Potential Well (Bound Electron) 19
4.3. Finite Potential Barrier (Tunnel Effect) 23
4.4. Electron in a Periodic Field of a Crystal (the Solid State) 26
Problems 33

CHAPTER 5
Energy Bands in Crystals 35
5.1. One-Dimensional Zone Schemes 35
5.2. One- and Two-Dimensional Brillouin Zones 40
*5.3. Three-Dimensional Brillouin Zones 44
*5.4. Wigner–Seitz Cells 44
*5.5. Translation Vectors and the Reciprocal Lattice 44
*5.6. Free Electron Bands 50
5.7. Band Structures for Some Metals and Semiconductors 53
5.8. Curves and Planes of Equal Energy 57
Problems 59

CHAPTER 6
Electrons in a Crystal 60
6.1. Fermi Energy and Fermi Surface 60
6.2. Fermi Distribution Function 61
6.3. Density of States 62
6.4. Population Density 64
6.5. Complete Density of States Function Within a Band 65
6.6. Consequences of the Band Model 66
6.7. Effective Mass 67
6.8. Conclusion 70
Problems 70
Suggestions for Further Reading (Part I) 71

PART II
Electrical Properties of Materials 73

CHAPTER 7
Electrical Conduction in Metals and Alloys 75
7.1. Introduction 75
7.2. Survey 76
7.3. Conductivity—Classical Electron Theory 78
7.4. Conductivity—Quantum Mechanical Considerations 80
7.5. Experimental Results and Their Interpretation 84
7.5.1. Pure Metals 84
7.5.2. Alloys 86
7.5.3. Ordering 87
7.5.4. Thermoelectric Phenomena 88
7.6. Superconductivity 89
7.6.1. Experimental Results 90
*7.6.2. Theory 93
Problems 96

CHAPTER 8
Semiconductors 98
8.1. Band Structure 98
8.2. Intrinsic Semiconductors 100

8.3. Extrinsic Semiconductors 104
8.3.1. Donors and Acceptors 104
8.3.2. Band Structure 106
8.3.3. Temperature Dependence of the Number of Carriers 106
8.3.4. Conductivity 108
8.3.5. Fermi Energy 109
*8.4. Effective Mass 109
8.5. Hall Effect 110
8.6. Compound Semiconductors 112
8.7. Semiconductor Devices 113
8.7.1. Metal–Semiconductor Contacts 113
8.7.2. Rectifying Contacts (Schottky Barrier Contacts) 114
8.7.3. Ohmic Contacts (Metallizations) 118
8.7.4. *p–n* Rectifier (Diode) 119
8.7.5. Zener Diode 121
8.7.6. Solar Cell (Photodiode) 123
*8.7.7. Avalanche Photodiode 125
*8.7.8. Tunnel Diode 126
8.7.9. Transistors 127
*8.7.10. Quantum Semiconductor Devices 136
8.7.11. Semiconductor Device Fabrication 139
*8.7.12. Digital Circuits and Memory Devices 145
Problems 152

CHAPTER 9
Electrical Conduction in Polymers, Ceramics, and Amorphous Materials 155
9.1. Conducting Polymers and Organic Metals 155
9.2. Ionic Conduction 162
9.3. Conduction in Metal Oxides 165
9.4. Amorphous Materials (Metallic Glasses) 167
9.4.1. Xerography 171
Problems 173
Suggestions for Further Reading (Part II) 173

PART III
Optical Properties of Materials 175

CHAPTER 10
The Optical Constants 177
10.1. Introduction 177
10.2. Index of Refraction, n 178
10.3. Damping Constant, k 178
10.4. Characteristic Penetration Depth, W, and Absorbance, α 182
10.5. Reflectivity, R, and Transmissivity, T 182
10.6. Hagen–Rubens Relation 184
Problems 185

CHAPTER 11
Atomistic Theory of the Optical Properties 186

11.1. Survey 186
11.2. Free Electrons Without Damping 188
11.3. Free Electrons With Damping (Classical Free Electron Theory of Metals) 191
11.4. Special Cases 194
11.5. Reflectivity 195
11.6. Bound Electrons (Classical Electron Theory of Dielectric Materials) 196
*11.7. Discussion of the Lorentz Equations for Special Cases 199
11.7.1. High Frequencies 199
11.7.2. Small Damping 200
11.7.3. Absorption Near ν_0 200
11.7.4. More Than One Oscillator 200
11.8. Contributions of Free Electrons and Harmonic Oscillators to the Optical Constants 201
Problems 202

CHAPTER 12
Quantum Mechanical Treatment of the Optical Properties 204

12.1. Introduction 204
12.2. Absorption of Light by Interband and Intraband Transitions 204
12.3. Optical Spectra of Materials 208
*12.4. Dispersion 208
Problems 213

CHAPTER 13
Applications 214

13.1. Measurement of the Optical Properties 214
*13.1.1. Kramers–Kronig Analysis (Dispersion Relations) 215
*13.1.2. Spectroscopic Ellipsometry 215
*13.1.3. Differential Reflectometry 218
13.2. Optical Spectra of Pure Metals 220
13.2.1. Reflection Spectra 220
*13.2.2. Plasma Oscillations 225
13.3. Optical Spectra of Alloys 226
*13.4. Ordering 230
*13.5. Corrosion 232
13.6. Semiconductors 232
13.7. Insulators (Dielectric Materials and Glass Fibers) 237
13.8. Lasers 239
13.8.1. Principles 239
13.8.2. Helium–Neon Laser 243
13.8.3. Carbon Dioxide Laser 243
13.8.4. Semiconductor Laser 243
13.8.5. Direct Versus Indirect Band-Gap Semiconductor Lasers 246
13.8.6. Wavelength of Emitted Light 247

13.8.7. Threshold Current Density 248
13.8.8. Homojunction Versus Heterojunction Lasers 249
13.8.9. Laser Modulation 250
13.8.10. Laser Amplifier 251
13.8.11. Quantum Well Lasers 252
13.8.12. Light-Emitting Diode (LED) 253
13.9. Integrated Optoelectronics 254
13.9.1. Passive Waveguides 254
13.9.2. Electro-Optical Waveguides (EOW) 256
13.9.3. Optical Modulators and Switches 257
13.9.4. Coupling and Device Integration 258
13.9.5. Energy Losses 260
13.10. Optical Storage Devices 261
13.11. The Optical Computer 263
Problems 266
Suggestions for Further Reading (Part III) 267

PART IV
Magnetic Properties of Materials 269

CHAPTER 14
Foundations of Magnetism 271
14.1. Introduction 271
14.2. Basic Concepts in Magnetism 272
*14.3. Units 274
Problems 275

CHAPTER 15
Magnetic Phenomena and Their Interpretation—Classical Approach 276
15.1. Overview 276
15.1.1. Diamagnetism 276
15.1.2. Paramagnetism 278
15.1.3. Ferromagnetism 281
15.1.4. Antiferromagnetism 286
15.1.5. Ferrimagnetism 288
15.2. Langevin Theory of Diamagnetism 291
*15.3. Langevin Theory of (Electron Orbit) Paramagnetism 293
*15.4. Molecular Field Theory 297
Problems 299

CHAPTER 16
Quantum Mechanical Considerations 302
16.1. Paramagnetism and Diamagnetism 302
16.2. Ferromagnetism and Antiferromagnetism 307
Problems 311

CHAPTER 17
Applications 312
17.1. Introduction 312
17.2. Electrical Steels (Soft Magnetic Materials) 312
17.2.1. Core Losses 313
17.2.2. Grain Orientation 315
17.2.3. Composition of Core Materials 317
17.2.4. Amorphous Ferromagnetics 317
17.3. Permanent Magnets (Hard Magnetic Materials) 318
17.4. Magnetic Recording 321
17.5. Magnetic Memories 323
Problems 325
Suggestions for Further Reading (Part IV) 325

PART V
Thermal Properties of Materials 327

CHAPTER 18
Introduction 329

CHAPTER 19
Fundamentals of Thermal Properties 332
19.1. Heat, Work, and Energy 332
19.2. Heat Capacity, C' 333
19.3. Specific Heat Capacity, c 334
19.4. Molar Heat Capacity, C_v 334
19.5. Thermal Conductivity, K 336
19.6. The Ideal Gas Equation 336
19.7. Kinetic Energy of Gases 337
Problems 339

CHAPTER 20
Heat Capacity 340
20.1. Classical (Atomistic) Theory of Heat Capacity 340
20.2. Quantum Mechanical Considerations—The Phonon 342
20.2.1. Einstein Model 342
20.2.2. Debye Model 345
20.3. Electronic Contribution to the Heat Capacity 346
Problems 350

CHAPTER 21
Thermal Conduction 351
21.1. Thermal Conduction in Metals and Alloys—Classical Approach 351
21.2. Thermal Conduction in Metals and Alloys—Quantum Mechanical Considerations 353
21.3. Thermal Conduction in Dielectric Materials 354
Problems 356

CHAPTER 22
Thermal Expansion 358
Problems 360
Suggestions for Further Reading (Part V) 360

Appendices 361
App. 1. Periodic Disturbances 363
App. 2. Euler Equations 367
App. 3. Summary of Quantum Number Characteristics 368
App. 4. Tables 370
App. 5. About Solving Problems 380

Index 387

PART I

FUNDAMENTALS OF ELECTRON THEORY

CHAPTER 1

Introduction

The understanding of the behavior of electrons in solids is one of the keys to understanding materials. The electron theory of solids is capable of explaining optical, magnetic, thermal, as well as electrical properties of materials. In other words, the electron theory provides important fundamentals for a technology which is often considered to be the basis for modern civilization. A few examples will illustrate this. Magnetic materials are used in electric generators, motors, loudspeakers, transformers, tape recorders, and tapes. Optical properties of materials are utilized in lasers, optical communication, windows, lenses, optical coatings, solar collectors, and reflectors. Thermal properties play a role in refrigeration and heating devices and in heat shields for spacecraft. Some materials are extremely good electrical conductors, such as silver and copper; others are good insulators, such as porcelain or quartz. Semiconductors are generally poor conductors at room temperature. However, if traces of certain elements are added, the electrical conductivity increases.

Since the invention of the transistor in the late 1940s, the electronics industry has grown to an annual sales level of about one trillion dollars. From the very beginning, materials and materials research have been the lifeblood of the electronics industry.

For the understanding of the electronic properties of materials three approaches have been developed during the past hundred years or so which differ considerably in their philosophy and their level of sophistication. In the last century, a phenomenological description of the experimental observation was widely used. The laws which were eventually discovered were empirically derived. This "continuum theory" considered only macroscopic quantities and interrelated experimental data. No assumptions were made about the structure of matter when the equations were formulated. The conclusions which can be drawn from the empirical laws still have validity, at least as long as no oversimplifications are made during their interpretation. Ohm's law, the Maxwell equations, Newton's law, or the Hagen–Rubens equation may serve as examples.

A refinement in understanding the properties of materials was accomplished at the turn of this century by introducing atomistic principles into the description of matter. The "classical electron theory" postulated that free electrons in metals drift as a response to an external force and interact with certain lattice atoms. Paul Drude was the principal proponent of this approach. He developed several fundamental equations which are still widely utilized today. We will make extensive use of the Drude equations in subsequent parts of this book.

A further refinement was accomplished at the beginning of this century by quantum theory. This approach was able to explain important experimental observations which could not be readily interpreted by classical means. It was realized that Newtonian mechanics become inaccurate when they are applied to systems with atomic dimensions, i.e., when attempts are made to explain the interactions of electrons with solids. Quantum theory, however, lacks vivid visualization of the phenomena which it describes. Thus, a considerable effort needs to be undertaken to comprehend its basic concepts; but mastering its principles leads to a much deeper understanding of the electronic properties of materials.

The first part of the present book introduces the reader to the fundamentals of quantum theory. Upon completion of this part the reader should be comfortable with terms such as Fermi energy, density of states, Fermi distribution function, band structure, Brillouin zones, effective mass of electrons, uncertainty principle, and quantization of energy levels. These concepts will be needed in the following parts of the book.

It is assumed that the reader has taken courses in freshman physics, chemistry, and differential equations. From these courses the reader should be familiar with the necessary mathematics and relevant equations and definitions, such as:

Newton's law: force equals mass times acceleration ($F = ma$); (1.1)

Kinetic energy: $E_{\text{kin}} = \frac{1}{2}mv^2$ (v is the particle velocity); (1.2)

Momentum: $p = mv$; (1.3)

Combining (1.2) and (1.3) yields $E_{\text{kin}} = \dfrac{p^2}{2m}$; (1.4)

Speed of light: $c = \nu\lambda$ (ν = frequency of the light wave, and λ its wavelength); (1.5)

Velocity of a wave: $v = \nu\lambda$; (1.6)

Angular frequency: $\omega = 2\pi\nu$; (1.7)

Einstein's mass–energy equivalence: $E = mc^2$. (1.8)

It would be further helpful if the reader has taken an introductory course in materials science or a course in crystallography in order to be familiar with terms such as lattice constant, Miller's indices, X-ray diffraction, Bragg's law, etc. Regardless, these concepts are briefly summarized in this text whenever they are needed. In order to keep the book as self-contained as possible, some fundamentals in mathematics and physics are summarized in the Appendices.

CHAPTER 2

The Wave-Particle Duality

This book is mainly concerned with the interactions of electrons with matter. Thus, the question "What is an electron?" is quite in order. Now, to our knowledge, nobody has so far seen an electron, even by using the most sophisticated equipment. We experience merely the *actions* of electrons, e.g., on a television screen or in an electron microscope. In each of these instances, the electrons seem to manifest themselves in quite a different way, i.e., in the first case as a particle and in the latter case as an electron wave. Accordingly, we shall use, in this book, the terms "wave" and "particle" as convenient means to describe the different aspects of the properties of electrons. This "duality" of the manifestations of electrons should not overly concern us. The reader has probably been exposed to a similar discussion when the properties of light have been introduced to him.

We perceive light intuitively as a wave (specifically, an electromagnetic wave) which travels in undulations from a given source to a point of observation. The color of the light is related to its wavelength, λ, or to its frequency, ν, i.e., its number of vibrations per second. Many crucial experiments, such as diffraction, interference, or dispersion clearly confirm the wavelike nature of light. Nevertheless, at least since the discovery of the photoelectric effect in 1887 by Hertz, and its interpretation in 1905 by Einstein, do we know that light also has a particle nature. (The photoelectric effect describes the emission of electrons from a metallic surface after it has been illuminated by light of appropriately high energy, e.g., by blue light.) Interestingly enough, Newton, about 300 years ago, was a strong proponent of the particle concept of light. His original ideas, however, were in need of some refinement, which was eventually provided in 1901 by quantum theory. We know today (based on Planck's famous hypothesis) that a certain minimal energy of light, i.e., at least one *light quantum*, called a *photon*, with the energy

$$E = \nu h = \omega \hbar, \tag{2.1}$$

needs to impinge on a metal in order that a negatively charged electron may overcome its binding energy to its positively charged nucleus, and can escape into free space. (This is true regardless of the *intensity* of the light.) In (2.1) h is the Planck constant whose numerical value is given in Appendix 4. Frequently, the reduced Planck constant

$$\hbar = \frac{h}{2\pi} \tag{2.2}$$

is utilized in conjunction with the angular frequency, $\omega = 2\pi\nu$ (1.7). In short, the wave-particle duality of *light* (or more generally, of electromagnetic radiation) had been firmly established at the beginning of this century (1924).

On the other hand, the wave-particle duality of *electrons* needed more time until it was fully recognized. The particle property of electrons, having a rest mass m_0 and charge e, was discovered in 1897 by Thomson in an experiment in which he observed the deviation of a cathode ray by electric and magnetic fields. The proof of the free mobility of electrons in a metal was accomplished by measurements of inertia effects by Tolman. In 1924 de Broglie, who believed in a unified creation of the universe, introduced the idea that electrons should also possess a wave-particle duality. In other words, he suggested, based on the hypothesis of a general reciprocity of physical laws, the wave nature of electrons. He connected the wavelength, λ, of an electron wave and the momentum, p, of the particle by the relation

$$\lambda p = h. \tag{2.3}$$

This equation can be derived by combining equivalents to the photonic equations $E = \nu h$ (2.1), $E = mc^2$ (1.8), $p = mc$ (1.3), and $c = \lambda\nu$ (1.5).

In 1926, Schrödinger gave this idea of de Broglie a mathematical form. In 1927, Davisson and Germer and, independently in 1928, G.P. Thomson discovered electron diffraction by a crystal which finally proved the wave nature of electrons.

What is a wave? A wave is a "disturbance" which is periodic in position *and* time. (In contrast to this, a *vibration* is a disturbance which is only periodic in position *or* time.[1]) Waves are characterized by a velocity, v, a frequency, ν, and a wavelength, λ, which are interrelated by

$$v = \nu\lambda. \tag{2.4}$$

Quite often, however, the wavelength is replaced by its inverse quantity

[1] A summary of the equations which govern waves and vibrations is given in Appendix 1.

(multiplied by 2π), i.e., λ is replaced by the *wave number*

$$k = \frac{2\pi}{\lambda}. \tag{2.5}$$

Concomitantly, the frequency ν is replaced by the angular frequency $\omega = 2\pi\nu$ (1.7). Equation (2.4) then becomes

$$v = \frac{\omega}{k}. \tag{2.6}$$

One of the simplest waveforms is mathematically expressed by a sine (or a cosine) function. This simple disturbance is called a "harmonic wave." (We restrict our discussion below to harmonic waves since a mathematical manipulation, called a Fourier transformation, can substitute for any odd type of waveform by a series of harmonic waves, each having a different frequency.)

The properties of electrons will be described in the following by a harmonic wave, i.e., by a wave function Ψ (which contains, as outlined above, a time- and a space-dependent component):

$$\Psi = \sin(kx - \omega t). \tag{2.7}$$

This wave function does not represent, as far as we know, any physical waves or other physical quantities. It should be understood merely as a mathematical description of a particle (the electron) which enables us to calculate its actual behavior in a convenient way. This thought probably sounds unfamiliar to a beginner in quantum physics. However, by repeated exposure, one can become accustomed to this kind of thought.

The wave-particle duality may be better understood by realizing that the electron can be represented by a *combination* of several wave trains having slightly different frequencies between ω and $\omega + \Delta\omega$ and different wave numbers between k and $k + \Delta k$. Let us study this, assuming at first only *two* waves, which will be written as above:

$$\Psi_1 = \sin[kx - \omega t] \tag{2.7}$$

and

$$\Psi_2 = \sin[(k + \Delta k)x - (\omega + \Delta\omega)t]. \tag{2.8}$$

Superposition of Ψ_1 and Ψ_2 yields a new wave Ψ. With $\sin\alpha + \sin\beta = 2\cos\frac{1}{2}(\alpha - \beta)\cdot\sin\frac{1}{2}(\alpha + \beta)$ we obtain

$$\Psi_1 + \Psi_2 = \Psi = \underbrace{2\cos\left(\frac{\Delta\omega}{2}t - \frac{\Delta k}{2}x\right)}_{\substack{\text{Modulated}\\ \text{amplitude}}} \cdot \underbrace{\sin\left[\left(k + \frac{\Delta k}{2}\right)x - \left(\omega + \frac{\Delta\omega}{2}\right)t\right]}_{\text{Sine wave}}. \tag{2.9}$$

Equation (2.9) describes a sine wave (having a frequency intermediate between ω and $\omega + \Delta\omega$) whose amplitude is slowly modulated by a cosine

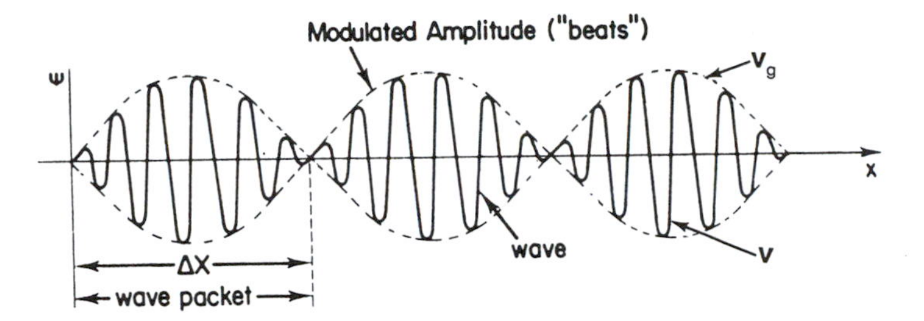

Figure 2.1. Combination of two waves of slightly different frequencies. ΔX is the distance over which the particle can be found.

function. (This familiar effect in acoustics can be heard in the form of "beats" when two strings of a piano have a slightly different pitch. The beats become less rapid the smaller the difference in frequency, $\Delta\omega$, between the two strings until they finally cease once both strings have the same pitch, (2.9).) Each of the "beats" represents a "wave packet" (Fig. 2.1). The wave packet becomes "larger" the slower the beats, i.e., the smaller $\Delta\omega$. The extreme conditions are as follows:

(a) No variation in ω and k (i.e., $\Delta\omega = 0$ and $\Delta k = 0$), which yields an "infinitely long" wave packet, i.e., a monochromatic wave. This corresponds to the wave picture of an electron (see Fig. 2.2).
(b) Alternately, $\Delta\omega$ and Δk could be assumed to be very large. This yields small wave packets. If a large number of different waves are considered (rather than two waves Ψ_1 and Ψ_2), filling the frequencies between ω and $\Delta\omega$, the string of wave packets shown in Fig. 2.1 reduces to *one* wave packet only. The electron is then represented by a particle.

Different velocities need to be distinguished:

(a) The velocity of the *matter wave* is called the wave velocity or "*phase velocity*," v. As we saw above, the matter wave is a monochromatic wave (or a stream of particles of equal velocity whose frequency, ω, wavelength,

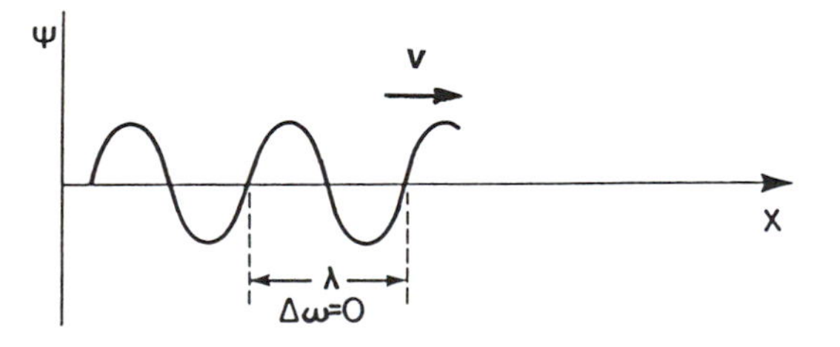

Figure 2.2. Monochromatic matter wave ($\Delta\omega$ and $\Delta k = 0$). The wave has constant amplitude. The matter wave travels with the phase velocity, v.

λ, momentum, p, or energy, E, can be exactly determined (Fig. 2.2)). The location of the particles, however, is undetermined. From the second part of (2.9) (marked "sine wave"), we deduce

$$v = \frac{x}{t} = \frac{\omega + \Delta\omega/2}{k + \Delta k/2} = \frac{\omega'}{k'}, \tag{2.6a}$$

which is a restatement of (2.6). We obtain the velocity of a matter wave that has a frequency $\omega + \Delta\omega/2$ and a wave number $k + \Delta k/2$. The phase velocity varies for different wavelengths (a phenomenon which is called "dispersion," and which the reader knows from the rainbow colors which emerge from a prism when white light impinges on it).

(b) We mentioned above that a *particle* can be understood to be "composed of" a group of waves or a "wave packet." Each individual wave has a slightly different frequency, ranging in value between ω and $\omega + \Delta\omega$. Appropriately, the velocity of a particle is called "*group velocity*," v_g. The "envelope" in Fig. 2.1 propagates with the group velocity, v_g. From the left part of (2.9) (marked "modulated amplitude") we obtain this group velocity

$$v_g = \frac{x}{t} = \frac{\Delta\omega}{\Delta k} = \frac{d\omega}{dk}. \tag{2.10}$$

Equation (2.10) is the velocity of a "*pulse wave*," i.e., of a moving particle.

The location X of a particle is known precisely, whereas the frequency is not. This is due to the fact that a wave packet can be thought to "consist" of several wave functions $\Psi_1, \Psi_2, \ldots, \Psi_n$, with different frequencies between ω and $\omega + \Delta\omega$. Another way of looking at it is to perform a Fourier analysis of a pulse wave (Fig. 2.3) which results in a series of sine and cosine functions (waves) which have different wavelengths. The better the location, ΔX, of a particle can be determined, the wider is the frequency range, $\Delta\omega$, of its waves. This is one form of Heisenberg's uncertainty principle,

$$\Delta p \cdot \Delta X \geqq h, \tag{2.11}$$

stating that the distance over which there is a finite probability of finding an electron, ΔX, and the range of momenta, Δp (or wavelengths (2.3)), of the electron wave is a constant. This means that both the location *and* frequency of an electron cannot be accurately determined at the same time.

A word of encouragement should be added at this point for those readers

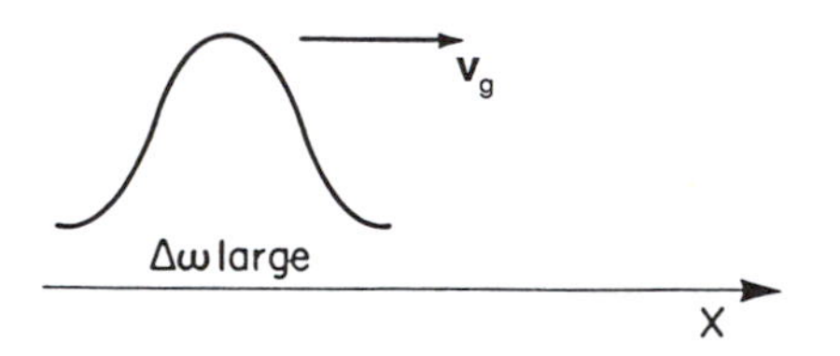

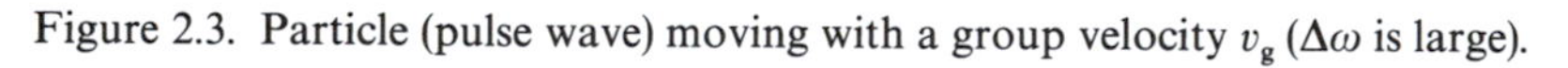
Figure 2.3. Particle (pulse wave) moving with a group velocity v_g ($\Delta\omega$ is large).

who (quite legitimately) might ask the question: What can I do with wave functions which supposedly have no equivalent in real life? For the interpretation of the wave functions, we will use in future chapters Born's postulate which states that the square of the wave function (or because Ψ is generally a complex function, the quantity $\Psi\Psi^*$) is the probability of finding a particle at a certain location. (Ψ^* is the complex conjugate quantity of Ψ.) In other words,

$$\Psi\Psi^* \, dx \, dy \, dz = \Psi\Psi^* \, d\tau \tag{2.12}$$

is the probability of finding an electron in the volume element $d\tau$. This makes it clear that in wave mechanics probability statements are often obtained, whereas in classical mechanics the location of a particle can be determined exactly. We will see in future chapters, however, that this does not affect the usefulness of our results.

Finally, the reader may ask the question: Is an electron wave the same as an electromagnetic wave? Most definitely not! *Electromagnetic waves* (radio waves, infrared radiation (heat), visible light, ultraviolet (UV) light, X-rays, or γ-rays) propagate by an interaction of electrical and magnetic disturbances. Detection devices for electromagnetic waves include the human eye, photomultiplier tubes, photographic films, heat sensitive devices, such as the skin, or antennas in conjunction with electrical circuits. For the detection of *electrons* (e.g., in an electron microscope or on a television screen) certain chemical compounds called "phosphors" are utilized. Materials which possess "phosphorescence" (see Section 13.8) include zinc sulfide, zinc–cadmium sulfide, tungstates, molybdates, salts of the rare earths, uranium compounds, or organic compounds. They vary in color and strength and in the length of emitting visible light.

At the end of this chapter, let us revisit the fundamental question which stood at the outset of our discussion concerning the wave-particle duality: Are particles and waves really two completely unrelated phenomena? Seen conceptually, they probably are. But consider (2.9) and its discussion. Both waves and particles are *mathematically* described essentially by the same equation, i.e., the former by setting $\Delta\omega$ and $\Delta k = 0$ and the latter for a large $\Delta\omega$ and Δk. Thus, waves and particles appear to be interrelated in a certain way. It is left to the reader to contemplate further on this idea.

Problems

1. Calculate the wavelength of an electron which has a kinetic energy of 4 eV.
2. What should be the energy of an electron so that the associated electron waves have a wavelength of 600 nm?
3. Since the visible region spans between approximately 400 nm and 700 nm, why can the electron wave mentioned in Problem 2 not be seen by the human eye? What kind of device is necessary to detect electron waves?

4. What is the energy of a light quantum (photon) which has a wavelength of 600 nm? Compare the energy with the electron wave energy calculated in Problem 2 and discuss the difference.

5. A tennis ball, having a mass of 50 g, travels with a velocity of 200 km/h. What is the equivalent wavelength of this "particle"? Compare your result with that obtained in Problem 1 above and discuss the difference.

6. Derive (2.9) by adding (2.7) and (2.8).

7. Derive (2.3) by combining (1.3), (1.5), (1.8), and (2.1).

8. *Computer problem.*
   (a) Insert numerical values of your choice into (2.9) and plot the result. In particular, vary $\Delta\omega$ and Δk.
   (b) Add more than two equations of the type of (2.7) and (2.8) by using different values of $\Delta\omega$ and plot the result. Does this indeed reduce the number of wave packets, as stated in the text?

CHAPTER 3

The Schrödinger Equation

We shall now make use of the conceptual ideas which we introduced in the previous chapter, i.e., we shall cast, in mathematical form, the description of an electron as a wave, as suggested by Schrödinger in 1926. All "derivations" of the Schrödinger equation start in one way or another from certain assumptions which cause the uninitiated reader to ask the legitimate question, "Why just in this way?" The answer to this question can naturally be given, but these explanations are relatively involved. In addition, the "derivations" of the Schrödinger equation do not further our understanding of quantum mechanics. It is, therefore, not intended to "derive" here the Schrödinger equation. We consider this relation as a fundamental equation for the description of wave properties of electrons, just as Newton's equations describe the matter properties of large particles.

3.1. The Time-Independent Schrödinger Equation

The *time-independent* Schrödinger equation will always be applied when the properties of atomic systems have to be calculated in stationary conditions, i.e., when the property of the surroundings of the electron does not change with time. This is the case for most of the applications which will be discussed in this text. Thus, we introduce, at first, this simpler form of the Schrödinger equation in which the potential energy V depends only on the location (and not, in addition, on the time). Therefore, the time-independent Schrödinger equation is an equation of a *vibration*. It has the following form:

$$\boxed{\nabla^2\psi + \frac{2m}{\hbar^2}(E - V)\psi = 0,} \tag{3.1}$$

where

$$\nabla^2\psi = \frac{\partial^2\psi}{\partial x^2} + \frac{\partial^2\psi}{\partial y^2} + \frac{\partial^2\psi}{\partial z^2}, \tag{3.2}$$

and m is the (rest) mass of the electron,[2] and

$$E = E_{\text{kin}} + V \tag{3.3}$$

is the total energy of the system.

In (3.1) we wrote for the wave function a lowercase ψ which we will use from now on, when we want to state explicitly that the wave function is only space dependent. Thus, we split from Ψ a time-dependent part:

$$\Psi(x, y, z, t) = \psi(x, y, z) \cdot e^{i\omega t}. \tag{3.4}$$

*3.2. The Time-Dependent Schrödinger Equation

The *time-dependent* Schrödinger equation is a *wave* equation because it contains derivatives of Ψ with respect to space *and* time (see below, (3.8)). One obtains this equation from (3.1) by eliminating the total energy,

$$E = \nu h = \omega \hbar, \tag{2.1}$$

where ω is obtained by differentiating (3.4) with respect to time:

$$\frac{\partial\Psi}{\partial t} = \psi i\omega e^{i\omega t} = \Psi i\omega. \tag{3.5}$$

This yields

$$\omega = -\frac{i}{\Psi}\frac{\partial\Psi}{\partial t}. \tag{3.6}$$

Combining (2.1) with (3.6) provides

$$E = -\frac{\hbar i}{\Psi}\frac{\partial\Psi}{\partial t}. \tag{3.7}$$

Finally, combining (3.1) with (3.7) yields

$$\boxed{\nabla^2\Psi - \frac{2mV}{\hbar^2}\Psi - \frac{2mi}{\hbar}\frac{\partial\Psi}{\partial t} = 0.} \tag{3.8}$$

It should be noted here that quantum mechanical equations can be obtained from classical equations by applying differential operators to the wave func-

[2] In most cases we shall denote the rest mass by m instead of m_0.

tion Ψ (Hamiltonian operators). They are

$$E \triangleq -\hbar i \frac{\partial}{\partial t} \tag{3.9}$$

and

$$\mathbf{p} \triangleq -\hbar i \nabla. \tag{3.10}$$

When these operators are applied to

$$E_{\text{total}} = E_{\text{kin}} + E_{\text{pot}} = \frac{p^2}{2m} + V \tag{3.11}$$

we obtain

$$-\hbar i \frac{\partial \Psi}{\partial t} = \frac{\hbar^2 i^2}{2m} \nabla^2 \Psi + V\Psi, \tag{3.12}$$

which yields, after rearranging, the time-dependent Schrödinger equation (3.8).

*3.3. Special Properties of Vibrational Problems

The solution to an equation for a vibration is determined, except for certain constants. These constants are calculated by using boundary or starting conditions

$$(\text{e.g., } \psi = 0 \text{ at } x = 0). \tag{3.13}$$

As we will see in Section 4.2, only certain vibrational forms are possible when boundary conditions are imposed. This is similar to the vibrational forms of a vibrating string, where the fixed ends cannot undergo vibrations. Vibrational problems, which are determined by boundary conditions, are called *boundary* or *eigenvalue problems.* It is a peculiarity of vibrational problems with boundary conditions that not all frequency values are possible and, therefore, because of

$$E = \nu h, \tag{3.14}$$

not all values for the energy are allowed (see next chapter). One calls the values, which are allowed, eigenvalues. The functions ψ, which belong to the eigenvalues and which are a solution of the vibration equation and, in addition, satisfy the boundary conditions, are called *eigenfunctions* of the differential equation.

In Section 2 we have related the product $\psi\psi^*$ (which is called the "norm") to the probability of finding a particle at a given location. The probability of finding a particle somewhere in space is one, or

$$\int \psi\psi^* \, d\tau = \int |\psi|^2 \, d\tau = 1. \tag{3.15}$$

Equation (3.15) is called the normalized eigenfunction.

Problems

1. Write a mathematical expression for a vibration (vibrating string, for example) and for a wave. (See Appendix 1.) Familiarize yourself with the way these differential equations are solved. What is a "trial solution?" What is a boundary condition?
2. Define the terms "vibration" and "wave."
3. What is the difference between a damped and an undamped vibration? Write the appropriate equations.
4. What is the complex conjugate function of:
   (a) $\hat{x} = a + bi$; and
   (b) $\Psi = 2Ai \sin \alpha x$.

CHAPTER 4

Solution of the Schrödinger Equation for Four Specific Problems

4.1. Free Electrons

At first we solve the Schrödinger equation for a simple but, nevertheless, very important case. We consider electrons which propagate freely, i.e., in a potential free space in the positive x-direction. In other words, it is assumed that no "wall," i.e., no potential barrier (V), restricts the propagation of the electron wave. The potential energy V is then zero and the Schrödinger equation (3.1) assumes the following form:

$$\frac{d^2\psi}{dx^2} + \frac{2m}{\hbar^2} E\psi = 0. \tag{4.1}$$

This is a differential equation for an undamped vibration[3] with spatial periodicity whose solution is known to be[3]

$$\psi(x) = Ae^{i\alpha x}, \tag{4.2}$$

where

$$\alpha = \sqrt{\frac{2m}{\hbar^2} E}. \tag{4.3}$$

(For our special case we do not write the second term in (A.5)[3]

$$u = Ae^{i\alpha x} + Be^{-i\alpha x} \tag{4.4}$$

because we stipulated above that the electron wave[3]

$$\Psi(x) = Ae^{i\alpha x} \cdot e^{i\omega t} \tag{4.5}$$

[3] See Appendix 1.

propagates only in the positive x-direction and not, in addition, in the negative x-direction.)

From (4.3), it follows that

$$\boxed{E = \frac{\hbar^2}{2m}\alpha^2.} \tag{4.6}$$

Since no boundary condition had to be considered for the calculation of the free flying electron, all values of the energy are "allowed," i.e., one obtains an energy continuum (Fig. 4.1). This statement seems to be trivial at this point. The difference to the bound electron case will become, however, evident in the next section.

Before we move ahead, let us combine equations (4.3), (2.3), and (1.4), i.e.,

$$\alpha = \sqrt{\frac{2mE}{\hbar^2}} = \frac{p}{\hbar} = \frac{2\pi}{\lambda} = k, \tag{4.7}$$

which yields

$$E = \frac{\hbar^2}{2m}k^2. \tag{4.8}$$

The term $2\pi/\lambda$ was defined in (2.5) to be the wave number k. Thus, α is here identical with k. We see from (4.7) that the quantity k is proportional to the momentum p and, because of $\mathbf{p} = m\mathbf{v}$, also proportional to the velocity of the electrons. Since both momentum and velocity are vectors, it follows that k is a vector too. Therefore, we actually should write k as a vector which has the components k_x, k_y, and k_z:

$$|\mathbf{k}| = \frac{2\pi}{\lambda}. \tag{4.9}$$

Since $\mathbf{k}$ is inversely proportional to the wavelength, λ, it is also called the "wave vector." We shall use the wave vector in the following sections frequently. The $\mathbf{k}$-vector describes the wave properties of an electron, just as one describes in classical mechanics the particle property of an electron with the momentum. As mentioned above, $\mathbf{k}$ and $\mathbf{p}$ are mutually proportional, as one can see from (4.8). The proportionality factor is $1/\hbar$.

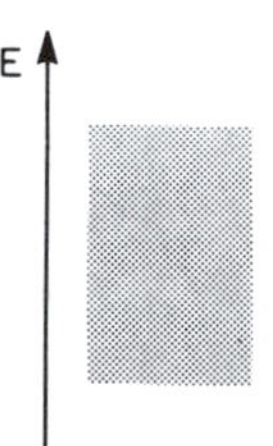

Figure 4.1. Energy continuum of a free electron (compare with Fig. 4.3).

4.2. Electron in a Potential Well (Bound Electron)

We now consider an electron which is bound to its atomic nucleus. For simplicity, we assume that the electron can move freely between two infinitely high potential barriers (Fig. 4.2). The potential barriers do not allow the electron to escape from this potential well, which means that $\psi = 0$ for $x \leqq 0$ and $x \geqq a$. We first treat the one-dimensional case just as in Section 4.1, i.e., we assume that the electron propagates only along the x-axis. However, because the electron is reflected on the walls of the well, it can now propagate in the positive, as well as in the negative, x-direction. In this respect, the present problem is different from the preceding one. The potential energy inside the well is zero, as before, so that the Schrödinger equation for an electron in this region can be written, as before,

$$\frac{d^2\psi}{dx^2} + \frac{2m}{\hbar^2}E\psi = 0. \tag{4.10}$$

Because of the two propagation directions of the electron, the solution of (4.10) is

$$\psi = Ae^{i\alpha x} + Be^{-i\alpha x} \tag{4.11}$$

(see Appendix 1), where

$$\alpha = \sqrt{\frac{2m}{\hbar^2}E}. \tag{4.12}$$

We now determine the constants A and B by means of boundary conditions. We just mentioned that at $x \leqq 0$ and $x \geqq a$ the ψ function is zero. This boundary condition is similar to that known for a vibrating string which does not vibrate at the two points where it is clamped down. (See also Fig. 4.4(a).) Thus, for $x = 0$ we stipulate $\psi = 0$. Then we obtain from (4.11)

$$B = -A. \tag{4.13}$$

Similarly, we obtain $\psi = 0$ for $x = a$, which gives with (4.13)

$$0 = Ae^{i\alpha a} + Be^{-i\alpha a} = A(e^{i\alpha a} - e^{-i\alpha a}). \tag{4.14}$$

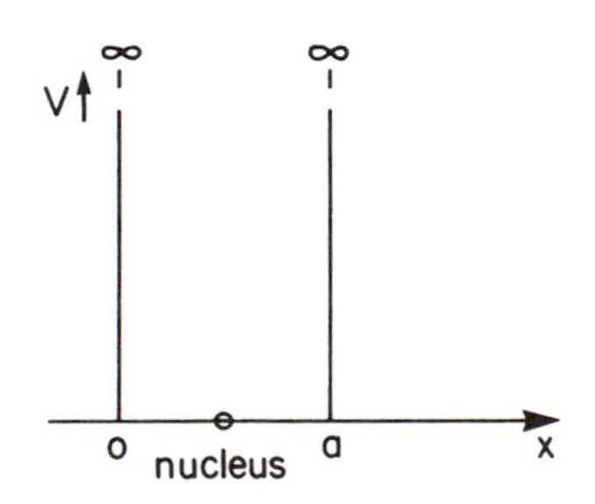

Figure 4.2. One-dimensional potential well. The walls consist of infinitely high potential barriers.

With (4.13) and Euler's equation

$$\sin \rho = \frac{1}{2i}(e^{i\rho} - e^{-i\rho}) \tag{4.15}$$

(see Appendix 2), we write

$$A[e^{i\alpha a} - e^{-i\alpha a}] = 2Ai \cdot \sin \alpha a = 0. \tag{4.16}$$

Equation (4.16) is only valid if $\sin \alpha a = 0$, i.e., if

$$\alpha a = n\pi, \qquad n = 0, 1, 2, 3, \ldots . \tag{4.17}$$

Substituting the value of α from (4.12) into (4.17) provides

$$\boxed{E_n = \frac{\hbar^2}{2m}\alpha^2 = \frac{\hbar^2 \pi^2}{2ma^2} n^2,} \qquad n = 1, 2, 3, \ldots . \tag{4.18}$$

(We exclude the solution $n = 0$ because for $n = 0$ it follows that $\psi = 0$ and thus $\psi\psi^* = 0$.) We notice immediately a striking difference from the case in Section 4.1. Because of the boundary conditions, only certain solutions of the Schrödinger equation exist, namely those for which n is an integer. In the present case the energy assumes only those values which are determined by (4.18). All other energies are not allowed. The allowed values are called "energy levels." They are shown in Fig. 4.3 for a one-dimensional case. Because of the fact that an electron of an isolated atom can assume only certain energy levels, it follows that the energies which are excited or absorbed also possess only discrete values. The result is called an "energy quantization."

We determine now the wave function ψ, and the probability ($\psi\psi^*$) for finding an electron within the potential well. According to (4.11), (4.13), and Euler's equation (4.15) we obtain within the well

$$\psi = 2Ai \cdot \sin \alpha x, \tag{4.19}$$

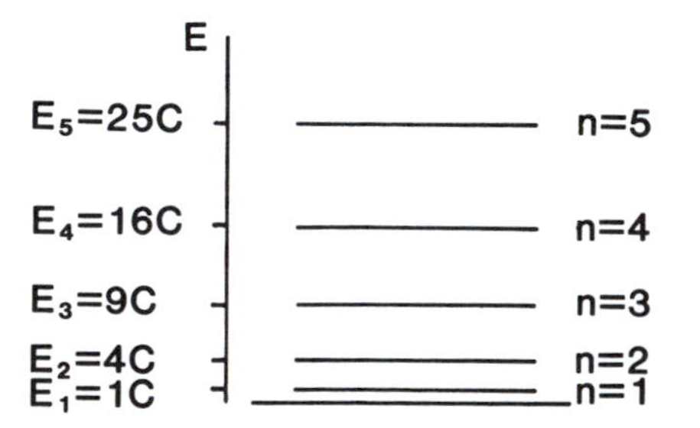

Figure 4.3. Allowed energy values of an electron which is bound to its atomic nucleus. E is the excitation energy in the present case. $C = \hbar^2\pi^2/2ma^2$, see (4.18). (E_1 is the zero point energy.)

and the complex conjugate of ψ

$$\psi^* = -2Ai \sin \alpha x. \tag{4.20}$$

The product $\psi\psi^*$ is then

$$\psi\psi^* = 4A^2 \sin^2 \alpha x. \tag{4.21}$$

This equation is rewritten by making use of (3.15):

$$\int_0^a \psi\psi^* \, dx = 4A^2 \int_0^a \sin^2(\alpha x) \, dx = \frac{4A^2}{\alpha}\left[-\tfrac{1}{2}\sin \alpha x \cos \alpha x + \frac{\alpha x}{2}\right]_0^a = 1. \tag{4.22}$$

Inserting the boundaries in (4.22) and using (4.17) provides

$$A = \sqrt{\frac{1}{2a}}. \tag{4.23}$$

Then one obtains from (4.19)

$$|\psi| = \sqrt{\frac{2}{a}} \sin \frac{n\pi}{a} x, \tag{4.24}$$

and from (4.21)

$$\psi\psi^* = \frac{2}{a} \sin^2 \frac{n\pi}{a} x. \tag{4.25}$$

Equations (4.24) and (4.25) are plotted for various n-values in Fig. 4.4. From Fig. 4.4(a), we see that standing electron waves are created between the walls of the potential well. Note that integer multiples of half a wavelength are equal to the length a of the potential well. The present case, in its mathematical treatment as well as in its result, is analogous to that of a vibrating string.

Of special interest is the behavior of the function $\psi\psi^*$, i.e., the probability

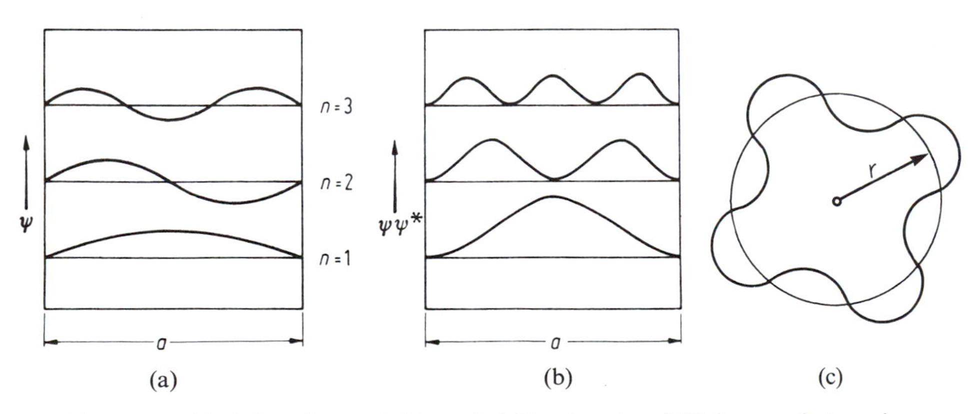

Figure 4.4. (a) ψ function and (b) probability function $\psi\psi^*$ for an electron in a potential well for different n-values. (c) Allowed electron orbit of an atom.

of finding the electron at a certain place within the well (Fig. 4.4(b)). In the classical case the electron travels back and forth between the walls. Its probability function is therefore equally distributed along the whole length of the well. In wave mechanics the deviation from the classical case is most pronounced for $n = 1$. In this case, $\psi\psi^*$ is largest in the middle of the well and vanishes at the boundary. For higher n-values, i.e., for higher energies, the wave mechanical values for $\psi\psi^*$ are approaching the classical value.

* The results which are obtained by considering an electron in a square well are similar to the ones which one receives when the wave mechanical properties of a hydrogen atom are calculated. As above, one considers an electron with charge $-e$ to be bound to its nucleus. The potential V in which the electron propagates is taken as the Coulombic potential $V = -e^2/r$. Since V is a function of the radius r, the Schrödinger equation is more conveniently expressed in polar coordinates. Of main interest are, again, the conditions under which solutions to this Schrödinger equation exist. The treatment leads, similarly as above, to discrete energy levels:

$$E = -\frac{me^4}{2\hbar^2}\frac{1}{n^2} = -13.6 \cdot \frac{1}{n^2}\,(\text{eV}). \tag{4.18a}$$

The main difference compared to the square well model is, however, that the energy is now proportional to $-1/n^2$ (and not as in (4.18) to n^2). This results in a "crowding" of energy levels at higher energies. The energy at the lowest level is called the ionization energy, which has to be supplied to remove an electron from its nucleus. Energy diagrams, as in Fig. 4.5, are common in spectroscopy. The origin of the energy scale is arbitrarily set at $n = \infty$ and the ionization energies are counted negative. Since we are mainly concerned with the solid state, the detailed calculation of the hydrogen atom is not treated here.

It was implied above that an electron can be considered to orbit around its nucleus. (Rutherford model, 1910). Similarly, as shown in Fig. 4.4(a), the electron waves associated with an orbiting electron have to be standing waves. If this were not the case, the wave would be out of phase with itself after one orbit. After a large number of orbits, all possible phases would be

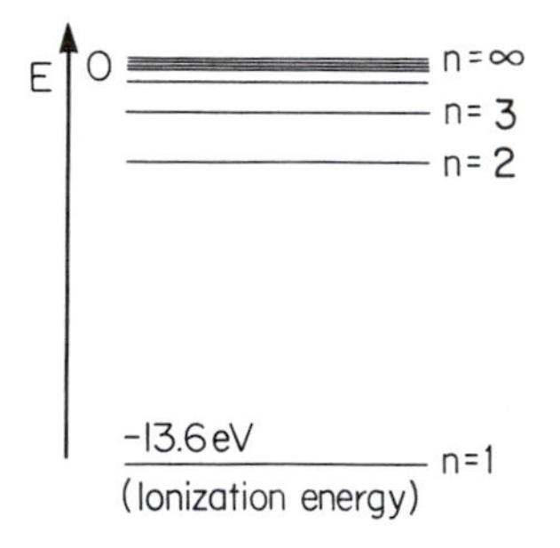

Figure 4.5. Energy levels of atomic hydrogen. E is the binding energy.

obtained and the wave would be annihilated by destructive interference. This can only be avoided if a radius is chosen so that the wave joins on itself (Fig. 4.4(c)). In this case the circumference $2\pi r$ of the orbit is an integer multiple n of the wavelength λ or

$$2\pi r = n\lambda$$

which yields

$$r = \frac{\lambda}{2\pi} n.$$

This means that only certain distinct orbits are allowed, which brings us back to the allowed energy levels which we discussed above. Actually, this model was proposed in 1913 by Niels Bohr.

So far we have considered the electron to be confined to a one-dimensional well. A similar calculation for a three-dimensional potential well ("electron in a box") leads to an equation which is analogous to (4.18):

$$E_n = \frac{\hbar^2 \pi^2}{2ma^2}(n_x^2 + n_y^2 + n_z^2). \tag{4.26}$$

The smallest allowed energy in a three-dimensional potential well is occupied by an electron if $n_x = n_y = n_z = 1$. For the next higher energy there are three different possibilities for combining the n-values; namely, $(n_x, n_y, n_z) = (1, 1, 2), (1, 2, 1)$, or $(2, 1, 1)$. One calls the states which have the same energy but different quantum numbers "degenerate" states. The example just given describes a threefold degenerate energy state.

4.3. Finite Potential Barrier (Tunnel Effect)

Let us assume that a free electron, propagating in the positive x-direction, encounters a potential barrier whose potential energy V_0 ("height" of the barrier) is larger than the total energy E of the electron, but is still finite (Fig. 4.6). For this case we have to write two Schrödinger equations which take into

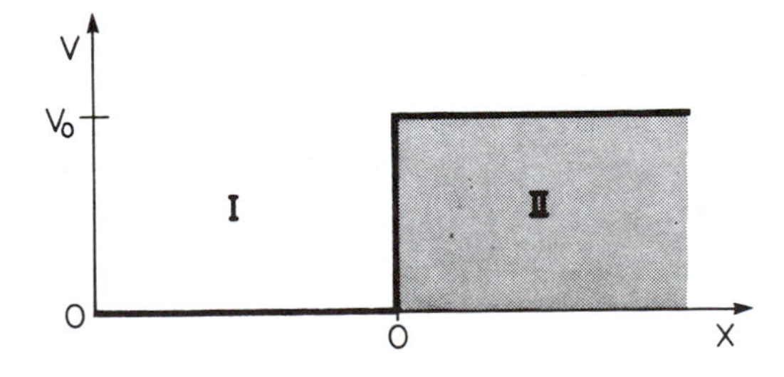

Figure 4.6. Finite potential barrier.

account the two different areas. In region I ($x < 0$) the electron is assumed to be free and we can write

$$\text{(I)} \quad \frac{d^2\psi}{dx^2} + \frac{2m}{\hbar^2} E\psi = 0. \tag{4.27}$$

Inside the potential barrier ($x > 0$) the Schrödinger equation reads

$$\text{(II)} \quad \frac{d^2\psi}{dx^2} + \frac{2m}{\hbar^2}(E - V_0)\psi = 0. \tag{4.28}$$

The solutions to these equations are as before (see Appendix 1):

$$\text{(I)} \quad \psi_{\text{I}} = Ae^{i\alpha x} + Be^{-i\alpha x}, \tag{4.29}$$

where

$$\alpha = \sqrt{\frac{2mE}{\hbar^2}} \tag{4.30}$$

and

$$\text{(II)} \quad \psi_{\text{II}} = Ce^{i\beta x} + De^{-i\beta x} \tag{4.31}$$

with

$$\beta = \sqrt{\frac{2m}{\hbar^2}(E - V_0)}. \tag{4.32}$$

A word of caution has to be inserted here. We stipulated above that V_0 is larger than E. As a consequence of this $(E - V_0)$ is negative and β becomes imaginary. Therefore, we define a new parameter

$$\gamma = i\beta. \tag{4.33}$$

This yields, for (4.32),

$$\gamma = \sqrt{\frac{2m}{\hbar^2}(V_0 - E)}, \tag{4.34}$$

i.e., a parameter γ which is prevented under the stated conditions from becoming imaginary. Rearranging (4.33)

$$\beta = \frac{\gamma}{i}, \tag{4.35}$$

and inserting (4.35) into (4.31) yields

$$\psi_{\text{II}} = Ce^{\gamma x} + De^{-\gamma x}. \tag{4.36}$$

Next, one of the constants C or D needs to be determined by means of a boundary condition:

For $x \to \infty$ it follows from (4.36) that

$$\psi_{\text{II}} = C \cdot \infty + D \cdot 0. \tag{4.37}$$

The consequence of (4.37) could be that ψ_{II} and therefore $\psi_{\text{II}}\psi_{\text{II}}^*$ are infinity.

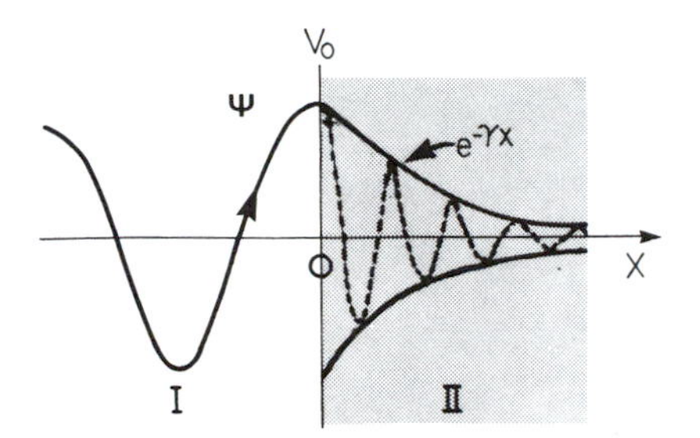

Figure 4.7. Electron wave meeting a potential barrier. (The electron wave in region II is $\Psi = De^{-\gamma x} \cdot e^{i\omega t}$.)

Since the probability $\psi\psi^*$ can never be larger than one (certainty), $\psi_{II} \to \infty$ is no solution. To avoid this, C has to go to zero:

$$C \to 0. \tag{4.38}$$

Then, (4.36) reduces to

$$\psi_{II} = De^{-\gamma x}, \tag{4.39}$$

which reveals that the ψ-function decreases in region II exponentially, as shown in Fig. 4.7. The decrease is faster the larger γ is chosen, i.e., for a large potential barrier, V_0. If the potential barrier is only moderately high and relatively narrow, the electron wave may continue on the opposite side of the barrier. This penetration of a potential barrier by an electron wave is called "tunneling" and has important applications in solid state physics (tunnel diode, tunnel electron microscope, field ion microscope). Tunneling is a quantum mechanical effect. In classical physics, the equivalent to (4.29), i.e., a wave running towards a wall and its reflection, is mainly observed.

*If an electron is confined to a square well (Section 4.2) with *finite* potential barriers, the penetration of the electron wave into these barriers has to be considered as shown in Fig. 4.8. To treat this case, some additional boundary conditions need to be taken into consideration:

(1) The functions ψ_I and ψ_{II} are continuous at $x = 0$. As a consequence, $\psi_I \equiv \psi_{II}$ at $x = 0$. This yields, with (4.29), (4.36), and (4.38),

$$Ae^{i\alpha x} + Be^{-i\alpha x} = De^{-\gamma x}.$$

With $x = 0$, we obtain

$$A + B = D. \tag{4.40}$$

(2) The slopes of the wave functions in regions I and II are continuous at $x = 0$, i.e., $(d\psi_I/dx) \equiv (d\psi_{II}/dx)$. This yields

$$Ai\alpha e^{i\alpha x} - Bi\alpha e^{-i\alpha x} = -\gamma De^{-\gamma x}. \tag{4.41}$$

With $x = 0$, one obtains

$$Ai\alpha - Bi\alpha = -\gamma D. \tag{4.42}$$

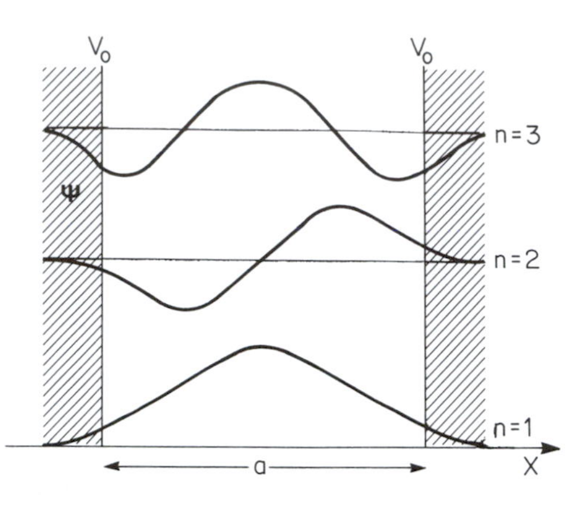

Figure 4.8. Square well with finite potential barriers. (The zero points on the vertical axis have been shifted for clarity.)

Inserting (4.40) yields

$$A = \frac{D}{2}\left(1 + i\frac{\gamma}{\alpha}\right) \tag{4.43}$$

and

$$B = \frac{D}{2}\left(1 - i\frac{\gamma}{\alpha}\right). \tag{4.43a}$$

From this, the ψ-functions in region I can be expressed in terms of a constant D. Figure 4.8 illustrates this behavior for various values of n which should be compared to Fig. 4.4(a).

4.4. Electron in a Periodic Field of a Crystal (the Solid State)

In the preceding sections we became acquainted with some special cases, namely, the completely free electron and the electron which is confined to a potential well. The goal of this section is to study the behavior of an electron in a crystal. We will see eventually that the extreme cases which we treated previously can be derived from this general case.

Our first task is to find a potential distribution which is suitable for a solid. From X-ray diffraction investigations it is known that the atoms in a crystal are arranged periodically. Thus, for the treatment of our problem a periodic repetition of the potential well of Fig. 4.2, i.e., a periodic arrangement of potential wells and potential barriers, is probably close to reality and is

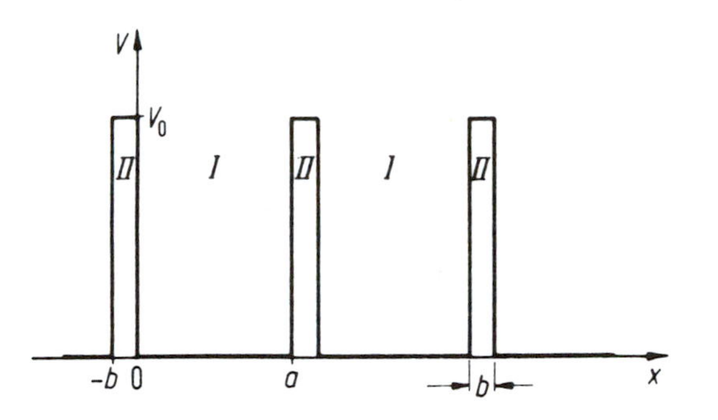

Figure 4.9. One-dimensional periodic potential distribution (simplified) (Kronig-Penney model).

also best suited for a calculation. Such a periodic potential is shown in Fig. 4.9 for the one-dimensional case.[4]

The potential distribution shows potential wells of length a which we call region I. These wells are separated by potential barriers of height V_0 and width b (region II), where V_0 is assumed to be larger than the energy E of the electron.

This model is certainly a coarse simplification of the actual potential distribution in a crystal. It does not take into consideration that the inner electrons are more strongly bound to the core, i.e., that the potential function of a point charge varies as $1/r$. It also does not consider that the individual potentials from each lattice site overlap. A potential distribution which takes these features into consideration is shown in Fig. 4.10. It is immediately evident, however, that the latter model is less suitable for a simple calculation than the one which is shown in Fig. 4.9. Thus, we utilize the model shown in Fig. 4.9.

We now write the Schrödinger equation for regions I and II:

$$\text{(I)} \qquad \frac{d^2\psi}{dx^2} + \frac{2m}{\hbar^2} E\psi = 0, \tag{4.44}$$

$$\text{(II)} \qquad \frac{d^2\psi}{dx^2} + \frac{2m}{\hbar^2}(E - V_0)\psi = 0. \tag{4.45}$$

For abbreviation we write, as before,

$$\alpha^2 = \frac{2m}{\hbar^2} E, \tag{4.46}$$

and

$$\gamma^2 = \frac{2m}{\hbar^2}(V_0 - E). \tag{4.47}$$

[4] R. De L. Kronig and W. G. Penney, *Proc. Roy. Soc. London*, **130**, 499 (1931).

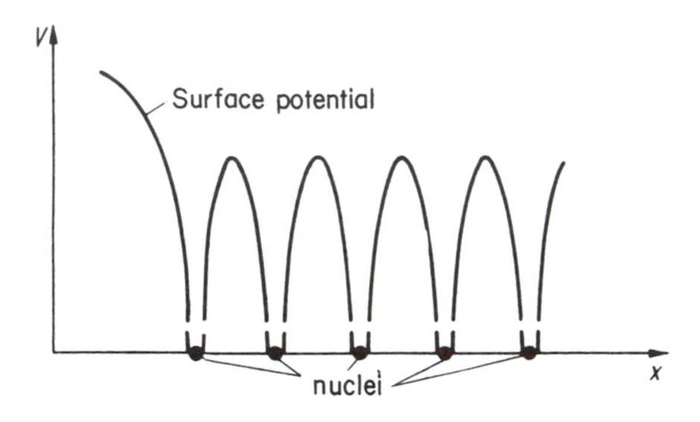

Figure 4.10. One-dimensional periodic potential distribution for a crystal (muffin tin potential).

(γ^2 is chosen in a way to avoid it from becoming imaginary, see Section 4.3.) Equations (4.44) and (4.45) need to be solved simultaneously, a task which can be achieved only with considerable mathematical effort. Bloch[5] showed that the solution of this type of equation has the following form:

$$\psi(x) = u(x) \cdot e^{ikx} \tag{4.48}$$

(Bloch function), where $u(x)$ is a periodic function which possesses the periodicity of the lattice in the x-direction. Therefore, $u(x)$ is no longer a constant (amplitude A) as in (4.2), but changes periodically with increasing x (modulated amplitude). Of course, $u(x)$ is different for various directions in the crystal lattice.

The reader who is basically interested in the results, and their implications for the electronic structure of crystals, may skip the mathematical treatment given below and refer directly to (4.67).

Differentiating the Bloch function (4.48) twice with respect to x provides

$$\frac{d^2\psi}{dx^2} = \left(\frac{d^2u}{dx^2} + \frac{du}{dx}2ik - k^2u\right)e^{ikx}. \tag{4.49}$$

We insert (4.49) into (4.44) and (4.45) and take into account the abbreviations (4.46) and (4.47):

$$\text{(I)} \quad \frac{d^2u}{dx^2} + 2ik\frac{du}{dx} - (k^2 - \alpha^2)u = 0, \tag{4.50}$$

$$\text{(II)} \quad \frac{d^2u}{dx^2} + 2ik\frac{du}{dx} - (k^2 + \gamma^2)u = 0. \tag{4.51}$$

Equations (4.50) and (4.51) have the form of an equation of a damped vibra-

[5] F. Bloch, *Z. Phys.* **52**, 555 (1928); **59**, 208 (1930).

tion. The solution[6] to (4.50) and (4.51) is

$$\text{(I)} \quad u = e^{-ikx}(Ae^{i\alpha x} + Be^{-i\alpha x}), \tag{4.55}$$

$$\text{(II)} \quad u = e^{-ikx}(Ce^{-\gamma x} + De^{\gamma x}). \tag{4.56}$$

We have four constants A, B, C, and D which we need to dispose of by means of four boundary conditions: The functions ψ and $d\psi/dx$ pass over continuously from region I into region II at the point $x = 0$. Equation I = Equation II for $x = 0$ yields

$$A + B = C + D. \tag{4.57}$$

(du/dx) for I = (du/dx) for II at $x = 0$ provides

$$A(i\alpha - ik) + B(-i\alpha - ik) = C(-\gamma - ik) + D(\gamma - ik). \tag{4.58}$$

Further, ψ and therefore u is continuous at the distance $(a + b)$. This means that equation I at $x = 0$ must be equal to equation II at $x = a + b$, or, more simply, equation I at $x = a$ is equal to equation II at $x = -b$ (see Fig. 4.9). This yields

$$Ae^{(i\alpha - ik)a} + Be^{(-i\alpha - ik)a} = Ce^{(ik+\gamma)b} + De^{(ik-\gamma)b}. \tag{4.59}$$

Finally, (du/dx) is periodic in $a + b$

$$Ai(\alpha - k)e^{ia(\alpha-k)} - Bi(\alpha + k)e^{-ia(\alpha+k)}$$
$$= -C(\gamma + ik)e^{(ik+\gamma)b} + D(\gamma - ik)e^{(ik-\gamma)b}. \tag{4.60}$$

The constants A, B, C, and D can be determined by means of these four equations which, when inserted in (4.55) and (4.56), provide values for u. This also means that solutions for the function ψ can be given by using (4.48). However, as in the preceding sections, the knowledge of the ψ function is not of primary interest. We are searching instead for a condition which tells us where solutions to the Schrödinger equation (4.44) and (4.45) exist. We recall that these limiting conditions were leading to the energy levels in Section 4.2. We proceed here in the same manner. We use the four equations (4.57)–(4.60) and eliminate the four constants A–D. (This can be done by simple algebraic manipulation or by forming the determinant out of the coefficients A–D and equating this determinant to zero.) The lengthy calculation provides, using

[6] Differential equation of a damped vibration for spatial periodicity (see Appendix 1)

$$\frac{d^2u}{dx^2} + D\frac{du}{dx} + Cu = 0. \tag{4.52}$$

Solution:

$$u = e^{-(D/2)x}(Ae^{i\delta x} + Be^{-i\delta x}), \tag{4.53}$$

where

$$\delta = \sqrt{C - \frac{D^2}{4}}. \tag{4.54}$$

Euler's equations,[7]

$$\frac{\gamma^2 - \alpha^2}{2\alpha\gamma} \sinh(\gamma b) \cdot \sin(\alpha a) + \cosh(\gamma b) \cos(\alpha a) = \cos k(a + b). \qquad (4.61)$$

For simplification of the discussion of this equation we make the following stipulation. The potential barriers in Fig. 4.9 will be of the kind such that b is very small and V_0 is very large. It is further assumed that the product $V_0 b$, i.e., the area of this potential barrier, remains finite. In other words, if V_0 grows, b diminishes accordingly. The product $V_0 b$ is called the potential barrier strength.

If V_0 is very large, then E in (4.47) can be considered to be small compared to V_0 and can therefore be neglected so that

$$\gamma = \sqrt{\frac{2m}{\hbar^2}} \sqrt{V_0}. \qquad (4.62)$$

Multiplication of (4.62) with b yields

$$\gamma b = \sqrt{\frac{2m}{\hbar^2}} \sqrt{(V_0 b) b}. \qquad (4.63)$$

Since $V_0 b$ has to remain finite (see above) and $b \to 0$ it follows that γb becomes very small. For a small γb we obtain (see tables of the hyperbolic functions)

$$\cosh(\gamma b) \approx 1 \qquad \text{and} \qquad \sinh(\gamma b) \approx \gamma b. \qquad (4.64)$$

Finally, one can neglect α^2 compared to γ^2 and b compared to a (see (4.46), (4.47), and Fig. 4.9) so that (4.61) reads as follows:

$$\frac{m}{\alpha\hbar^2} V_0 b \sin \alpha a + \cos \alpha a = \cos ka. \qquad (4.65)$$

With the abbreviation

$$P = \frac{m a V_0 b}{\hbar^2} \qquad (4.66)$$

we finally get from (4.65)

$$\boxed{P \frac{\sin \alpha a}{\alpha a} + \cos \alpha a = \cos ka.} \qquad (4.67)$$

This is the desired relation which provides the allowed solutions to the Schrödinger equations (4.44) and (4.45). We notice that the boundary conditions lead to an equation with trigonometric functions. Therefore, only certain values of α are possible. This in turn means that because of (4.46), only

[7] See Appendix 2.

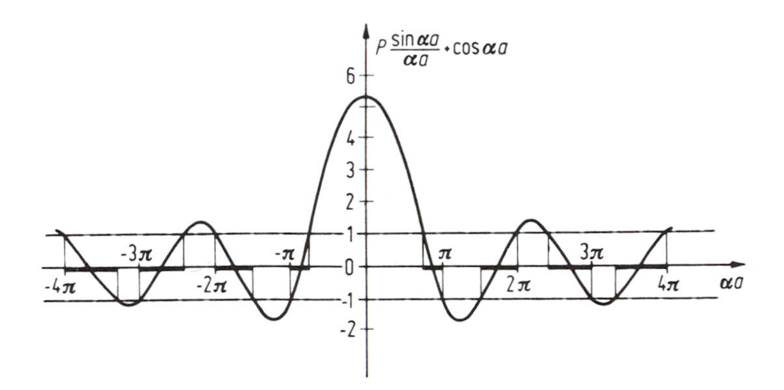

Figure 4.11. Function $P(\sin \alpha a/\alpha a) + \cos \alpha a$ versus αa. P was arbitrarily set to be $(3/2)\pi$.

certain values for the energy E are defined. This is quite similar to the case in Section 4.2. One can assess the situation best if one plots the function $P(\sin \alpha a/\alpha a) + \cos \alpha a$ versus αa which is done in Fig. 4.11 with $P = (3/2)\alpha$. It is of particular significance that the right-hand side of (4.67) allows only certain values of this function because $\cos ka$ is only defined between $+1$ and -1 (except for imaginary k-values). This is shown in Fig. 4.11, in which the allowed values of the function $P(\sin \alpha a/\alpha a) + \cos \alpha a$ are marked by heavy lines on the αa-axis.

We arrive herewith at the following very important result: Because αa is a function of the energy, the above-mentioned limitation means that an electron, which moves in a periodically varying potential field, can only occupy certain allowed energy zones. Energies outside of these allowed zones or "bands" are prohibited. One sees from Fig. 4.11 that with increasing values of αa (i.e., with increasing energy), the disallowed (or forbidden) bands become narrower. The size of the allowed and forbidden energy bands varies with the variation of P. Below, four special cases will be discussed.

(a) If the "potential barrier strength" $V_0 b$ (see Fig. 4.9) is large, then, according to (4.66), P is also large and the curve in Fig. 4.11 proceeds more steeply. The allowed bands are narrow.
(b) If the potential barrier strength and therefore P is small, the allowed bands become wider (see Fig. 4.12).
(c) If the potential barrier strength becomes smaller and smaller and finally disappears completely, P goes toward zero, and one obtains from (4.67)

$$\cos \alpha a = \cos ka \tag{4.68}$$

or $\alpha = k$. From this it follows, with (4.46), that

$$E = \frac{\hbar^2 k^2}{2m}.$$

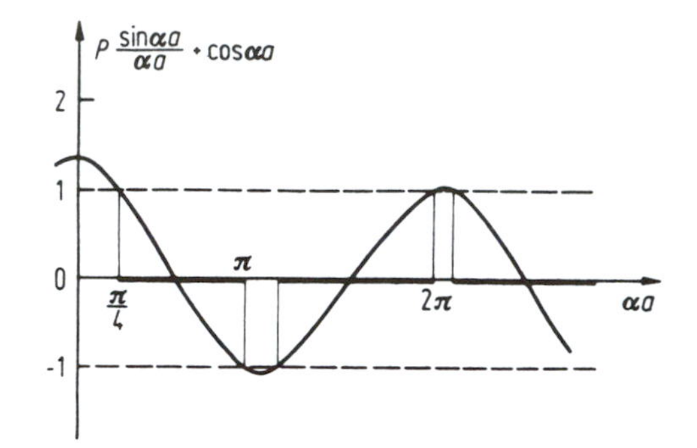

Figure 4.12. Function $P(\sin \alpha a/\alpha a) + \cos \alpha a$ with $P = \pi/10$.

This is the well-known equation (4.6) or (4.8) for free electrons which we derived in Section 4.1.

(d) If the potential barrier strength is very large, P approaches infinity. However, because the left-hand side of (4.67) has to stay within the limits ± 1, i.e., it has to remain finite, it follows that

$$\frac{\sin \alpha a}{\alpha a} \to 0,$$

i.e., $\sin \alpha a \to 0$. This is only possible if $\alpha a = n\pi$ or

$$\alpha^2 = \frac{n^2\pi^2}{a^2} \qquad \text{for} \quad n = 1, 2, 3, \ldots. \tag{4.69}$$

Combining (4.46) with (4.69) yields

$$E = \frac{\pi^2\hbar^2}{2ma^2} \cdot n^2,$$

which is the result of Section 4.2, equation (4.18).

We summarize (Fig. 4.13): If the electrons are strongly bound, i.e., if the potential barrier is very large, one obtains sharp energy levels. (Electron in

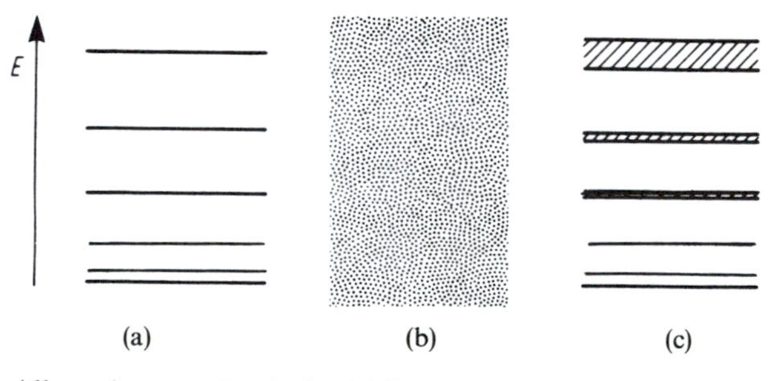

Figure 4.13. Allowed energy levels for (a) bound electrons, (b) free electrons, and (c) electrons in a solid.

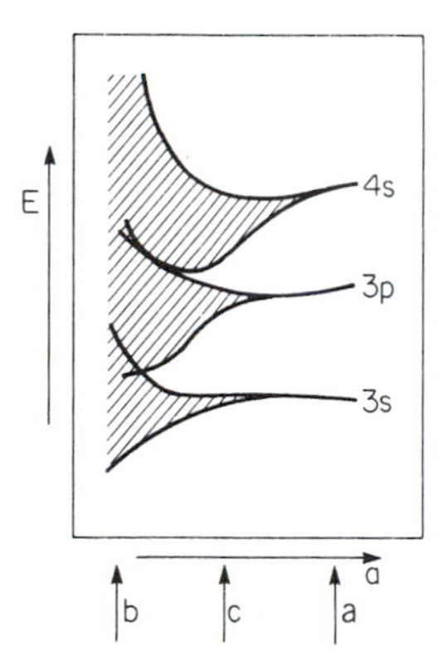

Figure 4.14. Widening of the sharp energy levels into bands and finally into a quasi-continuous energy region with decreasing atomic distance, a, for a metal (after calculations of Slater). The quantum numbers are explained in Appendix 3.

the potential field of *one* ion.) If the electron is not bound, one obtains a continuous energy region (free electrons). If the electron moves in a periodic potential field one receives energy bands (solid).

The widening of the energy levels into energy bands and the transition into a quasi-continuous energy region is shown in Fig. 4.14. This widening occurs because the atoms increasingly interact as their mutual distance decreases. The arrows a, b, and c refer to the three sketches of Fig. 4.13.

Problems

1. Describe the energy for:
   (a) a free electron;
   (b) a strongly bound electron; and
   (c) an electron in a periodic potential.
   Why do we get these different band schemes?

2. *Computer problem.* Plot $\psi\psi^*$ for an electron in a potential well. Vary n from 1 to ~ 100. What conclusions can be drawn from these graphs? (*Hint*: If for large values for n you see strange periodic structures, then you need to choose more data points!)

3. State the two Schrödinger equations for electrons in a periodic potential field (Kronig–Penney model). Use for their solutions, instead of the Bloch function, the trial solution

$$\psi(x) = Ae^{ikx}.$$

   Discuss the result. (*Hint*: For free electrons $V_0 = 0$.)

*4. When treating the Kronig–Penney model, we arrived at four equations for the constants A, B, C, and D. Confirm (4.61).

5. The differential equation for an undamped vibration is

$$a\frac{d^2u}{dx^2} + bu = 0 \tag{1}$$

whose solution is

$$u = Ae^{ikx} + Be^{-ikx}, \tag{2}$$

where

$$k = \sqrt{b/a}. \tag{3}$$

Prove that (2) is indeed a solution of (1).

6. Calculate the "ionization energy" for atomic hydrogen. Note that (4.18a) is written in cgs units. In SI units (4.18a) needs to be replaced by[8]

$$E = -\frac{me^4}{2(4\pi\varepsilon_0\hbar)^2}\frac{1}{n^2}, \tag{4.18b}$$

where $\varepsilon_0 = 8.854 \times 10^{-12}$ [F/m] is the "permittivity of free space." Convince yourself that the numerical value, as well as the unit of the ionization energy, is identical for both (4.18a) and (4.18b).

7. Derive (4.18a) in a semiclassical way by assuming that the centripetal force of an electron mv^2/r is counterbalanced by the Coulombic attraction force e^2/r^2 between the nucleus and the orbiting electron. Use Bohr's postulate which states that the angular momentum $L = mvr$ (v = linear electron velocity and r = radius of the orbiting electron) is a multiple integer of Planck's constant (i.e., $n \cdot \hbar$). (*Hint*: The kinetic energy of the electron is $E = \frac{1}{2}mv^2$.)

8. *Computer problem.* Plot equation (4.67) and vary values for P.

9. *Computer problem.* Plot equation (4.39) for various values for D and γ.

---

[8] See Appendix 4.

CHAPTER 5

Energy Bands in Crystals

5.1. One-Dimensional Zone Schemes

We are now in a position to make additional important statements which contribute considerably to the understanding of the properties of crystals. For this we plot the energy versus the momentum of the electrons or, because of (4.8), versus the wave vector **k**. As before we first discuss the one-dimensional case.

The relation between E and k_x is particularly simple in the case of free electrons, as can be seen from (4.8),

$$k_x = \text{const. } E^{1/2}. \tag{5.1}$$

The plot of E versus k_x is a parabola (Fig. 5.1).

We return now to (4.68) which we obtained from (4.67) for $P = 0$ (free electrons). Because the cosine function is periodic in 2π, (4.68) should be written in the more general form

$$\cos \alpha a = \cos k_x a \equiv \cos(k_x a + n2\pi), \tag{5.2}$$

where $n = 0, \pm 1, \pm 2, \ldots$. This gives

$$\alpha a = k_x a + n2\pi. \tag{5.3}$$

Combining (4.8),

$$\alpha = \sqrt{\frac{2m}{\hbar^2}} E^{1/2},$$

with (5.3) yields

$$k_x + n\frac{2\pi}{a} = \sqrt{\frac{2m}{\hbar^2}} E^{1/2}. \tag{5.4}$$

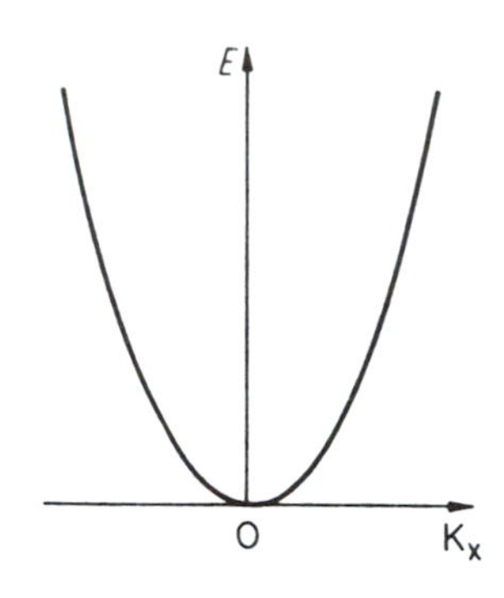

Figure 5.1. Electron energy E versus the wave vector k_x for free electrons.

We see from (5.4) that in the general case the parabola, shown in Fig. 5.1, is repeated periodically in intervals of $n \cdot 2\pi/a$ (Fig. 5.2). The energy is thus a periodic function of k_x with the periodicity $2\pi/a$.

We noted, when discussing Fig. 4.11, that if an electron propagates in a periodic potential we always observe discontinuities of the energies when $\cos k_x a$ has a maximum or a minimum, i.e., when $\cos k_x a = \pm 1$. This is only the case when

$$k_x a = n\pi, \qquad n = \pm 1, \pm 2, \pm 3, \ldots, \tag{5.5}$$

or

$$k_x = n \cdot \frac{\pi}{a}. \tag{5.6}$$

At these singularities a deviation from the parabolic E versus k_x curve

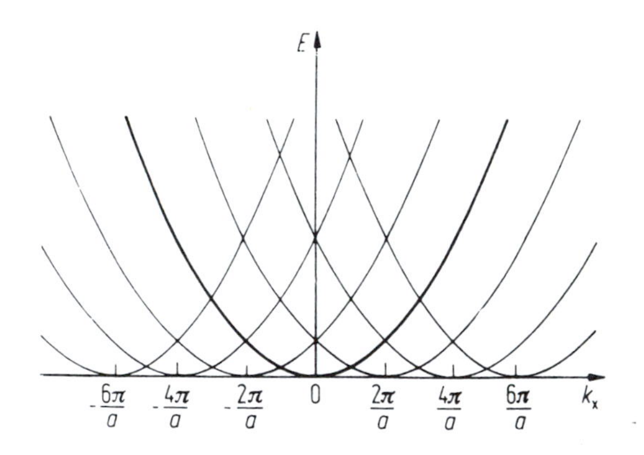

Figure 5.2. Periodic repetition of Fig. 5.1 at the points $k_x = n \cdot 2\pi/a$.

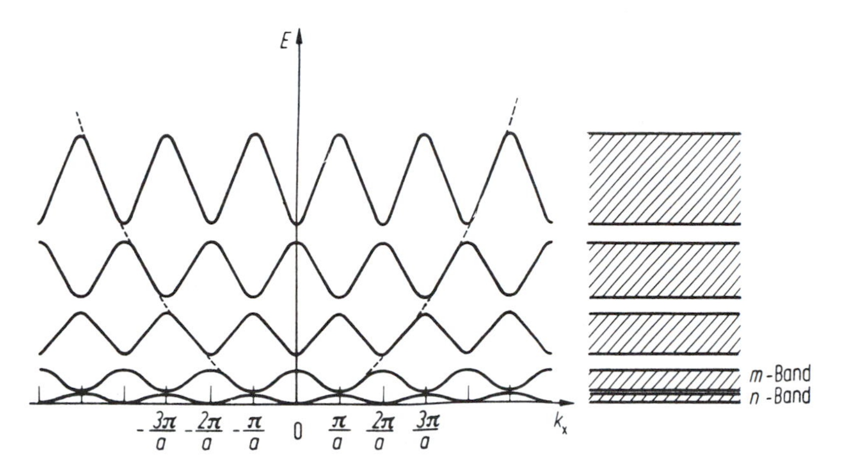

Figure 5.3. Periodic zone scheme.

occurs and the branches of the individual parabolas merge into the neighboring ones.[9] This is shown in Fig. 5.3.

The aforementioned consideration leads to a very important result. The electrons in a crystal behave, for most k_x values, like free electrons except when k_x approaches the value $n \cdot \pi/a$.

Besides this "periodic zone scheme" (Fig. 5.3), two further zone schemes are common. In the future we will use mostly the "reduced zone scheme" (Fig. 5.4) which is a section of Fig. 5.3 between the limits $\pm\pi/a$. In the "extended zone

[9] If two energy functions with equal symmetry cross, the quantum mechanical "noncrossing rule" requires that the eigenfunctions be split, so that they do not cross.

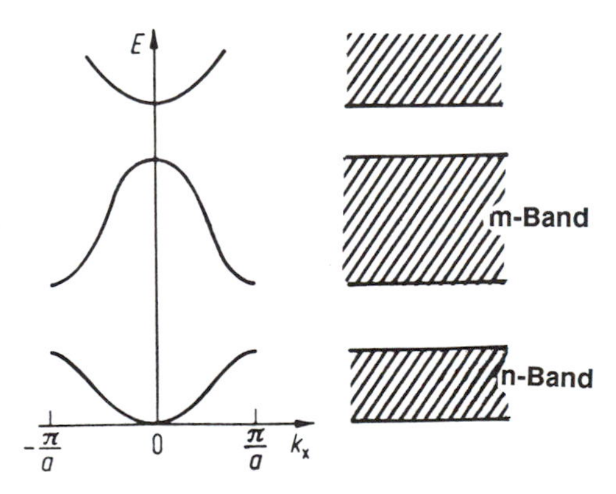

Figure 5.4. Reduced zone scheme. (This is a section of Fig. 5.3 between $-\pi/a$ and $+\pi/a$.)

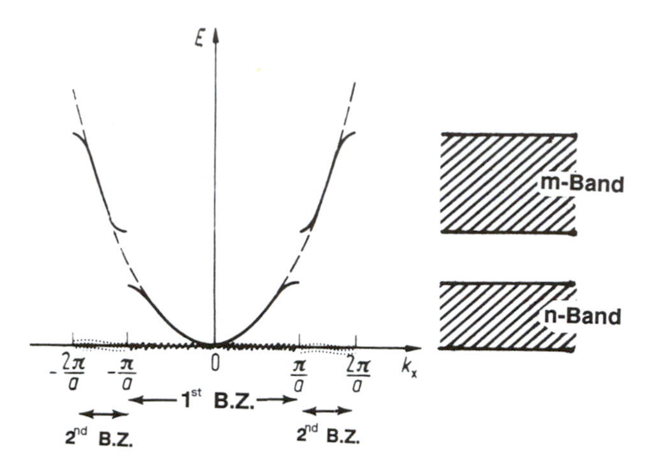

Figure 5.5. Extended zone scheme. The first and second Brillouin zones (BZ) are shown, see Section 5.2.

scheme" (Fig. 5.5), the deviations from the free electron parabola at the critical points $k_x = n \cdot \pi/a$ are particularly easy to identify.

Occasionally, it is useful to plot *free* electrons in a reduced zone scheme. In doing so, one considers the width of the forbidden bands to be reduced until the energy gap between the individual branches disappears completely. This leads to the "free electron bands" which are shown in Fig. 5.6 for a special case. The well-known band character disappears for free electrons, however, and one obtains a continuous energy region as explained in Section 4.1. As before, the shape of the individual branches in Fig. 5.6 is due to the $2\pi/a$ periodicity, as a comparison with Fig. 5.2 shows. From (5.4), it follows that

$$E = \frac{\hbar^2}{2m}\left(k_x + n\frac{2\pi}{a}\right)^2, \qquad n = 0, \pm 1, \pm 2, \ldots. \tag{5.7}$$

By inserting different n-values in (5.7), one can calculate the shape of the branches of the free electron bands. A few examples might illustrate this:

$\mathbf{n = 0}$ yields $E = \dfrac{\hbar^2}{2m} k_x^2$ (parabola with 0 as origin);

$\mathbf{n = -1}$ yields $E = \dfrac{\hbar^2}{2m}\left(k_x - \dfrac{2\pi}{a}\right)^2 \left(\text{parabola with } \dfrac{2\pi}{a} \text{ as origin}\right)$;

$$\text{for } k_x = 0 \text{ follows } E = 4\frac{\pi^2\hbar^2}{2ma^2};$$

$$\text{for } k_x = \frac{\pi}{a} \text{ follows } E = 1\frac{\pi^2\hbar^2}{2ma^2}.$$

The calculated data are depicted in Fig. 5.6.

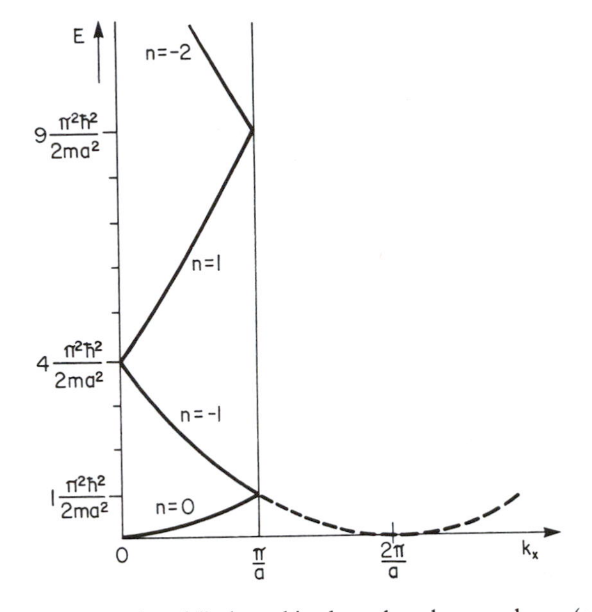

Figure 5.6. "Free electron bands" plotted in the reduced zone scheme (cubic primitive crystal structure).

One important question has remained essentially unanswered: What do these E versus $|\mathbf{k}|$ curves really mean? Simply stated, they relate the energy of an electron to its **k**-vector, i.e., with its momentum. We shall eventually learn to appreciate complete band diagrams in later chapters, in which we will draw from these diagrams important conclusions about the electronic properties of materials. In Figs. 5.3 and 5.5 the individual allowed energy regions and the disallowed energy regions, called band gaps, are clearly seen. We call the allowed bands, for the time being, the n-band, or the m-band, and so forth. In semiconductor physics (see Chapter 8) we will call one of these bands the *valence band*, and the next higher one the *conduction band*.

An additional item needs to be mentioned: It is quite common to use the word "band" for both the allowed energy regions, such as the n-band or the m-band, as well as for the individual branches within a band as seen, for example, in Fig. 5.6. As a rule this does not cause any confusion.

Finally, we need to stress one more point: The wave vector **k** is inversely proportional to the wavelength of the electrons (see equation (4.9)). Thus, **k** has the unit of a reciprocal length and is therefore defined in "reciprocal space." The reader might recall from a course in crystallography that each crystal structure has two lattices associated with it, one of them being the crystal (or real) lattice and the other the reciprocal lattice. We will show in Section 5.5 how these two lattices are related. The following may suffice for the moment: each lattice plane in real space can be represented by a vector which is normal to this plane and whose length is made proportional to the reciprocal of the interplanar distance. The tips of all such vectors from sets of parallel lattice planes form the points in a reciprocal lattice. An X-ray diffraction pattern is a map of such a reciprocal lattice.

5.2. One- and Two-Dimensional Brillouin Zones

Let us again inspect Fig. 5.5. We noticed there that the energy versus k_x-curve, between the boundaries $-\pi/a$ and $+\pi/a$, corresponds to the first electron band, which we arbitrarily labeled as *n-band*. This region in k-space between $-\pi/a$ and $+\pi/a$ is called the first *Brillouin zone* (BZ). Accordingly, the area between π/a and $2\pi/a$, and also between $-\pi/a$ and $-2\pi/a$, which corresponds to the m-band, is called the second Brillouin zone. In other words, the lowest band shown in Fig. 5.5 corresponds to the first Brillouin zone, the next higher band corresponds to the second Brillouin zone, and so on. Now, we learned above that the individual branches in an extended zone scheme (Fig. 5.5) are $2\pi/a$ periodic, i.e., they can be shifted by $2\pi/a$ to the left or to the right. We make use of this concept and shift the branch of the second Brillouin zone on the positive side of the $E-(k_x)$ diagram in Fig. 5.5 by $2\pi/a$ to the left, and likewise the left band of the second Brillouin zone by $2\pi/a$ to the right. A reduced zone scheme as shown in Fig. 5.4 is the result. Actually, we projected the second Brillouin zone into the first Brillouin zone. The same can be done with the third Brillouin zone, etc. This has very important implications: we do not need to plot E versus k-curves for *all* Brillouin zones; the relevant information is, because of the $2\pi/a$ periodicity, already contained in the first Brillouin zone, i.e., in a reduced zone scheme.

We now consider the behavior of an electron in the potential of a two-dimensional lattice. The electron movement in two dimensions can be described as before by the wave vector $\mathbf{k}$ which has the components k_x and k_y which are parallel to the x- and y-axes, in reciprocal space. Points in the k_x–k_y coordinate system form a two-dimensional reciprocal lattice (see Fig. 5.7). One obtains, in the two-dimensional case, a two-dimensional field of allowed energy regions which corresponds to the allowed energy bands, i.e., one obtains two-dimensional Brillouin zones.

We shall illustrate the construction of the Brillouin zones for a two-dimensional reciprocal lattice (Fig. 5.7). For the first zone one constructs the perpendicular bisectors on the shortest lattice vectors, G_1. The area which is enclosed by these four "Bragg planes" is the first Brillouin zone. For the following zones the bisectors of the next shortest lattice vectors are constructed. It is essential that for the zones of higher order the extended limiting lines of the zones of lower order have to be used as additional limiting lines. The first four Brillouin zones are shown in Fig. 5.8. Note that all the zones have the same

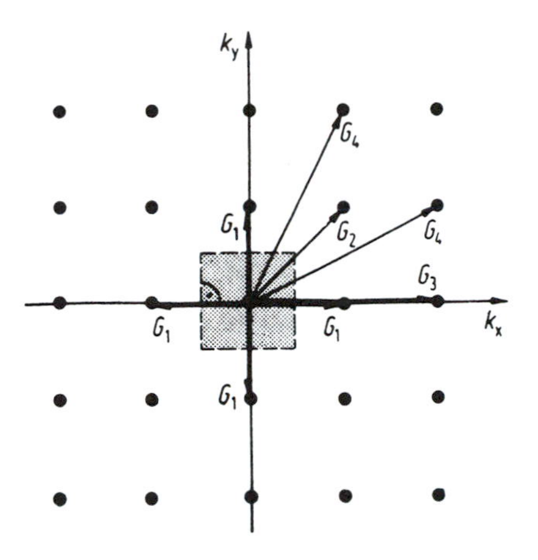

Figure 5.7. Four shortest lattice vectors in a k_x–k_y coordinate system and the first Brillouin zone in a two-dimensional reciprocal lattice.

area. The first four shortest lattice vectors G_1 through G_4 are drawn in Fig. 5.7.

The significance of the Brillouin zones will become evident in later sections, when the energy bands of solids are discussed. A few words of explanation should be mentioned here, however. The Brillouin zones are useful if one wants to calculate the behavior of an electron which may travel in a specific

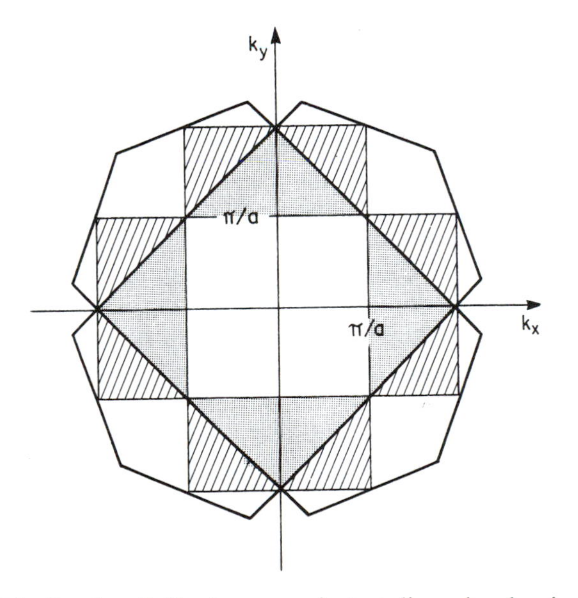

Figure 5.8. The first four Brillouin zones of a two-dimensional reciprocal lattice.

direction in reciprocal space. For example, if in a two-dimensional lattice, an electron travels at 45° to the k_x-axis, the boundary of the Brillouin zone is reached, according to Fig. 5.8, for

$$k_{\text{crit}} = \frac{\pi}{a}\sqrt{2}. \tag{5.8}$$

In contrast to this, the boundary of a Brillouin zone is reached at

$$k_{\text{crit}} = \frac{\pi}{a} \tag{5.9}$$

when an electron moves parallel to the k_x- or k_y-axes. The largest energies which electrons can assume in these two examples are therefore

$$E_{\max} = \frac{\pi^2\hbar^2}{a^2 m}, \tag{5.8a}$$

(at 45°, i.e., in the (110) direction) and

$$E_{\max} = \frac{1}{2}\left(\frac{\pi^2\hbar^2}{a^2 m}\right), \tag{5.9a}$$

along the k_x-axis, i.e., in the (100) direction (see (4.8)). Once this maximal energy has been reached, the electron waves form standing waves (or equivalently, the electrons are reflected back into the Brillouin zone).

The consequence of (5.8) and (5.9) is an overlapping of energy bands which can be seen when the bands are drawn in different directions in k-space (Fig. 5.27). We will learn later that these considerations can be utilized to determine the difference between metals, semiconductors, and insulators.

The occurrence of allowed and forbidden energy zones for the electron movement in a crystal can also be illustrated in a completely different way. This will be done briefly here because of its immediate intuitive power. We consider an electron wave which propagates in a lattice at an angle θ to a set of parallel lattice planes (Fig. 5.9). At a certain angle of incidence, constructive interference between rays 1′ and 2′ occurs. These rays have been diffracted on the lattice atoms. It has been shown by Bragg that each ray which is diffracted in this way can be considered as being reflected by a mirror parallel to the lattice planes. At this critical angle the transmitted rays will be weakened considerably. This is always the case when the path difference $2a\sin\theta$ is an integer multiple of the electron wavelength λ, i.e., when

$$2a\sin\theta = n\lambda, \qquad n = 1, 2, 3, \ldots \tag{5.10}$$

(Bragg relation). With (4.9) one obtains, from (5.10),

$$2a\sin\theta = n\frac{2\pi}{k}$$

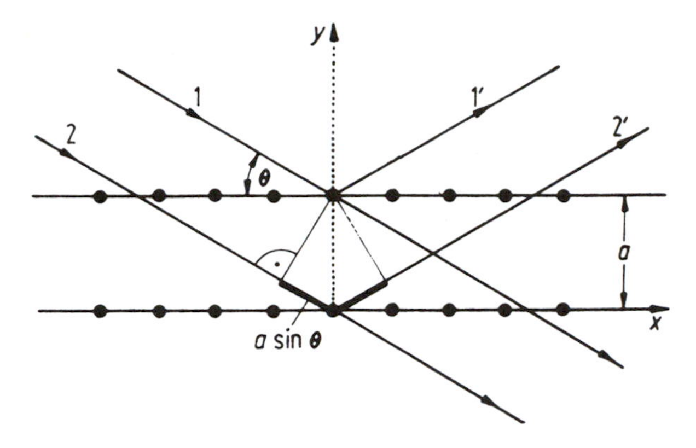

Figure 5.9. Bragg reflection of an electron wave in a lattice. The critical angle of incidence is θ.

and therefore

$$k_{\text{crit}} = n\frac{\pi}{a\sin\theta}. \tag{5.11}$$

For an electron movement parallel to a coordinate axis, i.e., for perpendicular incidence, (5.9) follows from (5.11); if $\theta = 45°$, one obtains (5.8).

Equation (5.11) leads to the result that for increasing electron energies a critical **k**-value is finally reached; for which "reflection" of the electron wave at the lattice planes occurs. At this critical **k**-value the transmission of an electron beam through the lattice is prevented. Then, the incident and the Bragg-reflected electron wave form a standing wave. The critical **k**-value is inversely proportional to the lattice constant a, as seen in (5.11). As a consequence, critical **k**- (or energy) values for which "reflection" occurs are reached first (i.e., they are reached for smaller k-values) for planes with small Miller indices (Fig. 5.10).

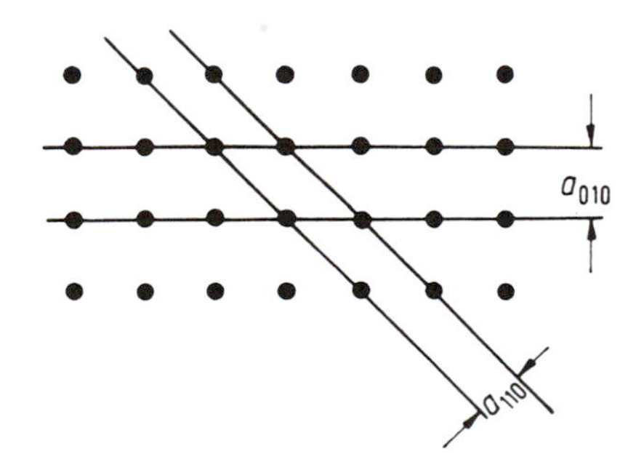

Figure 5.10. Lattice planes of a square lattice.

*5.3. Three-Dimensional Brillouin Zones

In the previous section, the physical significance of the Brillouin zones was discussed. It was shown that at the boundaries of these zones the electron waves were Bragg-reflected by the crystal. The wave vector $|\mathbf{k}| = 2\pi/\lambda$ was seen to have the unit of a reciprocal length and is therefore defined in the reciprocal lattice. We will now attempt to construct three-dimensional Brillouin zones for two important crystal structures, namely, the face-centered cubic (fcc) and the body-centered cubic (bcc) crystals. Since the Brillouin zones for these structures have some important features in common with the so-called Wigner–Seitz cells, it is appropriate to discuss, at first, the Wigner–Seitz cells and also certain features of the reciprocal lattice before we return to the Brillouin zones at the end of Section 5.5.

*5.4. Wigner–Seitz Cells

Crystals have symmetrical properties. Therefore, a crystal can be described as an accumulation of "unit cells." The smaller such a unit cell, i.e., the fewer atoms it contains, the simpler its description. The smallest possible cell is called a "primitive unit cell." Frequently, however, a larger, nonprimitive unit cell is used, which might have the advantage that the symmetry can be better recognized. Body-centered cubic and face-centered cubic are characteristic representatives of such "conventional" unit cells.[10]

[10] A lattice is a regular periodic arrangement of points in space; it is, consequently, a mathematical abstraction. All crystal structures can be traced to one of the 14 types of Bravais lattices (see textbooks on crystallography).

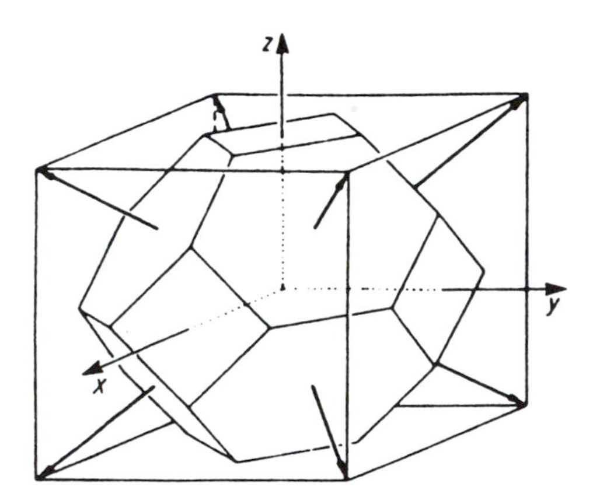

Figure 5.11. Wigner–Seitz cell for the body-centered structure.

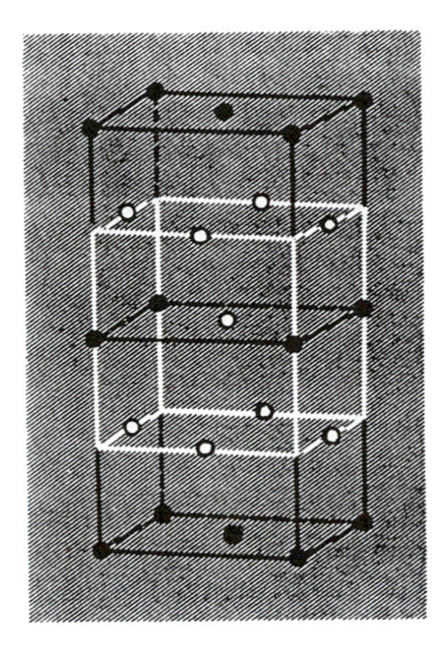

Figure 5.12. Conventional unit cell of the fcc structure. In the cell which is marked black, the atoms are situated on the corners and faces of the cubes. In the white cell, the atoms are at the centers of the edges and the center of the cell.

The Wigner–Seitz cell is a special type of primitive unit cell which shows the cubic symmetry of the cubic cells. For its construction, one bisects the vectors from a given atom to its nearest neighbors and places a plane perpendicular to these vectors at the bisecting points. This is shown in Fig. 5.11 for the bcc lattice.

In the fcc lattice, the atoms are arranged on the corners and faces of a cube,

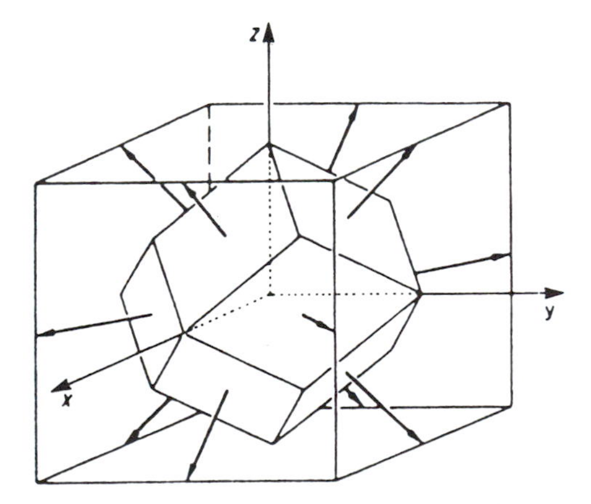

Figure 5.13. Wigner–Seitz cell for the fcc structure. It is constructed from the white cell which is marked in Fig. 5.12.

which is equivalent to the center points of the edges (Fig. 5.12). The Wigner–Seitz cell for this structure is shown in Fig. 5.13.

*5.5. Translation Vectors and the Reciprocal Lattice

In Fig. 5.14(a) the fundamental vectors $\mathbf{t}_1$, $\mathbf{t}_2$, $\mathbf{t}_3$ are inserted in a unit cell of a cubic primitive lattice. By combination of these "primitive vectors" a translation vector,

$$\mathbf{R} = n_1\mathbf{t}_1 + n_2\mathbf{t}_2 + n_3\mathbf{t}_3, \tag{5.12}$$

can be defined. Using this translation vector it is possible to reach, from a given lattice point, any other equivalent lattice point. For this, the factors n_1, n_2, n_3 have to be integers. In Fig. 5.14(b) the fundamental vectors $\mathbf{t}_1$, $\mathbf{t}_2$, $\mathbf{t}_3$ are shown in a conventional unit cell of a bcc lattice.

Similarly, as above, we now introduce for the reciprocal lattice three vectors, $\mathbf{b}_1$, $\mathbf{b}_2$, $\mathbf{b}_3$, and a translation vector

$$\mathbf{G} = 2\pi(h_1\mathbf{b}_1 + h_2\mathbf{b}_2 + h_3\mathbf{b}_3), \tag{5.13}$$

where h_1, h_2, and h_3 are, again, integers. (The factor 2π is introduced for convenience. In X-ray crystallography, this factor is omitted.)

The real and reciprocal lattices are related by a definition which states that the scalar product of the vectors $\mathbf{t}_1$ and $\mathbf{b}_1$ should be unity, whereas the scalar products of $\mathbf{b}_1$ and $\mathbf{t}_2$ or $\mathbf{b}_1$ and $\mathbf{t}_3$ are zero:

$$\mathbf{b}_1 \cdot \mathbf{t}_1 = 1, \tag{5.14}$$

$$\mathbf{b}_1 \cdot \mathbf{t}_2 = 0, \tag{5.15}$$

$$\mathbf{b}_1 \cdot \mathbf{t}_3 = 0. \tag{5.16}$$

Equivalent equations are defined for $\mathbf{b}_2$ and $\mathbf{b}_3$. These nine equations can be combined by using the Kronecker-Delta symbol,

$$\mathbf{b}_n \cdot \mathbf{t}_m = \delta_{nm}, \tag{5.17}$$

where $\delta_{nm} = 1$ for $n = m$ and $\delta_{nm} = 0$ for $n \neq m$. Equation (5.17) is from now on our *definition* for the three vectors $\mathbf{b}_n$, which are reciprocal to the vectors $\mathbf{t}_m$. From (5.15) and (5.16) it follows[11] that $\mathbf{b}_1$ is perpendicular to $\mathbf{t}_2$ and to $\mathbf{t}_3$ which means that $\mathbf{t}_2$ and $\mathbf{t}_3$ form a plane perpendicular to the vector $\mathbf{b}_1$ (Fig. 5.15). We therefore write[12]

$$\mathbf{b}_1 = \text{const. } \mathbf{t}_2 \times \mathbf{t}_3. \tag{5.18}$$

[11] The *scalar product* of two vectors $\mathbf{a}$ and $\mathbf{b}$ is $\mathbf{a}\cdot\mathbf{b} = ab\cos(\mathbf{ab})$. If $\mathbf{i}$, $\mathbf{j}$, and $\mathbf{l}$ are mutually perpendicular unit vectors, then we can write $\mathbf{i}\cdot\mathbf{j} = \mathbf{j}\cdot\mathbf{l} = \mathbf{l}\cdot\mathbf{i} = 0$ and $\mathbf{i}\cdot\mathbf{i} = \mathbf{jj} = \mathbf{ll} = 1$.

[12] The *vector product* of two vectors $\mathbf{a}$ and $\mathbf{b}$ is a vector which stands perpendicular to the plane formed by $\mathbf{a}$ and $\mathbf{b}$. It is $\mathbf{i} \times \mathbf{i} = \mathbf{j} \times \mathbf{j} = \mathbf{l} \times \mathbf{l} = 0$ and $\mathbf{i} \times \mathbf{j} = \mathbf{l}$ and $\mathbf{j} \times \mathbf{i} = -\mathbf{l}$.

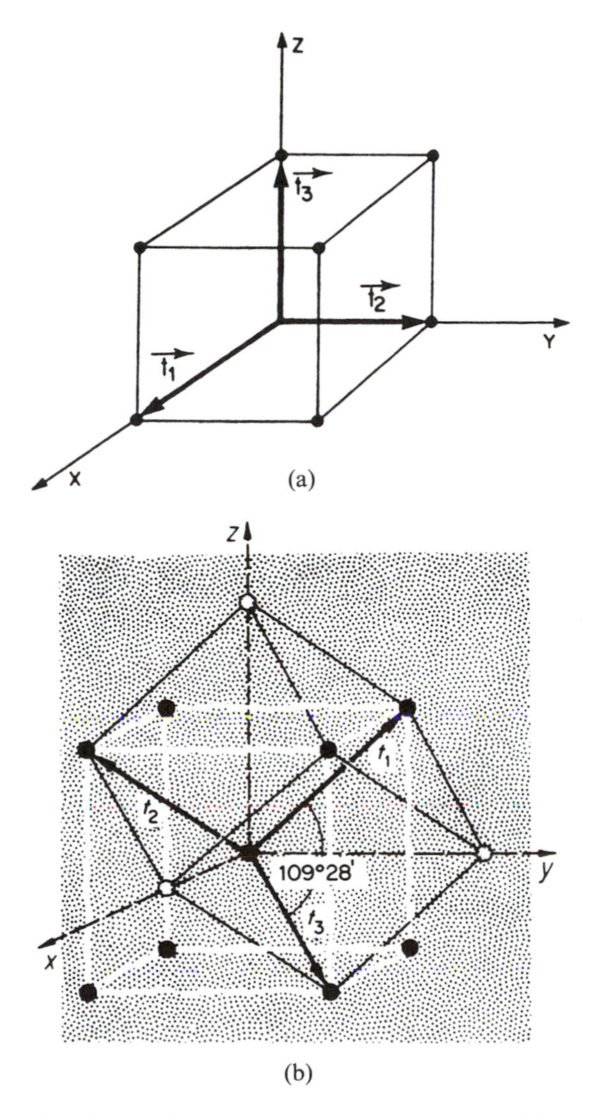

Figure 5.14. (a) Fundamental lattice vectors $\mathbf{t}_1$, $\mathbf{t}_2$, $\mathbf{t}_3$ in a cubic primitive lattice. (b) Fundamental lattice vectors in a conventional (white) and primitive, noncubic unit cell (black) of a bcc lattice. The axes of the primitive (noncubic) unit cell form angles of 109° 28′.

To evaluate the constant, we form the scalar product of $\mathbf{t}_1$ and $\mathbf{b}_1$ (5.18) and make use of (5.14):

$$\mathbf{b}_1 \cdot \mathbf{t}_1 = \text{const. } \mathbf{t}_1 \cdot \mathbf{t}_2 \times \mathbf{t}_3 = 1. \tag{5.19}$$

This yields

$$\text{const.} = \frac{1}{\mathbf{t}_1 \cdot \mathbf{t}_2 \times \mathbf{t}_3}. \tag{5.20}$$

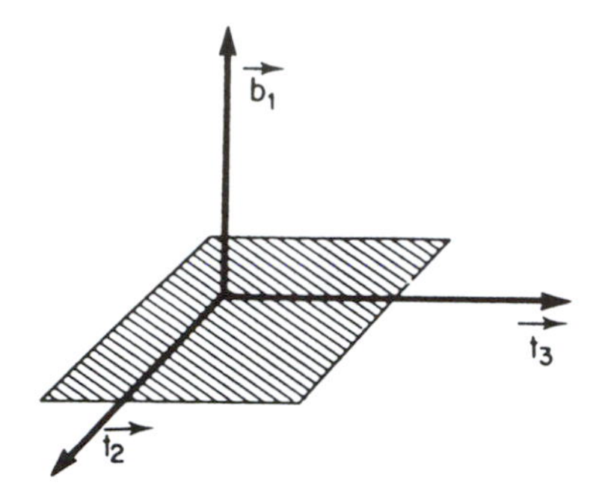

Figure 5.15. Plane formed by $\mathbf{t}_2$ and $\mathbf{t}_3$ with perpendicular vector $\mathbf{b}_1$.

Combining (5.18) with (5.20) gives

$$\mathbf{b}_1 = \frac{\mathbf{t}_2 \times \mathbf{t}_3}{\mathbf{t}_1 \cdot \mathbf{t}_2 \times \mathbf{t}_3}. \tag{5.21}$$

Equivalent equations can be obtained for $\mathbf{b}_2$ and $\mathbf{b}_3$:

$$\mathbf{b}_2 = \frac{\mathbf{t}_3 \times \mathbf{t}_1}{\mathbf{t}_1 \cdot \mathbf{t}_2 \times \mathbf{t}_3}, \tag{5.22}$$

$$\mathbf{b}_3 = \frac{\mathbf{t}_1 \times \mathbf{t}_2}{\mathbf{t}_1 \cdot \mathbf{t}_2 \times \mathbf{t}_3}. \tag{5.23}$$

Equations (5.21)–(5.23) are the transformation equations which express the fundamental vectors $\mathbf{b}_1$, $\mathbf{b}_2$, and $\mathbf{b}_3$ of the reciprocal lattice in terms of real lattice vectors.

As an example of how these transformations are performed, we calculate now the reciprocal lattice of a bcc crystal. The real crystal may have the lattice constant "a." We express the lattice vectors $\mathbf{t}_1$, $\mathbf{t}_2$, $\mathbf{t}_3$ in terms of the unit vectors, **i**, **j**, **l** in the x, y, z coordinate system (see Fig. 5.14(b)):

$$\mathbf{t}_1 = \frac{a}{2}(-\mathbf{i} + \mathbf{j} + \mathbf{l}), \tag{5.24}$$

or, abbreviated,

$$\mathbf{t}_1 = \frac{a}{2}(\bar{1}11) \tag{5.25}$$

and

$$\mathbf{t}_2 = \frac{a}{2}(1\bar{1}1), \tag{5.26}$$

$$\mathbf{t}_3 = \frac{a}{2}(11\bar{1}). \tag{5.27}$$

To calculate $\mathbf{b}_1$, using (5.21), we form at first the vector product[13]

$$t_2 \times t_3 = \frac{a^2}{4}\begin{vmatrix} \mathbf{i} & \mathbf{j} & \mathbf{l} \\ 1 & -1 & 1 \\ 1 & 1 & -1 \end{vmatrix} = \frac{a^2}{4}(\mathbf{i} + \mathbf{j} + \mathbf{l} + \mathbf{l} - \mathbf{i} + \mathbf{j}) = \frac{a^2}{4}(2\mathbf{j} + 2\mathbf{l})$$

$$= \frac{a^2}{2}(\mathbf{j} + \mathbf{l}) \tag{5.28}$$

and the scalar[14] product

$$\mathbf{t}_1 \cdot \mathbf{t}_2 \times \mathbf{t}_3 = \frac{a^3}{4}(-\mathbf{i} + \mathbf{j} + \mathbf{l}) \cdot (0 + \mathbf{j} + \mathbf{l}) = \frac{a^3}{4}(0 + 1 + 1) = \frac{a^3}{2}. \tag{5.29}$$

Combining (5.21) with (5.28) and (5.29) yields

$$\mathbf{b}_1 = \frac{\frac{a^2}{2}(\mathbf{j} + \mathbf{l})}{\frac{a^3}{2}} = \frac{1}{a}(\mathbf{j} + \mathbf{l}), \tag{5.30}$$

or

$$\mathbf{b}_1 = \frac{1}{a}(011). \tag{5.31}$$

Similar calculations yield

$$\mathbf{b}_2 = \frac{1}{a}(101), \tag{5.32}$$

$$\mathbf{b}_3 = \frac{1}{a}(110). \tag{5.33}$$

In Fig. 5.16, the vectors $\mathbf{b}_1$, $\mathbf{b}_2$, $\mathbf{b}_3$ are inserted into a cube of length $2/a$. We note immediately an important result. The end points of the reciprocal lattice vectors of a bcc crystal are at the center of the edges of a cube. This means that points of the reciprocal lattice of the bcc structure are identical to the lattice points in a real lattice of the fcc structure. Conversely, the reciprocal lattice points of the fcc structure and the real lattice points of the bcc structure are identical.

In Section 5.2, we constructed two-dimensional Brillouin zones by drawing perpendicular bisectors on the shortest lattice vectors. Similarly, a three-dimensional Brillouin zone can be obtained by bisecting all lattice vectors $\mathbf{b}$

---

[13] $\mathbf{a} \times \mathbf{b} = \begin{vmatrix} \mathbf{i} & \mathbf{j} & \mathbf{l} \\ a_x & a_y & a_z \\ b_x & b_y & b_z \end{vmatrix}.$

[14] $\mathbf{a} \cdot \mathbf{b} = a_x b_x + a_y b_y + a_z b_z.$

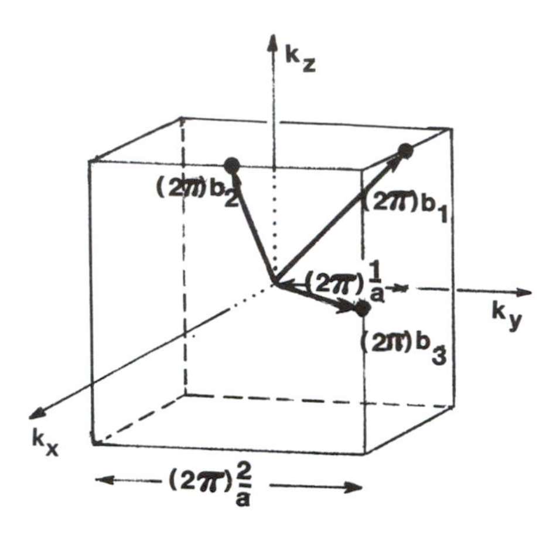

Figure 5.16. Lattice vectors in reciprocal space of a bcc crystal. The primitive vectors in the reciprocal lattice are (because of (5.13)) larger by a factor of 2π. The lattice constant of the cube then becomes $2\pi \cdot 2/a$.

and placing planes perpendicular on these points. As has been shown in Section 5.4, this construction is identical for a Wigner–Seitz cell. A comparison of the fundamental lattice vectors **b** and **t** gives the striking result that the Wigner–Seitz cell for an fcc crystal and the first Brillouin zone for a bcc crystal are identical in shape. The same is true for the Wigner–Seitz cell for bcc and the first Brillouin zone for fcc. Thus, a Brillouin zone can be defined as a Wigner–Seitz cell in the reciprocal lattice.

From (5.31) it can again be seen that the reciprocal lattice vector has the unit of a reciprocal length.

*5.6. Free Electron Bands

We mentioned in Section 5.1 that, because of the $E(\mathbf{k})$ periodicity, all information pertaining to the electronic properties of materials is contained in the first Brillouin zone. In other words, the energy $E_{\mathbf{k}'}$ for $\mathbf{k}'$ outside the first zone is identical to the energy $E_{\mathbf{k}}$ within the first zone if a suitable translation vector $\mathbf{G}$ can be found so that a wave vector $\mathbf{k}'$ becomes

$$\mathbf{k}' = \mathbf{k} + \mathbf{G}. \tag{5.34}$$

We have already used this feature in Section 5.1, where we plotted one-dimensional energy bands in the form of a reduced zone scheme. We proceed now to three-dimensional zone pictures. We might correctly expect that the energy bands are not alike in different directions in **k**-space. This can be demonstrated by using the "free electron bands" which we introduced in Fig. 5.6. We explain the details using the bcc crystal structure as an example.

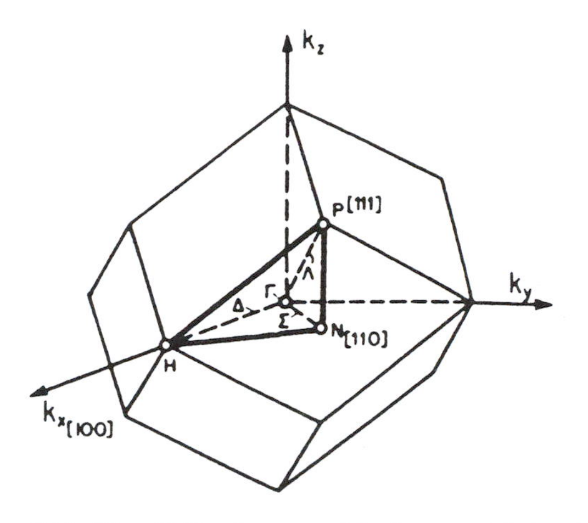

Figure 5.17. First Brillouin zone of the bcc crystal structure.

In three dimensions the equation analogous to (5.7) reads

$$E_{\mathbf{k}} = \frac{\hbar^2}{2m}(\mathbf{k} + \mathbf{G})^2. \tag{5.35}$$

In Fig. 5.17 three important directions in **k**-space are inserted into the first Brillouin zone of a bcc lattice. They are the [100] direction from the origin (Γ) to point H, the [110] direction from Γ to N, and the [111] direction from Γ to P. These directions are commonly labeled by the symbols Δ, Σ, and Λ, respectively. Figure 5.18 depicts the bands, calculated by using (5.35), for these distinct directions in **k**-space. The sequence of the individual subgraphs is established by convention and can be followed using Fig. 5.17.

We now show how some of these bands are calculated for a simple case. We select the Γ–H direction as an example. We vary the modulus of the vector $\mathbf{k}_{\Gamma H} \equiv \mathbf{k}_x$ between 0 and $2\pi/a$, the latter being the boundary of the Brillouin zone (see Fig. 5.16). For this direction, (5.35) becomes

$$E = \frac{\hbar^2}{2m}\left(\frac{2\pi}{a}x\mathbf{i} + \mathbf{G}\right)^2, \tag{5.36}$$

whereas x may take values between 0 and 1. To start with, let $\mathbf{G}$ be 0. Then (5.36) reads

$$E = \frac{\hbar^2}{2m}\left(\frac{2\pi}{a}\right)^2 (x\mathbf{i})^2 \equiv Cx^2. \tag{5.37}$$

This yields the well-known parabolic $E(\mathbf{k})$-dependence. The curve which represents (5.37) is labeled (000) in Fig. 5.18.

Now we let $h_1 = 0$, $h_2 = -1$, and $h_3 = 0$. Then we obtain by using (5.13)

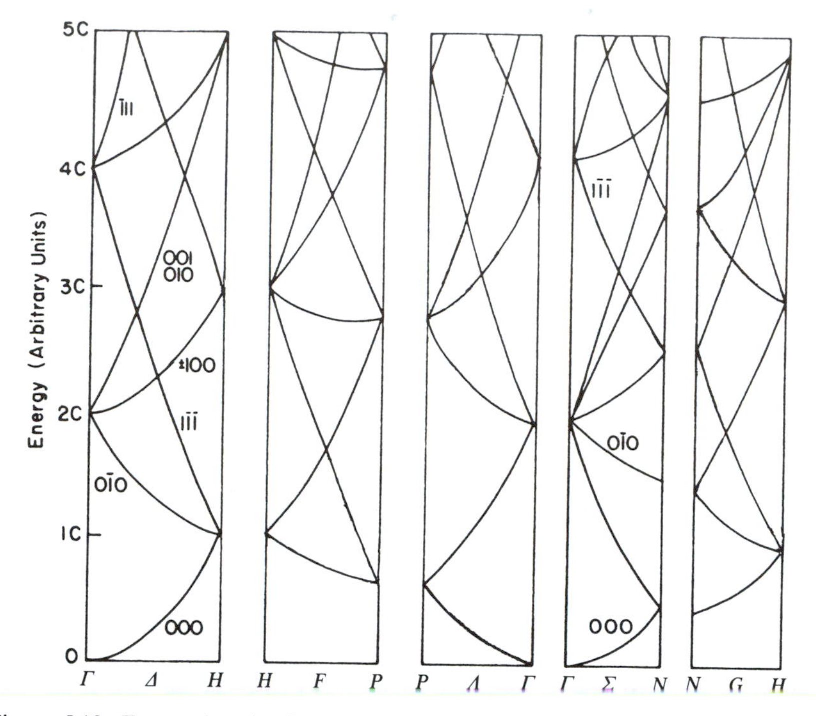

Figure 5.18. Energy bands of the free electrons for the bcc structure. The numbers given on the branches are the respective h_i values (see the calculation in the text). Compare to Fig. 5.6. $C = \hbar^2 2\pi^2/ma^2$, see (5.37).

and (5.32),

$$\mathbf{G} = -\frac{2\pi}{a}(\mathbf{i} + \mathbf{l}). \tag{5.38}$$

Combining (5.36) with (5.38) provides

$$\begin{aligned} E &= \frac{\hbar^2}{2m}\left[\frac{2\pi x}{a}\mathbf{i} - \frac{2\pi}{a}(\mathbf{i} + \mathbf{l})\right]^2 = C[\mathbf{i}(x-1) - \mathbf{l}]^2 \\ &= C[(x-1)^2 + 1] = C(x^2 - 2x + 2), \end{aligned} \tag{5.39}$$

which yields for

$$x = 0 \rightarrow E = 2C$$

and for

$$x = 1 \rightarrow E = 1C.$$

We obtain the band labeled $(0\bar{1}0)$ in Fig. 5.18. Similarly, all bands in Fig. 5.18 can be calculated by variation of the h values, **k**-directions and by using (5.35).

The free electron bands are very useful for the following reason: by comparing them with the band structures of actual materials, an assessment is

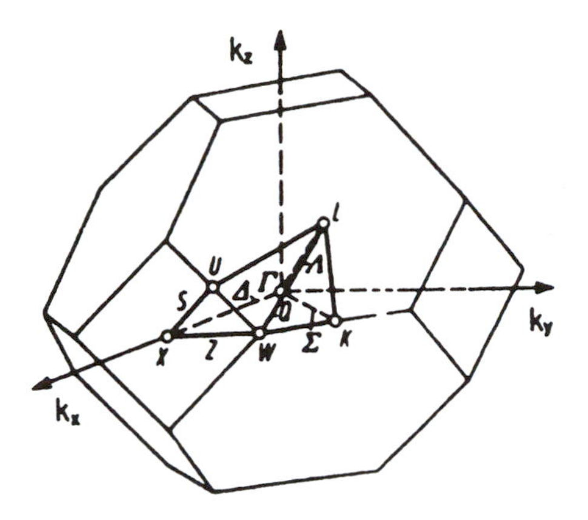

Figure 5.19. First Brillouin zone of the fcc structure.

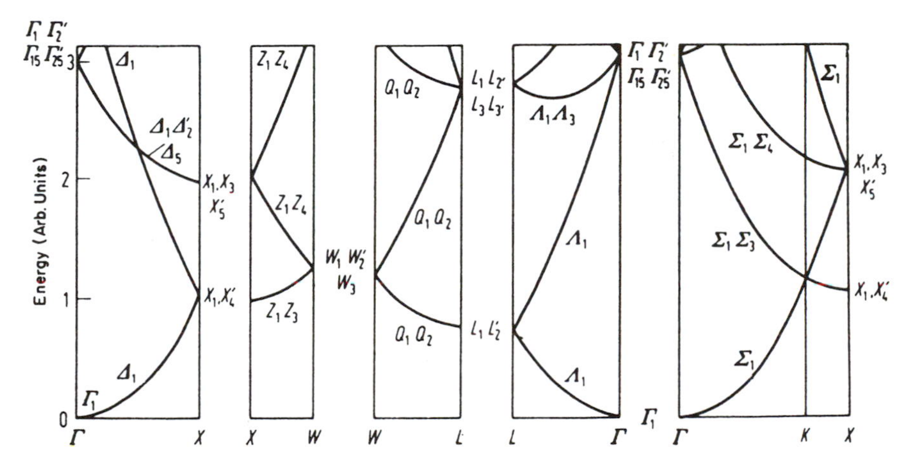

Figure 5.20. Free electron bands of the fcc structure.

possible if and to what degree the electrons in that material can be considered to be free.

In Figs. 5.19 and 5.20 the first Brillouin zone and the free electron bands of an fcc structure are shown.

5.7. Band Structures for Some Metals and Semiconductors

Those readers who have skipped Sections 5.3 through 5.6 need to familiarize themselves with the (three-dimensional) first Brillouin zone for the face centered cubic (fcc) crystal structure (Fig. 5.19). The (100), the (110), and the (111)

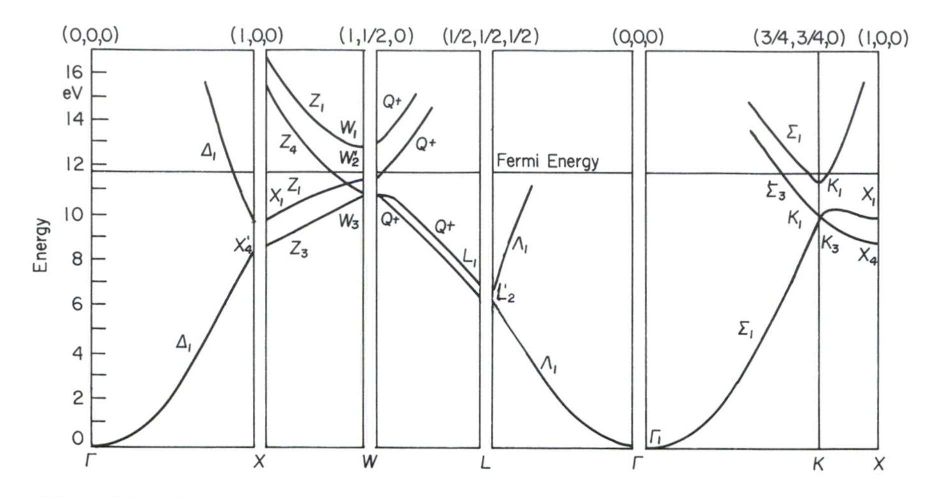

Figure 5.21. Energy bands for aluminum. Adapted from B. Segal, *Phys. Rev.* **124**, 1797 (1961). (The meaning of the Fermi energy will be explained in Section 6.1.)

directions in k-space are indicated by the letters $\Gamma - X$, $\Gamma - K$, and $\Gamma - L$, respectively. Other directions in **k**-space are likewise seen.

We inspect now some calculated energy-band structures. They should resemble the one shown in Fig. 5.4. In the present case, however, they are depicted for more than one direction in **k**-space. Additionally, they are displayed in the *positive* **k**-direction only, similarly as in Figs. 5.6 or 5.20.

We start with the band diagram for aluminum, Fig. 5.21. We recognize immediately the characteristic parabola-shaped bands in the $k_x(\Gamma - X)$ direction as seen before in Fig. 5.4. Similar parabola-shaped bands can be detected in the $\Gamma - K$ and the $\Gamma - L$ directions. The band diagram for aluminum looks quite similar to the free electron bands shown in Fig. 5.20. This suggests that the electrons in aluminum behave essentially free-electronlike (which is indeed the case).

We also detect in Fig. 5.21 some band gaps, for example, between the X_4' and the X_1 symmetry points, or between W_3 and W_2'. Note, however, that the individual energy bands overlap in different directions in k-space, so that as a whole, no band gap exists. (This is in marked difference to the band diagram of a semiconductor, as we shall see in a moment.) The lower, parabola-shaped bands are associated with the aluminum 3-s electrons (see Appendix 3). These bands are therefore called "3-s bands". The origin of the energy scale is positioned for convenience in the lower end of this s-band.

Next, we discuss the band structure for copper, Fig. 5.22. We notice in the lower half of this diagram closely spaced and flat running bands. Calculations show that these can be attributed to the 3d-bands of copper (see Appendix 3). They superimpose the 4s-bands (which are heavily marked in Fig. 5.22). The band which starts at Γ is, at first, s-electronlike, and becomes d-electronlike

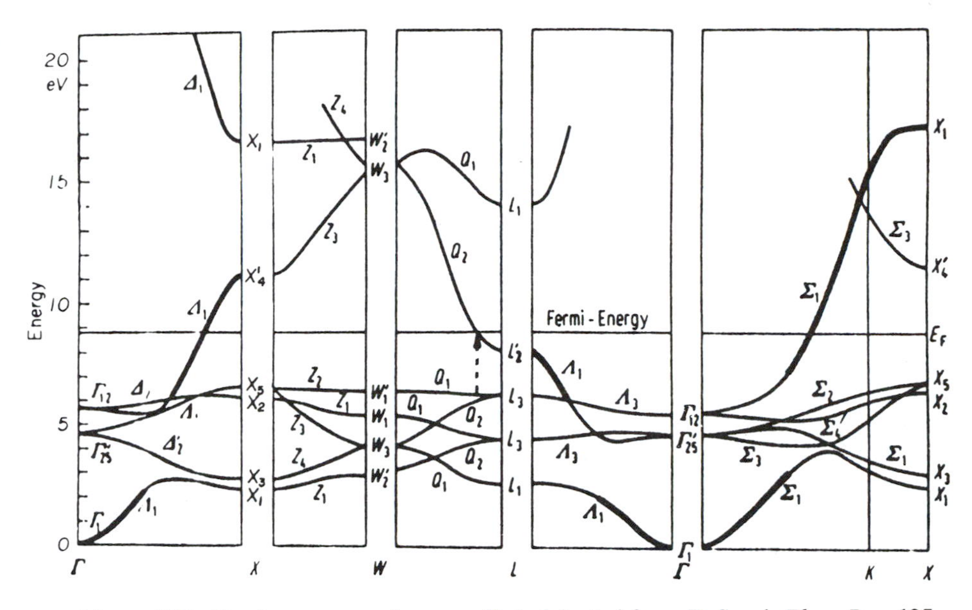

Figure 5.22. Band structure of copper (fcc). Adapted from B. Segal, *Phys. Rev.* **125**, 109 (1962). The calculation was made using the *l*-dependent potential. (For the definition of the Fermi energy, see Section 6.1.)

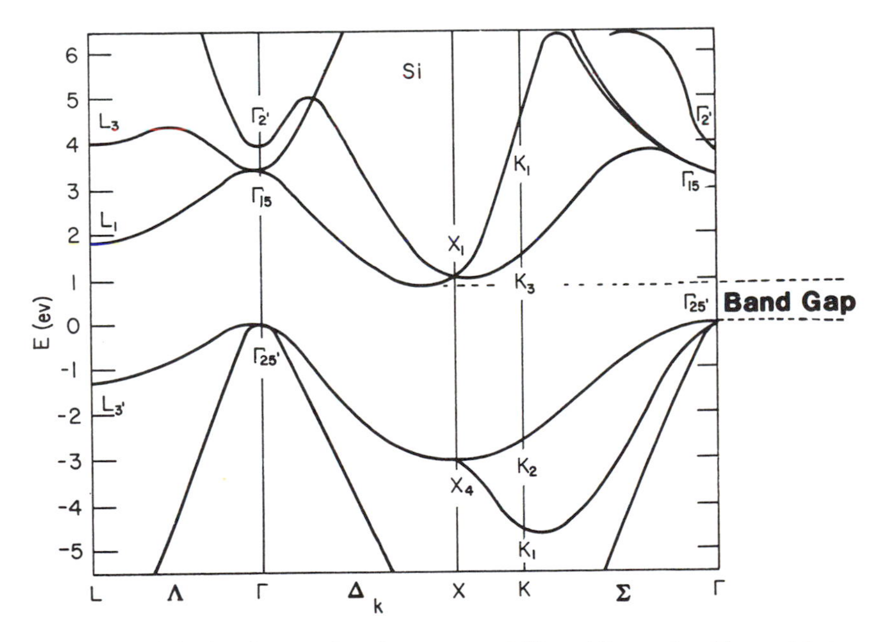

Figure 5.23. Calculated energy band structure of silicon (diamond-cubic crystal structure). Adapted from M.L. Cohen and T.K. Bergstresser, *Phys. Rev.* **14**, 789 (1966). See also J.R. Chelikowsky and M.L. Cohen, *Phys. Rev.* **B14**, 556 (1976).

while approaching point X. The first half of this band is continued at higher energies. It is likewise heavily marked. It can be seen, therefore, that the d-bands overlap the s-bands. Again, as for aluminum, no band gap exists if one takes all directions in **k**-space into consideration.

As a third example, the band structure of silicon is shown (Fig. 5.23). Of particular interest is the area between 0 and approximately 1 eV in which no energy bands are shown. This "energy gap," which is responsible for the well-known semiconductor properties, will be the subject of detailed discussion in a later chapter.

Finally, the band structure of gallium–arsenide is shown in Fig. 5.24. The so-called III–V semiconductor compounds, such as GaAs, may become of great technical importance, as we will discuss in Section 8.5. They have essentially the same crystal structure and the same total number of valence electrons as the element silicon. Again, a band gap is clearly seen.

It should be mentioned, in closing, that the band structures of actual solids, as shown in Figs. 5.21–5.24, are the result of extensive, computer-aided calculations, and that various investigators using different starting potentials arrive at slightly different band structures. Experimental investigations, such

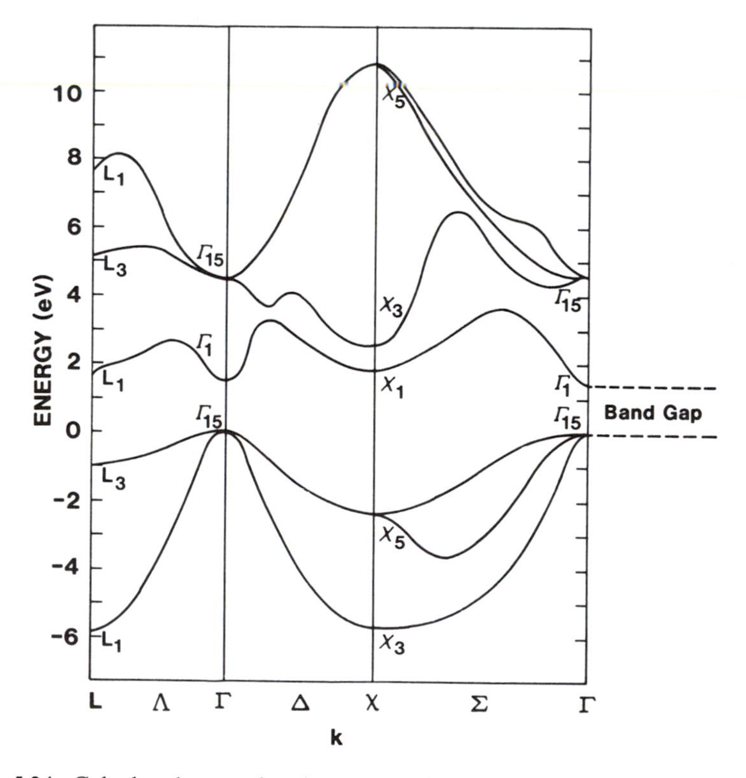

Figure 5.24. Calculated energy band structure of GaAs. Adapted from F. Herman and W.E. Spicer, *Phys. Rev.* **174**, 906 (1968).

as measurements of the frequency dependence of the optical properties, can decide which of the various calculated band structures are closest to reality.

5.8. Curves and Planes of Equal Energy

We conclude this chapter by discussing another interesting aspect of the energy versus wave vector relationship.

In one-dimensional **k**-"space" there is only *one* **k**-value which is connected with a given energy (see Fig. 5.1). In the two-dimensional case, i.e., when we

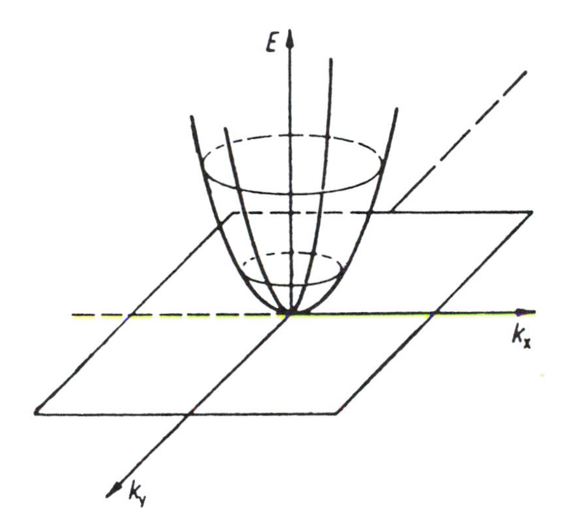

Figure 5.25. Electron energy E versus wave vector **k** (two-dimensional). This figure demonstrates various curves of equal energy for free electrons.

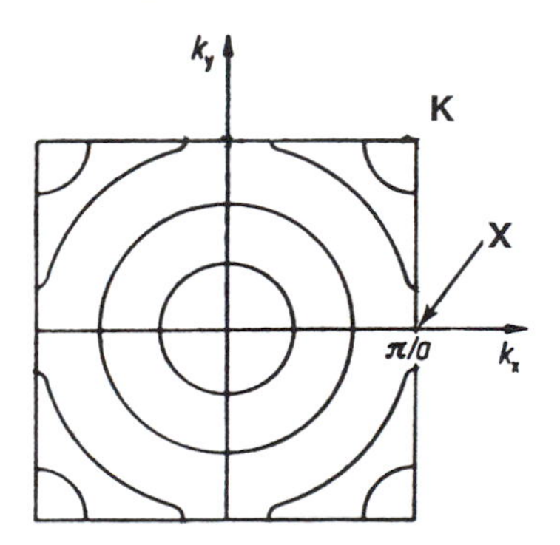

Figure 5.26. Curves of equal energy inserted into the first Brillouin zone for a two-dimensional square lattice.

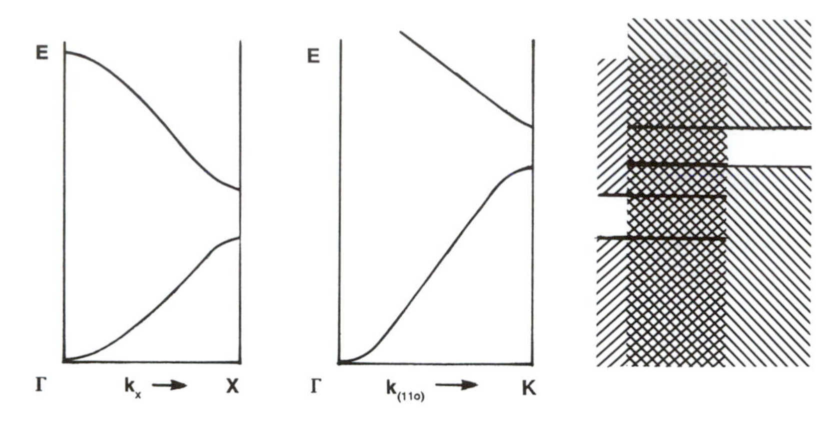

Figure 5.27. Overlapping of energy bands (compare with Fig. 5.19).

plot the electron energy over a k_x–k_y plane, more than one **k**-value can be assigned to a given energy. This leads to curves of equal energy, as shown in Fig. 5.25. For a two-dimensional square lattice and for small electron energies, the curves of equal energy are circles. However, if the energy of the electrons is approaching the energy of the boundary of a Brillouin zone, then a deviation from the circular form is known to occur. This is shown in Fig. 5.26, where curves of equal energy for a two-dimensional square lattice are inserted into the first Brillouin zone. It is of particular interest that the energy which belongs to point K in Fig. 5.26 is larger than the energy which belongs to point X (see (5.8a) and (5.9a)). Consequently, the curves of equal energy for

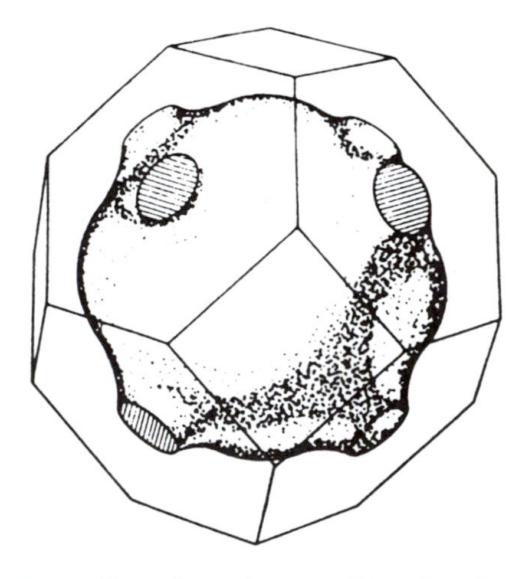

Figure 5.28. A particular surface of equal energy (Fermi surface, see Section 6.1) and the first Brillouin zone for copper. Adapted from A.B. Pippard, *Phil. Trans. Roy. Soc. London*, **A 250**, 325 (1957).

the first Brillouin zone may extend into the second zone. This leads to an overlapping of energy bands as schematically shown in Fig. 5.27, and in the band structures of Figs. 5.21–5.24. For copper and aluminum the band overlapping leads to quasi-continuous allowed energies (in different directions of **k**-space). For semiconductors the band overlapping is not complete, which results in the already-mentioned energy gap (Figs. 5.23 and 5.24).

In three-dimensional **k**-space one obtains *surfaces* of equal energy. For the free electron case and for a cubic lattice they are spheres. For a nonparabolic E-(**k**) behavior these surfaces become more involved. This is demonstrated in Fig. 5.28 for a special case.

Problems

1. What is the energy difference between the points L_2' and L_1 (upper) in the band diagram for copper?
2. How large is the "gap energy" for silicon? (*Hint*: Consult the band diagram for silicon.)
3. Calculate how much the kinetic energy of a free electron at the corner of the first Brillouin zone of a simple cubic lattice (three dimensions!) is larger than that of an electron at the midpoint of the face.
4. Construct the first four Brillouin zones for a simple cubic lattice in two dimensions.
5. Calculate the shape of the free electron bands for the cubic primitive crystal structure for $n = 1$ and $n = -2$ (see Fig. 5.6).
6. Calculate the free energy bands for a bcc structure in the $\mathbf{k}_x$-direction having the following values for $h_1/h_2/h_3$: (a) $1\bar{1}\bar{1}$; (b) 001; and (c) 010. Plot the bands in **k**-space. Compare with Fig. 5.18.
7. Calculate the main lattice vectors in reciprocal space of an fcc crystal.
8. Calculate the bands for the bcc structure in the 110 $[\Gamma - N]$ direction for: (a) (000); (b) $(0\bar{1}0)$; and (c) $(1\bar{1}\bar{1})$.
9. If $\mathbf{b}_1 \cdot \mathbf{t}_1 = 1$ is given (see equation (5.14)), does this mean that $\mathbf{b}_1$ is parallel to $\mathbf{t}_1$?

CHAPTER 6

Electrons in a Crystal

In the preceding chapters we considered essentially only *one* electron which was confined to the field of the atoms of a solid. This electron was in most cases an outer, i.e., a valence, electron. However, in a solid of one cubic centimeter at least 10^{22} electrons can be found. In this section we shall describe how these electrons are distributed among the available energy levels. It is impossible to calculate the exact place and the kinetic energy of each individual electron. We will see, however, that probability statements nevertheless give satisfying results.

6.1. Fermi Energy and Fermi Surface

The Fermi energy, E_F, is an important part of an electron band diagram. Many of the electronic properties of materials such as optical, electrical, or magnetic properties are related to the location of E_F within a band. The Fermi energies for aluminum and copper are shown in Figs. 5.21 and 5.22. Numerical values for the Fermi energies for some materials are given in Appendix 4. They range typically from 2 eV to 12 eV.

The Fermi energy is often defined as the "highest energy which the electrons assume at $T = 0$ K." However, this definition, even though convenient, can be misleading, particularly when dealing with semiconductors. Therefore, a more accurate definition of the Fermi energy will be given in Section 6.2. We will see there that at the Fermi energy the Fermi function, $F(E)$, equals $\frac{1}{2}$. An equation for the Fermi energy is given in (6.11).

In three-dimensional **k**-space the one-dimensional Fermi energy is replaced by a Fermi surface. The energy surface which is shown in Fig. 5.28 is the Fermi surface for copper.

6.2. Fermi Distribution Function

The distribution of the energies of a large amount of particles, and its change with increasing temperature can be calculated by means of statistical mechanics. The kinetic energy of an electron gas is governed by Fermi–Dirac statistics which states that the probability that a certain energy level is occupied by electrons is given by the Fermi function, $\mathrm{F}(E)$

$$\mathrm{F}(E) = \frac{1}{\exp\left(\dfrac{E - E_\mathrm{F}}{k_\mathrm{B} T}\right) + 1}. \tag{6.1}$$

If an energy level E is completely occupied by electrons, the Fermi distribution function $\mathrm{F}(E)$ equals 1; for an empty energy level one obtains $\mathrm{F}(E) = 0$. E_F is the Fermi energy which we introduced in Section 6.1, k_B is the Boltzmann constant, and T is the absolute temperature. In Fig. 6.1, the Fermi function is plotted versus the energy for $T = 0$ by using (6.1). One sees from this figure that at $T = 0$ all levels which have an energy smaller than E_F are completely filled with electrons, whereas higher energy states are empty.

The Fermi distribution function for higher temperatures ($T \neq 0$) is shown in Fig. 6.2. The decrease of the function $\mathrm{F}(E)$ from 1 to 0 at higher temperatures is "smeared out," i.e., it is extended to an energy interval $2\Delta E$. This decrease is heavily exaggerated in Fig. 6.2. ΔE at room temperature is in reality only about 1% of E_F.

At high energies ($E \gg E_\mathrm{F}$) the upper end of the Fermi distribution function can be approximated by the classical (Boltzmann) distribution function. This is best seen from (6.1) in which for large energies the exponential factor becomes significantly larger compared to 1. Then $\mathrm{F}(E)$ is approximately

$$\mathrm{F}(E) \approx \exp\left[-\left(\frac{E - E_\mathrm{F}}{k_\mathrm{B} T}\right)\right]. \tag{6.1a}$$

This expression is known to be the Boltzmann factor which gives, in classical

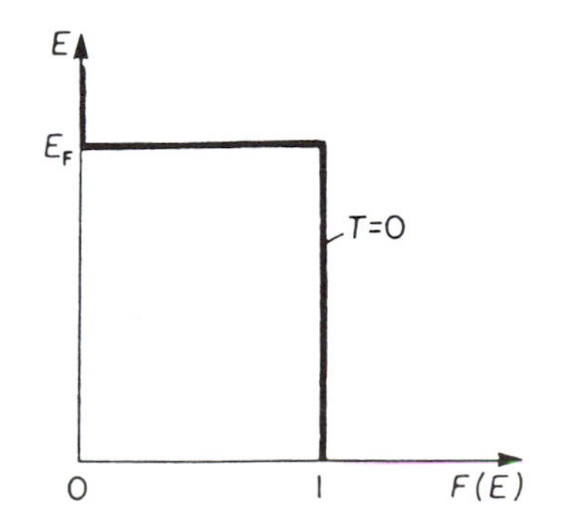

Figure 6.1. Fermi distribution function, $\mathrm{F}(E)$, versus energy, E, for $T = 0$.

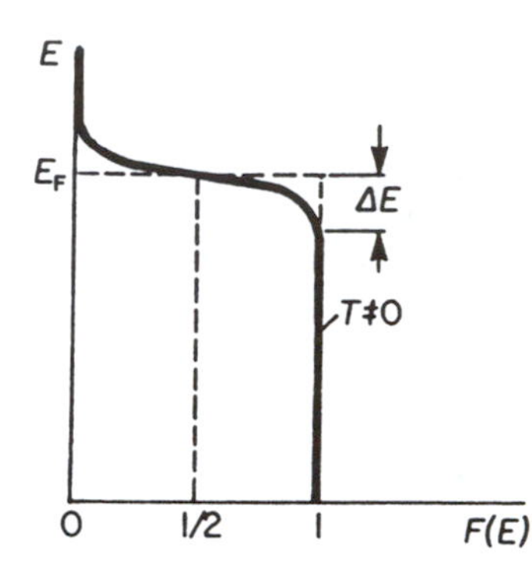

Figure 6.2. Fermi distribution function for $T \neq 0$.

thermodynamics, the probability that a given energy state is occupied. The F(E) curve for high energies is thus referred to as the "Boltzmann tail" of the Fermi distribution function.

Of particular interest is the value of the Fermi function F(E) at $E = E_F$ and $T \neq 0$. As can be seen from (6.1) and Fig. 6.2, F(E) is in this particular case $\frac{1}{2}$. This serves as a definition for the Fermi energy, as outlined in Section 6.1.

6.3. Density of States

We are now interested in the question of how energy levels are distributed over a band. We restrict our discussion for the moment to the lower part of the valence band (the s-band in copper, for example) because there the electrons can be considered to be essentially free due to their weak binding force to the nucleus. We assume that the free electrons (or the "electron gas") are confined into a square potential well from which they cannot escape. The dimensions of this potential well are thought to be identical to the dimensions of the crystal under consideration. Then our problem is similar to the case of *one* electron in a potential well of size a which we treated in Section 4.2. By using the appropriate boundary conditions the solution of the Schrödinger equation yields an equation which is analogous to (4.26)

$$E_n = \frac{\pi^2 \hbar^2}{2ma^2}(n_x^2 + n_y^2 + n_z^2), \tag{6.2}$$

where n_x, n_y, and n_z are the principal quantum numbers and a is now the *length*, etc., of the crystal. Now we pick an arbitrary set of quantum numbers n_x, n_y, n_z. To each such set we can find a specific energy level E_n, frequently called "energy state." An energy state can therefore be represented by a point in quantum number space (Fig. 6.3). In this space, n is the radius from the origin of the coordinate system to a point (n_x, n_y, n_z) where

$$n^2 = n_x^2 + n_y^2 + n_z^2. \tag{6.3}$$

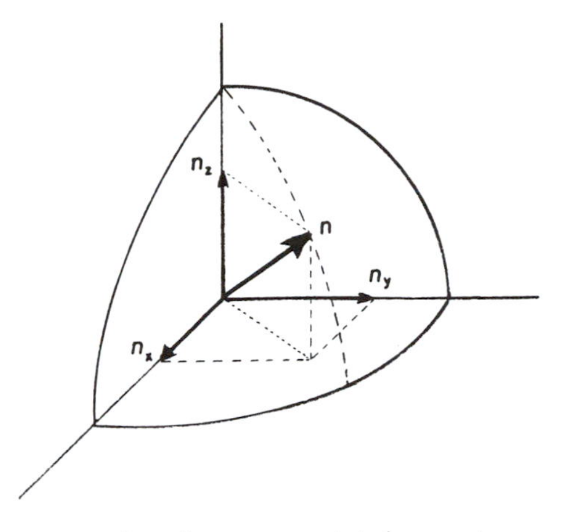

Figure 6.3. Representation of an energy state in quantum number space.

Equal values of the energy E_n are lying on the surface of a sphere with radius n. All points within the sphere therefore represent quantum states with energies smaller than E_n. The number of quantum states, η, with an energy equal to or smaller than E_n is proportional to the volume of the sphere. Since the quantum numbers are positive integers, the n-values can only be defined in the positive octant of the n-space. One-eighth of the volume of the sphere with radius n therefore gives the number of energy states, η, the energy of which is equal to or smaller than E_n. Thus, with (6.2) and (6.3), we obtain

$$\eta = \tfrac{1}{8} \cdot \tfrac{4}{3} \pi n^3 = \frac{\pi}{6} \left(\frac{2ma^2}{\pi^2 \hbar^2} \right)^{3/2} E^{3/2}. \tag{6.4}$$

Differentiation of η with respect to the energy E provides the *number of energy states per unit energy* in the energy interval dE, i.e., the density of the energy states, briefly called density of states $Z(E)$:

$$\frac{d\eta}{dE} = Z(E) = \frac{\pi}{4} \left(\frac{2ma^2}{\pi^2 \hbar^2} \right)^{3/2} E^{1/2} = \frac{V}{4\pi^2} \left(\frac{2m}{\hbar^2} \right)^{3/2} E^{1/2} \tag{6.5}$$

(a^3 is the volume, V, which the electrons can occupy).

The density of states plotted versus the energy gives, according to (6.5), a parabola. Figure 6.4 shows that at the lower end of the band considerably less energy levels (per unit energy) are available than at higher energies. One can compare the density of states concept with a high-rise apartment building in which the number of apartments per unit height (e.g., 8 feet) is counted. To stay within this analogy, only a very few apartments are thought to be available on the ground level. However, with increasing height of the building, the number of apartments per unit height becomes larger.

The area within the curve in Fig. 6.4 is, by definition, the number of states which have an energy equal to or smaller than E_n. Therefore, one obtains, for

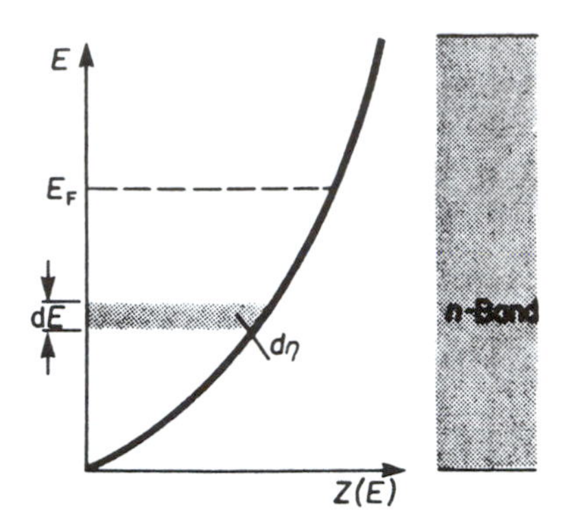

Figure 6.4. Density of states $Z(E)$ within a band. The electrons in this band are considered to be free.

an area element $d\eta$,

$$d\eta = Z(E) \cdot dE \tag{6.6}$$

as can be seen from (6.5) and Fig. 6.4.

6.4. Population Density

The number of electrons per unit energy, $N(E)$, within an energy interval dE can be calculated by multiplying the number of possible energy levels, $Z(E)$, with the probability for the occupation of these energy levels. We have to note, however, that because of the Pauli principle, each energy state can be occupied by one electron of positive spin and one of negative spin, i.e., each energy state can be occupied by two electrons. Therefore,

$$N(E) = 2 \cdot Z(E) \cdot F(E) \tag{6.7}$$

or, with (6.1) and (6.5),

$$N(E) = \frac{V}{2\pi^2}\left(\frac{2m}{\hbar^2}\right)^{3/2} E^{1/2} \frac{1}{\exp\left(\dfrac{E - E_F}{k_B T}\right) + 1}. \tag{6.8}$$

$N(E)$ is called the (electron) population density. We see immediately that for $T \to 0$ and $E < E_F$, the function $N(E)$ equals $2 \cdot Z(E)$ because $F(E)$ is unity in this case. For $T \neq 0$ and $E \simeq E_F$, the Fermi distribution function (6.1) causes a smearing out of $N(E)$ (Fig. 6.5).

The area within the curve in Fig. 6.5 represents the number of electrons, N^*, which has an energy equal to or smaller than the energy E_n. For an energy interval between E and $E + dE$, one obtains

$$dN^* = N(E)\,dE. \tag{6.9}$$

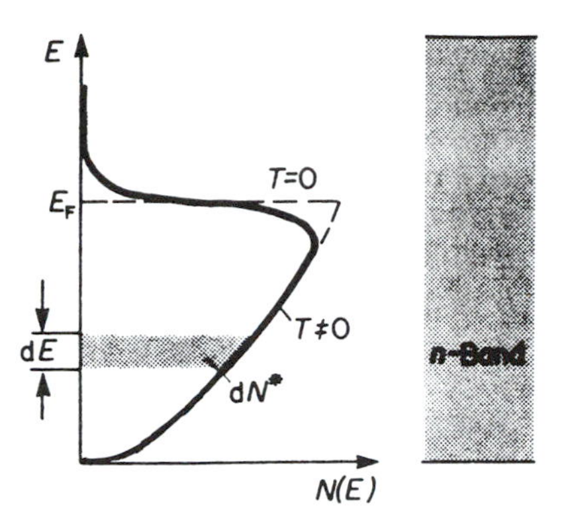

Figure 6.5. Population density $N(E)$ within a band for free electrons. dN^* is the number of electrons in the energy interval dE.

We are now in a position to calculate the Fermi energy by making use of (6.8) and (6.9). We consider the simple case $T \to 0$ and $E < E_F$ which yields $F(E) = 1$. Integration from the lower end of the band to the Fermi energy E_F provides

$$N^* = \int_0^{E_F} N(E)\,dE = \int_0^{E_F} \frac{V}{2\pi^2}\left(\frac{2m}{\hbar^2}\right)^{3/2} E^{1/2}\,dE = \frac{V}{3\pi^2}\left(\frac{2m}{\hbar^2}\right)^{3/2} E_F^{3/2}. \quad (6.10)$$

Rearranging (6.10) yields

$$E_F = \left(3\pi^2 \frac{N^*}{V}\right)^{2/3} \frac{\hbar^2}{2m}. \quad (6.11)$$

We define $N' = N^*/V$ as the number of electrons per unit volume. Then we obtain

$$E_F = (3\pi^2 N')^{2/3} \frac{\hbar^2}{2m}. \quad (6.11a)$$

It should be noted that N^* was calculated for simplicity for $T \to 0$ and $E < E_F$. This does not limit the applicability of (6.11), however, since the number of electrons does not change when the temperature is increased. In other words, integration from zero to infinity and using $T \neq 0$ would yield essentially the same result as above.

6.5. Complete Density of States Function Within a Band

We have seen in Section 6.3 that for the free electron case the density of states has a parabolic E versus $Z(E)$ relationship. In actual crystals, however, the density of states is modified by the energy conditions within the first Brillouin zone. Let us consider, for example, the curves of equal energy depicted in Fig.

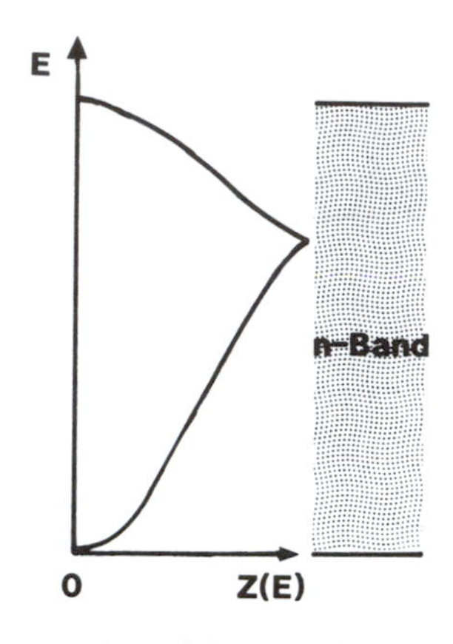

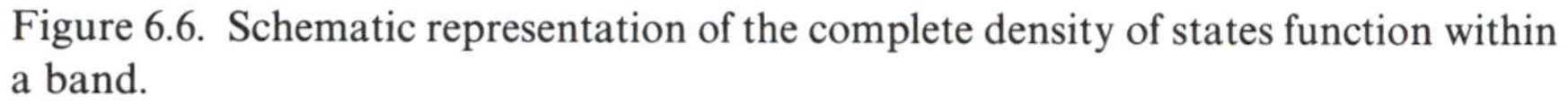

Figure 6.6. Schematic representation of the complete density of states function within a band.

5.26. For low energies, the equal energy curves are circles. Thus, the electrons behave free-electronlike for these low energies. The density of states curve is then, as before, a parabola. For larger energies, however, fewer energy states are available as is seen in Fig. 5.26. Thus, $Z(E)$ decreases with increasing E, until eventually the corners of the Brillouin zones are filled. At this point $Z(E)$ has dropped to zero. The largest number of energy states is thus found near the center of a band, as shown schematically in Fig. 6.6.

6.6. Consequences of the Band Model

We mentioned in Section 6.4 that because of the Pauli principle, each s-band of a crystal, consisting of N atoms, has space for $2N$ electrons, i.e., for two electrons per atom. If the highest filled s-band of a crystal is occupied by two electrons per atom, i.e., if the band is completely filled, we would expect that the electrons cannot drift through the crystal when an external electric field is applied (as it is similarly impossible to move a car in a completely occupied parking lot). An electron has to absorb energy in order to move. Keep in mind that, for a completely occupied band, higher energy levels are not allowed. (We exclude the possibility of electron jumps into higher bands.) Solids in which the highest filled band is completely occupied by electrons are, therefore, insulators (Fig. 6.7(a)).

In solids with one valence electron per atom (e.g., alkali metals) the valence band is essentially half-filled. An electron drift upon application of an external field is possible; the crystal shows metallic behavior (Fig. 6.7(b)).

Bivalent metals should be insulators according to this consideration, which is not the case. The reason for this lies in the fact that the upper bands partially overlap, which occurs due to the weak binding forces of the valence electrons on their atomic nuclei (see Fig. 5.27). If such an overlapping of bands

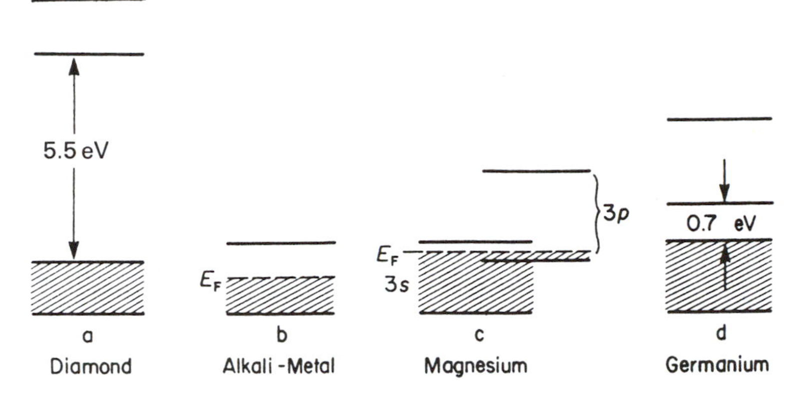

Figure 6.7. Simplified representation for energy bands for (a) insulators, (b) alkali metals, (c) bivalent metals, and (d) intrinsic semiconductors.

occurs, the valence electrons flow in the lower portion of the next higher band because the electrons tend to assume the lowest potential energy (Fig. 6.7(c)). As a result, bivalent solids may also possess partially filled bands. Thus, they are also conductors.

We shall see in Chapter 8 that the valence as well as the conduction bands of semiconductors can accommodate $4N$ electrons. Because germanium and silicon possess four valence electrons, the valence band is completely filled with electrons. Intrinsic semiconductors have a relatively narrow forbidden energy zone (Fig. 6.7(d)). A sufficiently large energy can, therefore, excite electrons from the completely filled valence band into the empty conduction band and thus provide some electron conduction.

This preliminary and very qualitative discussion on electronic conduction will be expanded substantially and the understanding will be deepened in Part II of this book.

6.7. Effective Mass

We implied in the previous sections that the mass of an electron in a solid is the same as the mass of a free electron. Experimentally-determined physical properties of solids, such as optical, thermal, or electrical properties indicate, however, that for some solids the mass is larger while for others it is slightly smaller than the free electron mass. This experimentally determined electron mass is usually called the effective mass m^*. The deviation of m^* from the free electron mass[15] m_0 can be easily appreciated by stating the ratio m^*/m_0 which

[15] We shall use the symbol m_0 only when we need to distinguish the free electron (rest) mass from the effective mass.

has values slightly above or below 1 (see Appendix 4). The cause for the deviation of the effective mass from the free electron mass is usually attributed to interactions between the drifting electrons and the atoms in a crystal. For example, an electron which is accelerated in an electric field might be slowed down slightly because of "collisions" with some atoms. The ratio m^*/m_0 is then larger than 1. On the other hand, the electron wave in another crystal might have just the right phase in order that the response to an external electric field is enhanced. In this case, m^*/m_0 is smaller than 1.

We shall now attempt to find an expression for the effective mass. For this, we shall compare the acceleration of an electron in an electric field calculated by classical as well as by quantum mechanical means. At first, we write an expression for the velocity of an electron in an energy band. We introduced in Chapter 2 the group velocity, i.e., the velocity with which a wave packet moves. Let ω be the angular frequency and $|\mathbf{k}| = 2\pi/\lambda$ the wave number of the electron wave. Then the group velocity is, according to (2.10),

$$v_g = \frac{d\omega}{dk} = \frac{d(2\pi\nu)}{dk} = \frac{d(2\pi E/h)}{dk} = \frac{1}{\hbar}\frac{dE}{dk}. \tag{6.12}$$

From this we calculate the acceleration

$$a = \frac{dv_g}{dt} = \frac{1}{\hbar}\frac{d^2E}{dk^2}\frac{dk}{dt}. \tag{6.13}$$

The relation between the energy E and the wave number $|\mathbf{k}|$ is known from the preceding sections. We now want to determine the factor dk/dt. Forming the first derivative of (4.8) ($p = \hbar k$) with respect to time yields

$$\frac{dp}{dt} = \hbar\frac{dk}{dt}. \tag{6.14}$$

Combining (6.14) with (6.13) yields

$$a = \frac{1}{\hbar^2}\frac{d^2E}{dk^2}\frac{dp}{dt} = \frac{1}{\hbar^2}\cdot\frac{d^2E}{dk^2}\cdot\frac{d(mv)}{dt} = \frac{1}{\hbar^2}\frac{d^2E}{dk^2}F, \tag{6.15}$$

where F is the force on the electron. The classical acceleration can be calculated from Newton's law (1.1)

$$a = \frac{F}{m}. \tag{6.16}$$

Comparing (6.15) with (6.16) yields the effective mass

$$m^* = \hbar^2\left(\frac{d^2E}{dk^2}\right)^{-1}. \tag{6.17}$$

We see from (6.17) that the effective mass is inversely proportional to the curvature of an electron band. Specifically, if the curvature of $E = f(k)$ at a given point in **k**-space is large, then the effective mass is small (and vice versa).

When inspecting band structures (Fig. 5.4 or Figs. 5.21–5.24) we notice some regions of high curvature. These regions might be found, particularly, near the center or near the boundary of a Brillouin zone. At these places, the effective mass is substantially reduced and may be as low as 1% of the free electron mass m_0. At points in **k**-space for which more than one electron band is found (Γ-point in Fig. 5.23, for example) more than one effective mass needs to be defined.

We shall demonstrate the **k**-dependence of the effective mass for a simple case and defer discussions about actual cases to Section 8.4. In Fig. 6.8(a) an ideal electron band within the first Brillouin zone is depicted. From this curve, both the first derivative and the reciprocal function of the second derivative, i.e., m^*, have been calculated. These functions are shown in Fig. 6.8(b) and (c). We notice in Fig. 6.8(c) that the effective mass of the electrons is small and positive near the center of the Brillouin zone and eventually increases for larger values of k_x. We likewise observe in Fig. 6.8(c) that electrons in the upper part of the given band have a negative effective mass. An electron with

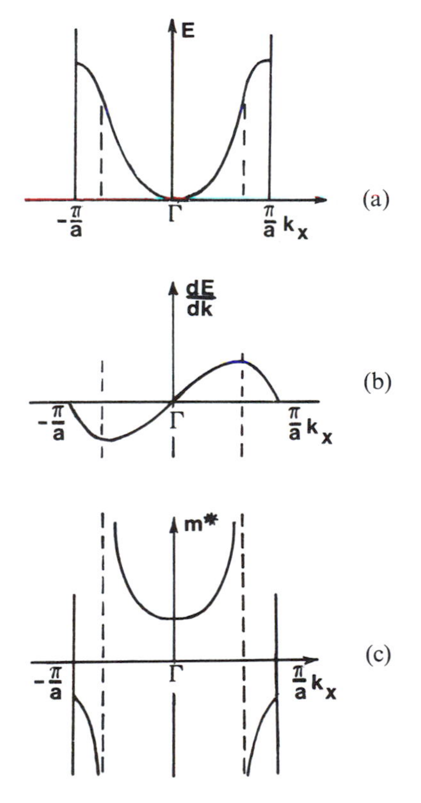

Figure 6.8. (a) Simple band structure, as shown in Fig. 5.4. (b) First derivative and (c) inverse function of the second derivative of the curve shown in (a).

a negative effective mass is called a "defect electron" or an "electron hole." (It is, however, common to ascribe to the hole a positive effective mass and a positive charge instead of a negative mass and a negative charge.) Electron holes play an important role in crystals whose valence bands are almost filled, e.g., in semiconductors. Solids which possess different properties in various directions (anisotropy) have a different m^* in each direction. The effective mass is a tensor in this case. An electron/hole pair is called an "exciton."

6.8. Conclusion

The first part of this book is intended to provide the reader with the necessary tools for a better understanding of the electronic properties of materials. We started our discussion by solving the Schrödinger equation for the free electron case, the bound electron case, and for electrons in a crystal. We learned that the distinct energy levels which are characteristic for isolated atoms widen into energy bands when the atoms are moved closer together and eventually form a solid. We also learned that the electron bands have "fine structure," i.e., they consist of individual "branches" in an energy versus momentum (actually **k**) diagram. We further learned that some of these energy bands are filled by electrons, and that the degree of this filling depends upon whether we consider a metal, a semiconductor, or an insulator. Finally, the degree to which electron energy levels are available within a band was found to be nonuniform. We discovered that the density of states is largest near the center of an electron band. All these relatively unfamiliar concepts will become more transparent to the reader when we apply them in the chapters to come.

Problems

1. What velocity has an electron near the Fermi surface of silver? ($E_F = 5.5$ eV).
2. Are there more electrons on the bottom or in the middle of the valence band of a metal? Explain.
3. At what temperature can we expect a 10% probability that electrons in silver have an energy which is 1% above the Fermi energy? ($E_F = 5.5$ eV).
4. Calculate the Fermi energy for silver assuming 6.1×10^{22} free electrons per cubic centimeter. (Assume the effective mass equals the free electron mass.)
5. Calculate the density of states of 1 m^3 of copper at the Fermi level ($m^* = m_0$, $E_F = 7$ eV). *Note*: Take 1 eV as energy interval. (Why?)
6. The density of states at the Fermi level (7 eV) was calculated for a certain metal to be about 10^{21} energy states per electron volt. Someone is asked to calculate the

number of electrons for this metal using the Fermi energy as the maximum kinetic energy which the electrons have. He argues that because of the Pauli principle, each energy state is occupied by two electrons. Consequently, there are 2×10^{21} electrons in that band.
(a) What is wrong with that argument?
(b) Why is the answer, after all, not too far from the correct numerical value?

7. Assuming the electrons to be free, calculate the total number of states below $E =$ 5 eV in a volume of 10^{-5} m^3.

8. (a) Calculate the number of free electrons per cubic centimeter in copper, assuming that the maximum energy of these electrons equals the Fermi energy ($m^* = m_0$).
(b) How does this result compare with that determined directly from the density of copper?
(c) How can we correct for the discrepancy?
(d) Does using the effective mass decrease the discrepancy?

9. What fraction of the 3*s*-electrons of sodium is found within an energy $k_B T$ below the Fermi level? (Take room temperature, i.e., $T = 300$ K.)

10. Calculate the Fermi distribution function for a metal at the Fermi level for $T \neq 0$.

11. Explain why, in a simple model, a bivalent material could be considered to be an insulator. Also explain why this simple argument is not true.

12. We stated in the text that the Fermi distribution function can be approximated by classical Boltzmann statistics if the exponential factor in the Fermi distribution function is significantly larger than one.
(a) Calculate $E - E_F = nk_B T$ for various values of n and state at which value for n,

$$\exp\left(\frac{E - E_F}{k_B T}\right)$$

can be considered to be "significantly larger" than 1 (assume $T = 300$ K). (*Hint*: Calculate the error in $F(E)$ for neglecting "1" in the denominator.)
(b) For what *energy* can we use Boltzmann statistics? (Assume $E_F = 5$ eV and $E - E_F = 4k_B T$.)

Suggestions for Further Reading (Part I)

N.W. Ashcroft and N.D. Mermin, *Solid State Physics*, Holt, Rinehart and Winston, New York (1976).
L.V. Azároff and J.J. Brophy, *Electronic Processes in Materials*, McGraw-Hill, New York (1963).
J.S. Blakemore, *Solid State Physics*, W.B. Saunders, Philadelphia, PA (1969).
R.H. Bube, *Electronic Properties of Crystalline Solids*, Academic Press, New York (1974).
R.H. Bube, *Electrons in Solids*, 2nd edn., Academic Press, New York (1988).
S. Datta, *Quantum Phenomena*, Vol. VIII of Modular Series on Solid State Devices, Addison-Wesley, Reading, MA (1989).

C. Kittel, *Solid State Physics*, Wiley, New York (1976).
N.F. Mott and H. Jones, *The Theory of the Properties of Metals and Alloys*, Dover, New York (1936).
M.A. Omar, *Elementary Solid State Physics*, Addison-Wesley, Reading, MA (1978).
R.F. Pierret, *Advanced Semiconductor Fundamentals*, Vol. VI of Modular Series on Solid State Devices, Addison-Wesley, Reading, MA (1987).
A.B. Pippard, *The Dynamics of Conduction Electrons*, Gordon and Breach, New York (1965).
H.A. Pohl, *Quantum Mechanics for Science and Engineering*, Prentice-Hall Series in Materials Science, Prentice-Hall, Englewood Cliffs, NJ.
R.M. Rose, L.A. Shepard, and J. Wulff, *The Structure and Properties of Materials*, Vol. IV, Electronic Properties, Wiley, New York (1966).
C.A. Wert and R.M. Thomson, *Physics of Solids*, McGraw-Hill, New York (1964).
P. Wikes, *Solid State Theory in Metallurgy*, Cambridge University Press, Cambridge (1973).

PART II

ELECTRICAL PROPERTIES OF MATERIALS

CHAPTER 7

Electrical Conduction in Metals and Alloys

7.1. Introduction

The first observations involving electrical phenomena probably began with the study of static electricity. Thales of Miletus, a Greek philosopher, discovered around 600 BC that a piece of amber, having been rubbed with a piece of cloth, attracted feathers and other light particles. Very appropriately, the word *electricity* was later coined by incorporating the Greek word *elektron*, which means *amber*.

It was apparently not before 2300 years later that man became again interested in electrical phenomena. Stephen Gray found in the early 1700s that some substances conduct electricity whereas others do not. In 1733 DuFay postulated the existence of two types of electricity which he termed *glass electricity* and *amber electricity* dependent on which material was rubbed. From then on a constant stream of well-known scientists contributed to our knowledge of electrical phenomena. Names such as Coulomb, Galvani, Volta, Oersted, Ampère, Ohm, Seebeck, Faraday, Henry, Maxwell, Thomson, and others, come to mind. What started 2600 years ago as a mysterious effect has been applied quite recently in an impressive technology which culminated in large-scale integration of electronic devices.

A satisfactory understanding of electrical phenomena on an atomistic basis was achieved by Drude at the turn of this century. A few decades later quantum mechanics refined our understanding. Both the classical as well as the quantum concepts of electrical phenomena will be covered in the chapters to come. Special emphasis is placed on the description of important applications.

7.2. Survey

The conductivity, σ, of different materials spans about twenty-five orders of magnitude (see Fig. 7.1). This is the largest known variation in a physical property.

It is generally accepted that in metals and alloys, the electrons, particularly the outer or valence electrons, play an important role in electrical conduction. Therefore, it seems most appropriate to make use of the electron theory which has been developed in the foregoing chapters. Before doing so, the reader is reminded of some fundamental equations of physics pertaining to electrical conduction. These laws have been extracted from experimental observations. Ohm's law,

$$V = RI, \tag{7.1}$$

relates the potential difference V (in volts) with the electrical resistance R (in ohms) and the electrical current I (in amps). Another form of Ohm's law,

$$j = \sigma\mathscr{E}, \tag{7.2}$$

links current density

$$j = \frac{I}{A}, \tag{7.2a}$$

i.e., the current per unit area (A/cm^2), with conductivity σ ($1/\Omega$ cm) and electric field strength[1]

$$\mathscr{E} = \frac{V}{L} \tag{7.3}$$

(V/cm). (In general, $\mathscr{E}$ and j are vectors. For our purpose, however, we need only their moduli.) The current density is frequently expressed by

$$j = Nve, \tag{7.4}$$

where N is the number of electrons (per unit volume), v their velocity, and e

[1] We use for the electric field strength a script $\mathscr{E}$ to distinguish it from the energy.

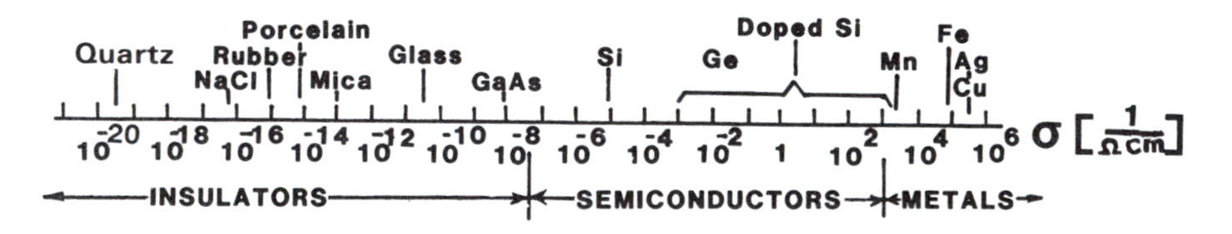

Figure 7.1. Room-temperature conductivity of various materials. (Superconductors, having conductivities of many orders of magnitude larger than copper, near 0 K, are not shown. The conductivity of semiconductors varies substantially with temperature and purity.) It is customary in engineering to use the centimeter as unit of length rather than the meter. We follow this practice.

their charge. The resistance of a conductor can be calculated from its physical dimensions by

$$R = \frac{L\rho}{A}, \tag{7.4a}$$

where L is the length of the conductor, A is its cross-sectional area, and ρ is the specific resistance, or resistivity (Ω cm). We define

$$\rho = \frac{1}{\sigma}. \tag{7.4b}$$

We discussed in Chapter 2 the existence of two alternatives to describe an electron. First, we may consider the electrons to have a particle nature. If this model is utilized, one can explain the resistivity by means of collisions of the drifting electrons with certain lattice atoms. The more collisions are encountered, the higher is the resistance. This concept qualitatively describes the increase in resistance with an increasing amount of lattice imperfections. It also explains the observed increase in resistance with increasing temperature: the thermal energy causes the lattice atoms to oscillate about their equilibrium positions (see Part V), thus increasing the probability for collisions with the drifting electrons.

Second, one may describe the electrons to have a wave nature. The matter waves may be thought to be scattered by lattice atoms. Scattering is the dissipation of radiation on small particles in all directions. The atoms absorb the energy of an incoming wave and thus become oscillators. These oscillators in turn re-emit the energy in the form of spherical waves. If two or more atoms are involved, the phase relationship between the individual re-emitted waves has to be taken into consideration. A calculation[2] shows that for a *periodic crystal structure* the individual waves in the forward direction are *in phase*, and thus interfere constructively. As a result, a wave which propagates through an ideal crystal (having periodically arranged atoms) does not suffer any change in intensity or direction. (Only its velocity is modified.) This mechanism is called *coherent scattering.*

If, however, the scattering centers are not periodically arranged (impurity atoms, vacancies, grain boundaries, thermal vibration of atoms, etc.) the scattered waves have no set phase relationship and the wave is said to be *incoherently scattered.* The energy of incoherently scattered waves is smaller in the forward direction.

The wave picture provides a deeper understanding of the electrical resistance in metals and alloys. In an ideal crystal, the electron waves are coherently scattered in the forward direction. As a consequence the electron wave passes without hindrance through the crystal. If, however, lattice defects, impurities, etc., are present, the matter waves are scattered and lose energy. This energy loss qualitatively explains the resistance. In the following two

[2] L. Brillouin, *Wave Propagation in Periodic Structures*, Dover, New York (1953).

sections we shall calculate the resistance using, at first, the particle and then the wave concept.

7.3. Conductivity—Classical Electron Theory

Our first approach towards an understanding of electrical conduction is to postulate, as Drude did, a free "electron gas" or "plasma," consisting of the valence electrons of the individual atoms in a crystal. We assume that in a monovalent metal, such as sodium, each atom contributes *one* electron to this plasma. The number of atoms, N_a, per cubic centimeter (and therefore the number of free electrons in a monovalent metal) can be obtained by applying

$$N_a = \frac{N_0 \delta}{M}, \tag{7.5}$$

where N_0 is the Avogadro constant, δ the density, and M the atomic mass of the element. One calculates about 10^{22} to 10^{23} atoms per cubic centimeter, i.e., 10^{22} to 10^{23} free electrons per cm^3 for a monovalent metal.

The electrons move randomly (in all possible directions) so that their individual velocities in the absence of an electric field cancel and no net velocity results. This situation changes when an electric field is applied. The electrons are then accelerated with a force $e\mathscr{E}$ towards the anode and a net drift of the electrons results, which can be expressed by a form of Newton's law ($F = ma$)

$$m\frac{dv}{dt} = e\mathscr{E}, \tag{7.6}$$

where e is the charge of the electrons and m is their mass. Equation (7.6) implies that as long as an electric field persists, the electrons are constantly accelerated. Equation (7.6) also suggests that after the field has been removed, the electrons keep drifting with constant velocity through the crystal. This is generally not observed, however, except for some materials at very low temperatures (superconductors). The *free electron model* needs, therefore, an adjustment to take into account the electrical resistance.

An electron, accelerated by an electric field, may be described to increase its drift velocity until it encounters a collision. At this time, the electron has acquired the velocity v_{max} which it may lose, all or in part, at the collision (Fig. 7.2(a)). Alternatively, and more appropriately, one may describe an electron motion to be counteracted by a "friction" force γv which opposes the electrostatic force $e\mathscr{E}$. We postulate that the resistance in metals and alloys is due to interactions of the drifting electrons with some lattice atoms, i.e., essentially with the imperfections in the crystal lattice (such as impurity atoms, vacancies, grain boundaries, dislocations, etc.). Thus, (7.6) is modified as follows:

$$m\frac{dv}{dt} + \gamma v = e\mathscr{E}, \tag{7.7}$$

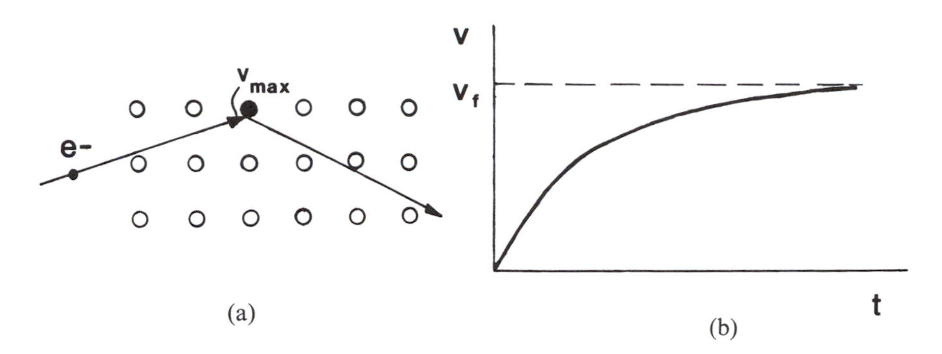

Figure 7.2. (a) Schematic representation of an electron collision with a lattice atom. (b) Velocity distribution of electrons due to an electrostatic force and a counteracting friction force. The electron eventually reaches the final velocity v_f.

where γ is a constant. The second term in (7.7) is a damping or *friction* force which contains the drift velocity, v, of the electrons. The electrons are thought to be accelerated until a final drift velocity v_f is reached (see Fig. 7.2(b)). At that time the electric field force and the friction force are equal in magnitude. In other words, the electrons are thought to move in a "viscous" medium.

For the steady state case ($v = v_f$) we obtain $dv/dt = 0$. Then (7.7) reduces to

$$\gamma v_f = e\mathscr{E}, \tag{7.8}$$

which yields

$$\gamma = \frac{e\mathscr{E}}{v_f}. \tag{7.9}$$

We insert (7.9) into (7.7) and obtain the complete equation for the drifting electrons under the influence of an electric field force and a friction force:

$$m\frac{dv}{dt} + \frac{e\mathscr{E}}{v_f}v = e\mathscr{E}. \tag{7.10}$$

The solution to this equation[3] is

$$v = v_f\left[1 - \exp\left(-\left(\frac{e\mathscr{E}}{mv_f}t\right)\right)\right]. \tag{7.11}$$

We note that the factor $mv_f/e\mathscr{E}$ in (7.11) has the unit of a time. It is customary to define this factor

$$\tau = \frac{mv_f}{e\mathscr{E}} \tag{7.12}$$

as a *relaxation time* (which can be interpreted as the average time between two

[3] The reader may convince himself of the correctness of this solution by inserting (7.11) and its first derivative by time into (7.10). Further, inserting $t \to \infty$ into (7.11) yields correctly $v = v_f$ (Fig. 7.2(b)). See also Problem 8.

consecutive collisions). Rearranging (7.12) yields

$$v_f = \frac{\tau e \mathscr{E}}{m}. \tag{7.13}$$

We make use of (7.4) which states that the current density j is proportional to the velocity of the drifting electrons and proportional to the number of free electrons N_f (per cm^3). This yields, with (7.2),

$$j = N_f v_f e = \sigma \mathscr{E}. \tag{7.14}$$

Combining (7.13) with (7.14) finally provides the sought-for equation for the conductivity

$$\boxed{\sigma = \frac{N_f e^2 \tau}{m}.} \tag{7.15}$$

Equation (7.15) teaches us that the conductivity is large for a large number of free electrons and for a large relaxation time. The latter is proportional to the mean free path between two consecutive collisions. The mean free path is defined to be

$$l = v\tau. \tag{7.15a}$$

7.4. Conductivity—Quantum Mechanical Considerations

It was stated above that the valence electrons perform, when in equilibrium, random motions with no preferential velocity in any direction. One can visualize this fact conveniently by plotting the velocities of the electrons in velocity space (Fig. 7.3(a)). The points inside a sphere (or inside a circle when

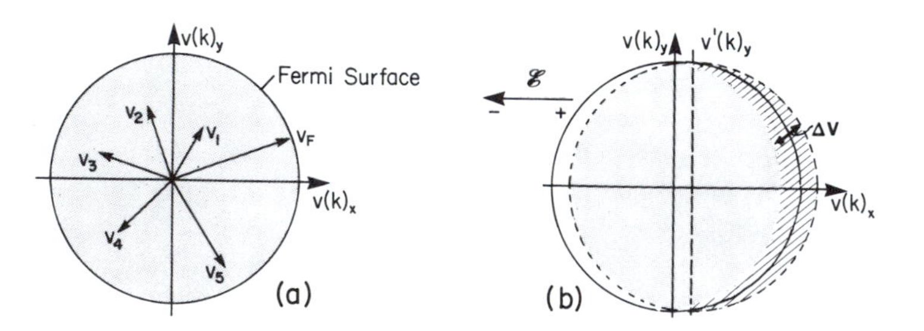

Figure 7.3. Velocity of electrons in two-dimensional velocity space. (a) Equilibrium and (b) when an electric field is applied. The shaded areas to the left and right of the $v(k)_y$-axis are of equal size. They cancel each other. The cross-hatched area remains uncompensated.

considering two dimensions) correspond to the endpoints of velocity vectors. The maximum velocity which the electrons are able to assume is the Fermi velocity v_F (i.e., the velocity of the electrons at the Fermi energy). The sphere having v_F as a radius represents, therefore, the Fermi surface. All points inside the Fermi sphere are occupied. As a consequence the velocity vectors cancel each other pairwise at equilibrium and no net velocity of the electrons results.

If an electric field is applied, the Fermi sphere is displaced opposite to the field direction, i.e., towards the positive end of the electric field, due to the net velocity gain of the electrons (Fig. 7.3(b) dashed circle). The great majority of the electron velocities still cancel each other pairwise (shaded area). However, some electrons remain uncompensated; their velocities $v + \Delta v$ are shown cross hatched in Fig. 7.3(b). These electrons cause the observed current. The Drude description of conduction thus needs a modification. In the *classical* picture one would assume that *all* electrons drift, under the influence of an electric field, with a modest velocity. Quantum mechanics, instead, teaches us that *only specific electrons* participate in conduction and that these electrons drift with a high velocity which is approximately the Fermi velocity v_F.

An additional point needs to be discussed and leads to an even deeper understanding. The largest energy which the electrons can assume in a metal at $T = 0$ is the Fermi energy E_F (Chapter 6). A large number of electrons actually possess this very energy since the density of states and thus the population density is highest around E_F (Fig. 7.4). Thus, only a little extra energy ΔE is needed to raise a substantial number of electrons from the Fermi level into slightly higher states. As a consequence, the energy (or the velocity) of electrons accelerated by the electric field $\mathscr{E}$ is only slightly larger than the Fermi energy E_F (or the Fermi velocity v_F) so that for all practical purposes the mean velocity can be approximated by the Fermi velocity v_F. We implied this fact already in our previous discussions.

We now calculate the conductivity by quantum mechanical means and apply, as before, Ohm's law $j = \sigma\mathscr{E}$, (7.2). The current density j is, as stated in

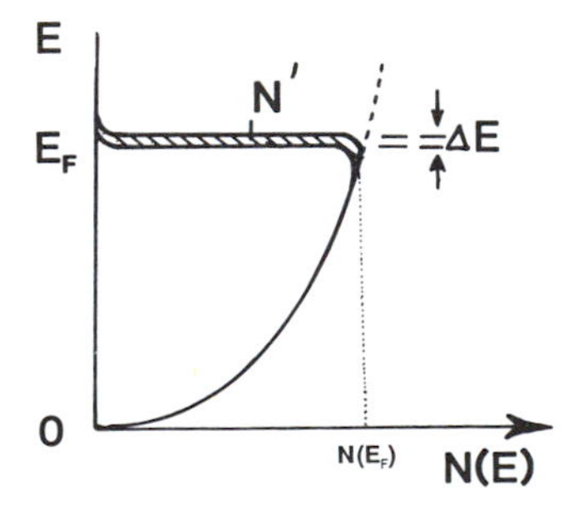

Figure 7.4. Population density $N(E)$ versus energy for free electrons (see Fig. 6.5) and displacement ΔE by an electric field. N' is the number of displaced electrons per unit volume (see (6.11a)) in the energy interval ΔE. $N(E)$ is, in the present case, also defined per unit volume, see (6.8).

(7.4), the product of the number of electrons, the electron velocity, and the electron charge. In our present case, we know that the velocity of the electrons which are responsible for the electron conduction is essentially the Fermi velocity v_F. Further, the number of electrons which need to be considered here is N', i.e., the number of displaced electrons per unit volume, as shown in Fig. 7.4. Thus, (7.4) needs to be modified to read

$$j = v_F e N'. \tag{7.16}$$

The number of electrons displaced by the electric field $\mathscr{E}$ is

$$N' = N(E_F)\Delta E \tag{7.17}$$

(see Fig. 7.4), which yields for the current density

$$j = v_F e N(E_F)\Delta E = v_F e N(E_F)\frac{dE}{dk}\Delta k. \tag{7.18}$$

The factor dE/dk is calculated by using the E versus $|\mathbf{k}|$ relationship known for free electrons (4.8), i.e.,

$$E = \frac{\hbar^2}{2m}k^2. \tag{7.19}$$

Taking the first derivative of (7.19) yields, with $k = p/\hbar$ (4.7),

$$\frac{dE}{dk} = \frac{\hbar^2}{m}k = \frac{\hbar^2 p}{m\hbar} = \frac{\hbar m v_F}{m} = \hbar v_F. \tag{7.20}$$

Inserting (7.20) into (7.18) yields

$$j = v_F^2 e N(E_F)\hbar\Delta k. \tag{7.21}$$

The displacement Δk of the Fermi sphere in k-space under the influence of an electric field can be calculated by using (7.6) and $p = \hbar k$ (4.7)

$$F = m\frac{dv}{dt} = \frac{d(mv)}{dt} = \frac{dp}{dt} = \hbar\frac{dk}{dt} = e\mathscr{E}, \tag{7.22}$$

which yields

$$dk = \frac{e\mathscr{E}}{\hbar}dt$$

or

$$\Delta k = \frac{e\mathscr{E}}{\hbar}\Delta t = \frac{e\mathscr{E}}{\hbar}\tau, \tag{7.23}$$

where τ is the time interval Δt between two "collisions" or the relaxation time (see Section 7.3). Inserting (7.23) into (7.21) yields

$$j = v_F^2 e^2 N(E_F)\mathscr{E}\tau. \tag{7.24}$$

One more consideration needs to be made. If the electric field vector points

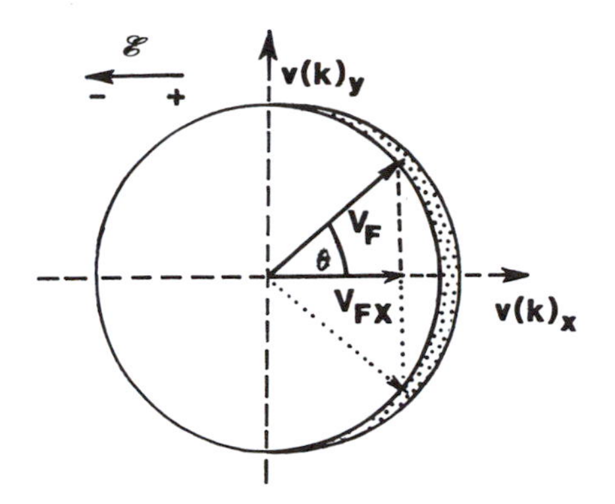

Figure 7.5. Two-dimensional velocity space.

in the negative $v(k)_x$ direction, then only the components of those velocities which are parallel to the positive $v(k)_x$ direction contribute to the electric current (Fig. 7.5). The $v(k)_y$ components cancel each other pairwise. In other words, only the projections of the velocities v_F on the positive $v(k)_x$-axis ($v_{Fx} = v_F \cos\theta$) contribute to the current. Thus, we have to sum up all contributions of the velocities in the first and fourth quadrant in Fig. 7.5 which yields

$$j = e^2 N(E_F)\mathscr{E}\tau \int_{-\pi/2}^{+\pi/2} (v_F \cos\theta)^2 \frac{d\theta}{\pi}$$

$$= e^2 N(E_F)\mathscr{E}\tau \frac{v_F^2}{\pi} \int_{-\pi/2}^{+\pi/2} \cos^2\theta \, d\theta$$

$$= e^2 N(E_F)\mathscr{E}\tau \frac{v_F^2}{\pi} \left[\tfrac{1}{4}\sin 2\theta + \frac{\theta}{2} \right]_{-\pi/2}^{+\pi/2},$$

$$j = \tfrac{1}{2} e^2 N(E_F)\mathscr{E}\tau v_F^2.$$

A similar calculation for a spherical Fermi surface yields

$$j = \tfrac{1}{3} e^2 N(E_F)\mathscr{E}\tau v_F^2. \tag{7.25}$$

Thus, the conductivity finally becomes, with $\sigma = j/\mathscr{E}$ (7.2),

$$\boxed{\sigma = \tfrac{1}{3} e^2 v_F^2 \tau N(E_F).} \tag{7.26}$$

This quantum mechanical equation reveals that the conductivity depends on the Fermi velocity, the relaxation time, and the population density (per unit volume). The latter is, as we know, proportional to the density of states. Equation (7.26) is more meaningful than the expression derived from the classical electron theory (7.15). Specifically, (7.26) contains the information that not *all* free electrons N_f are responsible for conduction, i.e., the conductivity in metals depends to a large extent on the population density of the

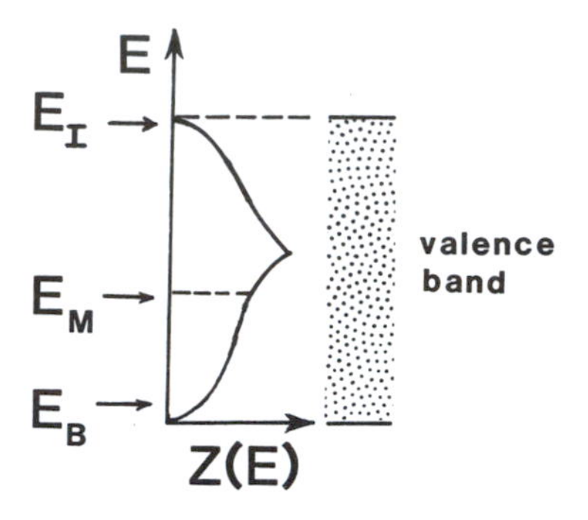

Figure 7.6. Schematic representation of the density of states (Fig. 6.6) and thus, with minor modifications, also the population density (6.7). Examples for highest electron energies for a monovalent metal (E_M), for a bivalent metal (E_B), and for an insulator (E_I) are indicated.

electrons near the Fermi surface. For example, monovalent metals (such as copper, silver, or gold) have partially filled valence bands, as shown in Figs. 5.22 or 6.7. Their electron population densities near their Fermi energy are high (Fig. 7.6), which results in a large conductivity according to (7.26). Bivalent metals, on the other hand, are distinguished by an overlapping of the upper bands and by a small electron concentration near the bottom of the valence band, as shown in Fig. 6.7(c). As a consequence the electron population near the Fermi energy is small (Fig. 7.6), which leads to a comparatively low conductivity. Finally, insulators and semiconductors have, under certain conditions, completely filled electron bands which results in a virtually zero population density near the top of the valence band (Fig. 7.6). Thus, the conductivity in these materials is extremely small.

7.5. Experimental Results and Their Interpretation

7.5.1. Pure Metals

The resistivity of a metal, such as copper, decreases linearly with decreasing temperature until it reaches a finite value (Fig. 7.7) according to the empirical equation

$$\rho_2 = \rho_1[1 + \alpha(T_2 - T_1)], \tag{7.27}$$

where α is the linear temperature coefficient of resistivity. We postulate that thermal energy causes lattice atoms to oscillate about their equilibrium positions, thus increasing the incoherent scattering of the electron waves. The residual resistivity ρ_{res} is interpreted to be due to imperfections in the crystal, such as impurities, vacancies, grain boundaries, or dislocations. The residual

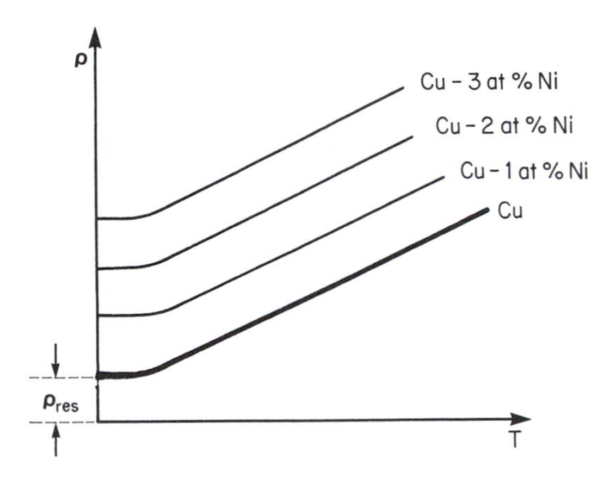

Figure 7.7. Schematic representation of the temperature dependence of the resistivity of copper and various copper–nickel alloys. ρ_{res} is the residual resistivity.

resistivity is essentially not temperature-dependent. According to Matthiessen's rule the resistivity arises from independent scattering processes which are additive, i.e.,

$$\rho = \rho_{th} + \rho_{imp} + \rho_{def} = \rho_{th} + \rho_{res}. \tag{7.28}$$

The thermally induced part of the resistivity ρ_{th} is called the *ideal* resistivity, whereas the resistivity which has its origin in impurities (ρ_{imp}) and defects (ρ_{def}) is summed up in the residual resistivity. The number of impurity atoms is generally constant in a given metal or alloy. The number of vacancies or grain boundaries, however, can be changed by various heat treatments. For example, if a metal is annealed at temperatures close to its melting point and then rapidly quenched into water at room temperature, its room temperature resistivity increases noticeably due to quenched-in vacancies. Frequently, this resistance increase diminishes during room temperature aging or annealing at slightly elevated temperatures due to the annihilation of some vacancies. Likewise, recrystallization, grain growth, and many other metallurgical processes change the resistivity of metals. As a consequence of this, and due to its simple measurement, the resistivity is one of the most widely studied properties in materials research.

It is interesting to compare the thermally induced change in conductivity in light of the quantum mechanical and classical models. The number of free electrons, N_f, essentially does not change with temperature. Likewise, $N(E)$ changes very little with T. However, the mean free path, and thus the relaxation time, decreases with increasing temperature (due to a large rate of collisions between the drifting electrons and the vibrating lattice atoms). This, in turn, decreases σ according to (7.15) and (7.26), in agreement with the observations in Fig. 7.7. Thus, both models accurately describe the temperature dependence of the resistivity.

7.5.2. Alloys

The resistivity of alloys increases with increasing amount of solute content (Fig. 7.7). The slopes of the individual ρ versus T lines remain, however, essentially constant. Small additions of solute cause a linear shift of the ρ versus T curves to higher resistivity values in accordance with Matthiessen's rule. This resistivity increase has its origin in several mechanisms. First, atoms of different size cause a variation in the lattice parameter and, thus, in electron scattering. Second, atoms having different valences introduce a local charge difference which also increases the scattering probability. Third, solutes which have a different electron concentration compared to the host element alter the position of the Fermi energy. This, in turn, changes the population density $N(E)$ according to (6.8) and thus the conductivity, see (7.26).

Various solute elements might alter the resistivity of the host material to different degrees. This is demonstrated in Fig. 7.8. Experiments have shown that the resistivity of *dilute single phase alloys* increases with the square of the valence difference between solute and solvent constituents (Linde's rule, Fig. 7.8(b)). Thus, the electron concentration of the solute element, i.e., the number of additional electrons the solute contributes, clearly plays a vital role in the resistance increase, as already mentioned above.

The isothermal resistivity of *concentrated* single phase alloys often has a maximum near 50% solute content, as shown in Fig. 7.9 (solid line). Specifically, the residual resistivity of these alloys depends, according to Nordheim's rule, on the fractional atomic compositions (X_A and X_B) of the constituents

$$\rho = X_A \rho_A + X_B \rho_B + C X_A X_B, \tag{7.29}$$

where C is a materials constant. Nordheim's rule holds strictly only for a few selected binary systems, because it does not take into consideration the changes in the density of states with composition. This is particularly true for alloys containing a transition metal.

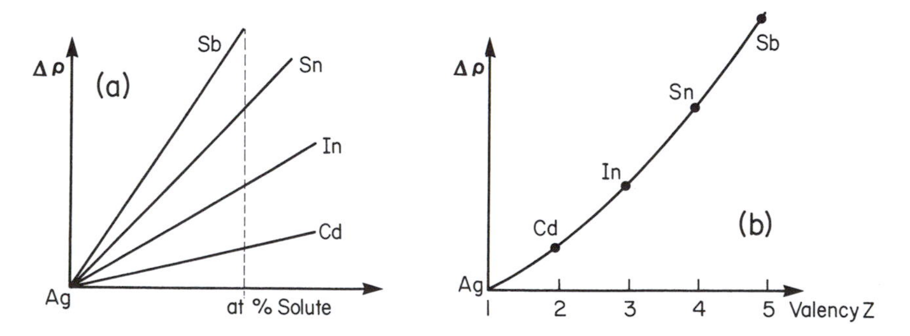

Figure 7.8. Resistivity change of various dilute silver alloys (schematic). Solvent and solute are all from the fifth period. (a) Resistivity change versus atomic % solute and (b) resistivity change due to 1 atomic % of solute.

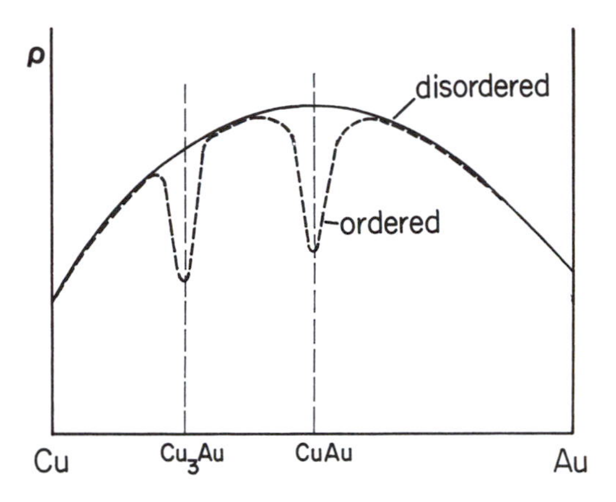

Figure 7.9. Schematic representation of the resistivity of ordered and disordered copper–gold alloys.

The resistivity of *two-phase alloys* is, in many instances, the sum of the resistivities of each of the components, taking the volume fractions of each phase into consideration. However, additional factors, such as the crystal structure and the kind of distribution of the phases in each other, have also to be considered. The concentration dependence of the resistivity of two-phase alloys does not exhibit a maximum, as in Fig. 7.9, but resembles instead a linear interpolation between the resistivities of the individual phases.

Some alloys (copper with small amounts of iron, for example) show a minimum in the resistivity at low temperatures. This anomaly is due to additional scattering of electrons by the magnetic moments of the solutes and is a deviation from the Matthiessen rule (Kondo effect).

7.5.3. Ordering

Solute atoms are generally randomly distributed in the solvent. Thus, the number of centers where incoherent scattering occurs increases proportionally with the number of substitutional atoms. If, however, the solute atoms are periodically arranged in the matrix, i.e., if, for example, in a 50/50 alloy the A and B atoms are alternately occupying successive lattice sites, then the electron waves are coherently scattered. This causes a decrease in resistivity (and an increase in the mean free path) (Fig. 7.9). Only selected alloys such as Cu_3Au, CuAu, Au_3Mn, etc., show a tendency towards long-range ordering.

The ordered state can be achieved by annealing an alloy of appropriate composition below the order–disorder transition temperature (about 395° C in Cu_3Au) or by slowly cooling from above the transition temperature. Annealing above this transition temperature destroys the ordering effect. In some alloys, however, such as in CuAu, the tendency towards ordering is so strong that even near the melting point some ordering remains. Long-range ordered alloys cause superlattice lines in X-ray patterns.

The disordered state can be retained at room temperature by quenching the alloy rapidly in ice brine from slightly above the transition temperature.

Some alloys, such as α-copper–aluminum, exhibit a much smaller resistance decrease due to annealing below a certain ordering temperature. This effect is called short-range ordering and has been found to be due to small domains in which the atoms are arranged in an ordered fashion. In the short-range ordered state the A–B interactions are slightly larger than the A–A or B–B interactions. (Short-range ordering can be identified by using small angle X-ray scattering. It causes small and broad intensity increases between the regular diffraction lines.[4])

7.5.4. Thermoelectric Phenomena

When two different types of conducting materials (e.g., metals) are brought into contact, electrons are transferred from the material with the higher Fermi energy (E_F) "down" into the material having a lower E_F. As a consequence, the material with the smaller E_F assumes a negative charge with respect to the material having a larger E_F. This results in a *contact potential* between the two materials (*Seebeck effect*).

The contact potential is temperature-dependent. Let us assume that two different types of wire (e.g., a copper and an iron wire) are connected at their ends to form a loop. One of the iron/copper contacts is brought to a higher temperature than the other iron/copper contact. Then a potential difference between these two "*thermocouples*" is observed which is

$$\Delta V = a(T - T_0) + b(T - T_0)^2, \tag{7.30}$$

where a and b are materials constants. The temperature dependence of ΔV (often called the thermoelectric power) of a thermocouple is

$$\Delta V/\Delta T = a + b(T - T_0). \tag{7.31}$$

For some combinations the constant b is essentially zero, so that the thermoelectric power can be considered to be a linear function of the temperature.

Thermocouples are frequently used as rigid, inexpensive, and fast probes for measuring temperatures even at remote locations. Several combinations of materials are commercially available which vary in sensitivity, temperature range, and price. Among them, copper and constantan (the latter alloy consisting of 55% Cu and 45% Ni) is quite popular because it is relatively sensitive and can be utilized between −180° C and +400° C.

A reversion of the Seebeck effect is the *Peltier* effect: An electric current which flows through a rod consisting of a metal combination of the type

---

[4] H. Warlimont, ed., *Order–Disorder Transformations in Alloys*, Springer-Verlag, Berlin (1974).

A/B/A causes a decrease in temperature on one junction and an increase in temperature on the other.

7.6. Superconductivity

Superconductors are materials whose resistivities become immeasurably small or actually become zero below a critical temperature T_c. The most sensitive measurements have shown that the resistance of these materials in the superconducting state is at least 10^{16} times smaller than their room temperature values. (See, in this context, Fig. 7.1.) So far, 27 elements, numerous alloys, ceramic materials (containing copper oxide), and organic compounds (based, e.g., on selenium or sulfur) have been discovered to possess superconductivity (see Table 7.1). Their critical T_c have values between 0.01 K and 125 K. Superconductors having a T_c above 77 K (boiling point of liquid nitrogen) are particularly interesting because they do not require liquid helium (boiling point 4 K) or liquid hydrogen (boiling point 20 K) for cooling. Some metals such as cesium become superconducting only if a large pressure is applied to them. The superconducting state has to be considered as a separate state such as the liquid, solid, or gaseous states. It has a higher degree of order—the entropy is zero. The superconducting transition is reversible.

A zero resistance combined with high current densities makes superconductors useful for strong electromagnets, e.g., in magnetic resonance imaging devices (used in medicine), high-energy particle accelerators, or electric power storage devices. (The latter can be appreciated by knowing that once an electrical current has been induced into a loop consisting of a superconducting wire, it continues to flow without significant decay for several weeks.)

Table 7.1. Critical Temperatures of Some Superconducting Materials.

Materials	T_c [K]	Remarks
Tungsten	0.01	—
Mercury	4.15	H.K. Onnes (1911)
Sulfur-based organic superconductor	8	S.S.P. Parkin et al. (1983)
V_3Si	17.1	J.K. Hulm (1953)
Nb_3Ge	23.2	(1973)
La–Ba–Cu–O	40	Bednorz and Müller (1986)
$YBa_2Cu_3O_{7-x}$ [a]	92	Wu, Chu, and others (1987)
$RBa_2Cu_3O_{7-x}$ [a]	~92	R = Gd, Dy, Ho, Er, Tm, Yb, Lu
$Bi_2Sr_2Ca_2Cu_3O_{10+\delta}$	113	Maeda et al. (1988)
$Th_2CaBa_2Cu_2O_{10+\delta}$	125	Hermann et al. (1988)

[a] The designation "1-2-3 compound" refers to the molar ratios of rare earth to alkaline earth to copper. (See chemical formula.)

Further potential applications are lossless power transmission lines, high-speed levitated trains, more compact and faster computers, or switching devices called cryotrons (based on the destruction of the superconducting state in a strong magnetic field). We shall cover the basic concepts for these applications in the following sections.

7.6.1. Experimental Results

When the temperature is lowered, the transition into the superconducting state is generally quite sharp for pure and structurally perfect elements (Fig. 7.10). A temperature range of less than 10^{-5} K has been observed in pure gallium. In alloys the transition may be spread, however, over a range of about 0.1 K. Ceramic superconductors generally display an even wider spread in transition temperatures.

The transition temperature T_c often varies with the atomic mass m_a according to

$$m_a^{\alpha} \cdot T_c = \text{const.}, \tag{7.32}$$

where α is a materials constant (Isotope effect). As an example, T_c for mercury varies from 4.185 K to 4.146 K when m_a changes from 199.5 to 203.4 atomic mass units.

Removal of the superconducting state does not only occur by raising the temperature, but also by subjecting the material to a magnetic field. The critical magnetic field strength, H_c, above which superconductivity is destroyed, depends upon the temperature to which the metal has been cooled. In general, the lower the sample temperature, the higher the critical field H_c (Fig. 7.11). One finds

$$H_c = H_0\left(1 - \frac{T^2}{T_c^2}\right), \tag{7.33}$$

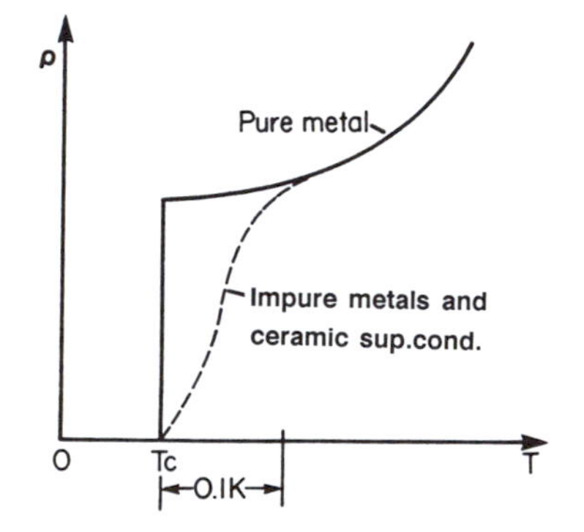

Figure 7.10. Schematic representation of the resistivity of pure and impure superconducting elements. T_c is the transition or critical temperature.

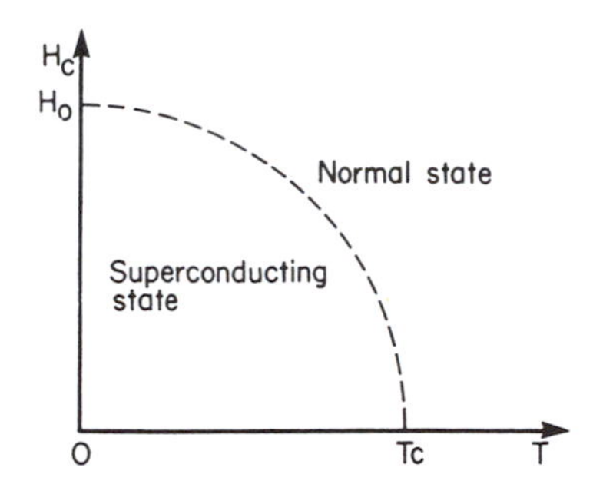

Figure 7.11. Dependence of critical field strength H_c at which superconductivity is destroyed, in relation to the temperature of the specimen.

where H_0 is the critical magnetic field strength at 0 K. Ceramic superconductors usually have a smaller H_c than metallic superconductors, i.e., they are more vulnerable to lose superconductivity by a moderate magnetic field.

We mentioned above that superconductors are utilized to produce electromagnets which have a high magnetic field strength and no loss in power due to heat dissipation. One limiting factor for ultrahigh field strengths, however, is that the magnetic field thus produced can reach H_c, so that the superconducting state is eventually destroyed by its own magnetic field.

In general, two classes of superconducting materials are distinguished. In *type* I *superconductors* the destruction of the superconducting state by a magnetic field, i.e., the transition between the superconducting and normal state, occurs sharply (Fig. 7.12). The critical field strength H_c is relatively low. Thus, type I superconductors are generally not used for coils for superconducting magnets. In *type* II *superconductors* the destruction of the superconducting state by a magnetic field is gradual. The superconducting properties are extended to a field H_{c2} which might be 100 times higher than H_{c1} (Fig. 7.13). Because of this stronger resistance against the magnetically induced destruction of the superconducting state, type II superconductors are mainly utilized for superconducting solenoids. Magnetic fields of several hundred

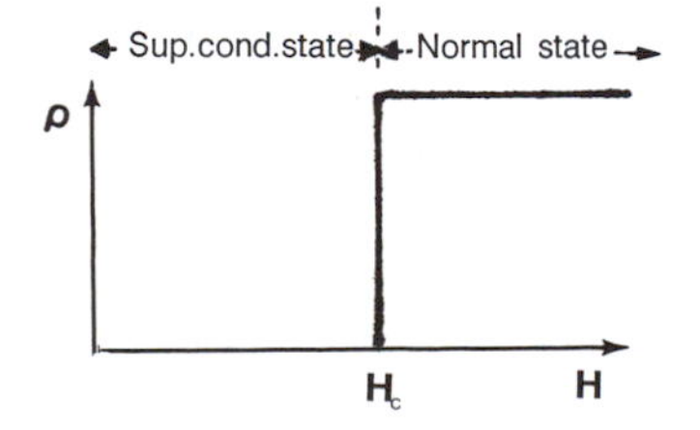

Figure 7.12. Schematic representation of the resistivity of a *type I* (or soft) superconductor when a magnetic field of field strength H is applied. These solids behave like normal conductors above H_c.

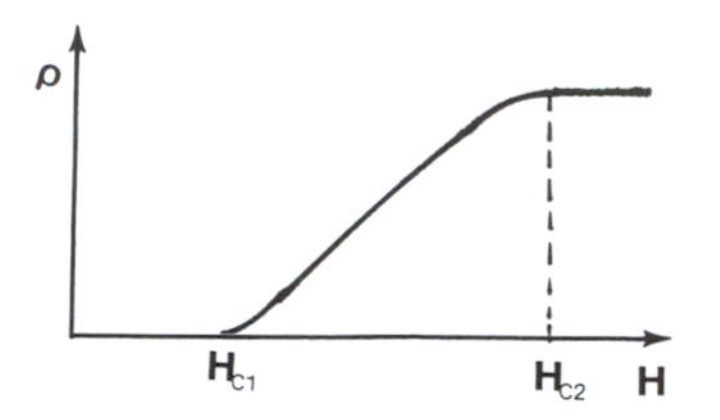

Figure 7.13. Schematic representation of the resistivity of a *type II* (or hard) superconductor. The region between H_{c1} and H_{c2} is called the *vortex state*. Above H_{c2}, the solid behaves like a normal conductor.

kiloGauss have been achieved with these materials. Among the type II superconductors are transition metals and alloys consisting of niobium, aluminum, silicon, or vanadium. Ceramic superconductors also belong to this group. (The terms "type I or type II superconductors" are often used likewise when the abrupt or gradual transition with respect to *temperature* is described, see Fig. 7.10).

The interval between H_{c1} and H_{c2} represents a state in which superconducting and normal conducting areas are mixed in the solid. Specifically, one observes small circular regions, called vortices, which are in the normal state. They are surrounded by large, superconducting regions.

It is noted in passing that superconducting materials have exceptional magnetic properties. For example, a permanent magnet levitates in mid-air above a piece of a superconducting material which is cooled below T_c. We shall return to the magnetic properties of superconductors in Section 15.1.1.

Ceramic superconductors seem to be characterized by two-dimensional sheets, a Cu–O nonstoichiometry (i.e., a limited amount of an oxygen deficiency, see Fig. 7.14), a reduced lattice parameter between the copper atoms, and a tetragonal (high temperature) to orthorhombic (room temperature) transition. Only the orthorhombic modification is superconducting. Further, ceramic superconductors appear to be antiferromagnetic (see Section 15.1.4). Thus the superconductivity is most likely connected to the entire lattice structure.

Despite the considerably higher transition temperatures, ceramic superconductors have not yet revolutionized new technologies, mainly because of their inherent brittleness, their incapability of carrying high current densities, and their environmental instability. These obstacles may be overcome eventually, e.g., by using bismuth-based materials which are capable of carrying high currents when cooled to about 20 K or by utilizing composite materials, i.e., by inserting the ingredient oxide powders into silver tubes and sintering them *after* plastic deformation (e.g., wire pulling). Other techniques employ depositions of ceramic superconducting films on ductile substrates. Additions of silver into some ceramic superconductors improve their environmental stability (by reducing the porosity of the material) without lowering T_c. In any event, the further development of superconducting materials should be followed with great anticipation.

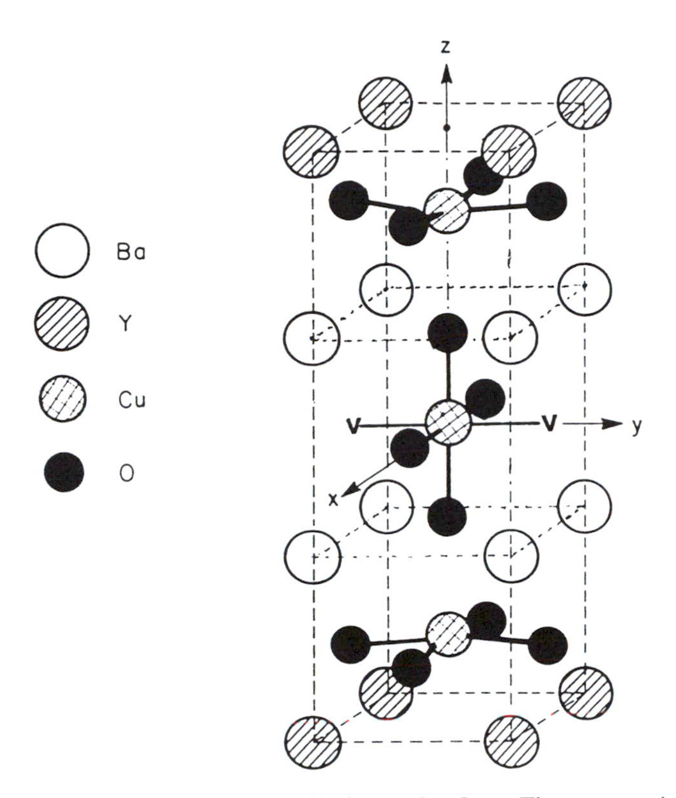

Figure 7.14. Room temperature unit cell of $YBa_2Cu_3O_{7-x}$. The structure is an orthorhombic layered perovskite ($BaTiO_3$), containing periodic oxygen vacancies. Two examples for oxygen vacancies are indicated by a "V." Adapted from M. Stavola, *Phys. Rev. B*, **36**, 850 (1987).

*7.6.2. Theory

Attempts to explain superconductivity have been made since its discovery in 1911. One of these theories makes use of the two-fluid model which postulates superelectrons which experience no scattering, have zero entropy (perfect order), and long coherence lengths, i.e., an area 1000 nm wide over which the superelectrons are spread. The London theory is semiphenomenological and dwells basically on the electrodynamic properties. The BCS theory (which was developed in 1957 by Bardeen, Cooper, and Schrieffer) is capable of explaining all superconduction properties reasonably well. The theory, however, is quite involved. Phenomenological descriptions of the concepts leading to this theory are probably simplifications of the actual mechanisms which govern superconduction and may thus provide temptations for misleading conclusions. (As so often in quantum mechanics, the mathematics is right—it is only our lack of imagination which holds us back from correctly interpreting the equations.) Nevertheless, a conceptual description of the BCS theory and its results is attempted.

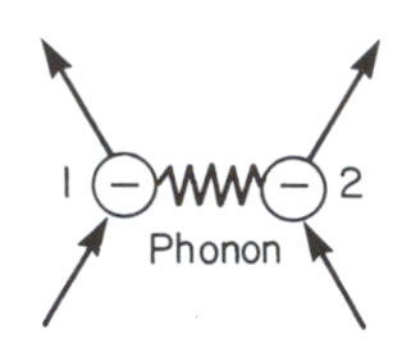

Figure 7.15. Schematic of a Cooper pair.

One key to the understanding of the BCS theory is accepting the existence of a pair of electrons (Cooper pair) which has a lower energy than two individual electrons. Imagine an electron in a metal at $T = 0$ K (no lattice vibrations). This electron perturbs the lattice slightly in its neighborhood. When such an electron drifts through a crystal the perturbation is only momentary and, after passing, a displaced ion reverts back into its original position. One can consider this ion to be held by springs in its lattice position so that after the electron has passed by, the ion does not simply return to its original site, but overshoots and eventually oscillates around its rest position. A *phonon* is created.[5] This phonon in turn interacts quickly with a second electron, which takes advantage of the deformation and lowers its energy. Electron 2 finally emits a phonon by itself which interacts with the first electron and so on. It is this passing back and forth of phonons which couples the two electrons together and brings them into a lower energy state (Fig. 7.15). One can visualize that all electrons on the Fermi surface having opposite momentum and opposite spin (i.e., $k\uparrow$ and $-k\downarrow$) form those Cooper pairs (Fig. 7.16), so that these electrons form a cloud of Cooper pairs which drift cooperatively through the crystal. Thus, the superconducting state is an *ordered* state of the *conduction electrons*. The scattering on the lattice atoms is eliminated, thus causing a zero resistance, as described similarly in Section 7.5.3 where we observed that ordering of the *atoms* in a crystal lattice reduces the resistivity.

[5] A *phonon* is a lattice vibration quantum. We will describe the properties of phonons in Chapter 20.

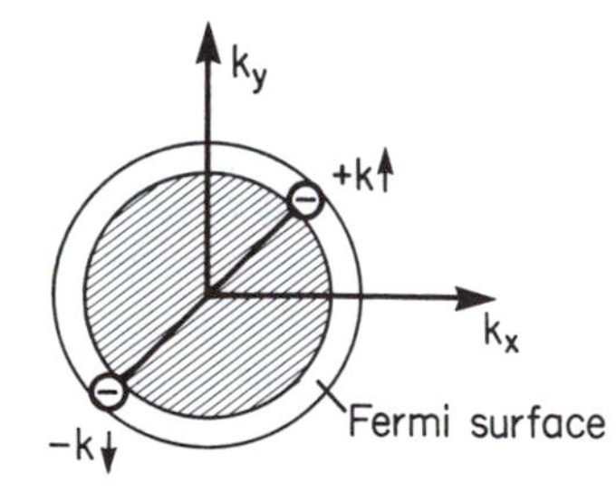

Figure 7.16. Fermi sphere, Fermi surface, and Cooper pair in a metal.

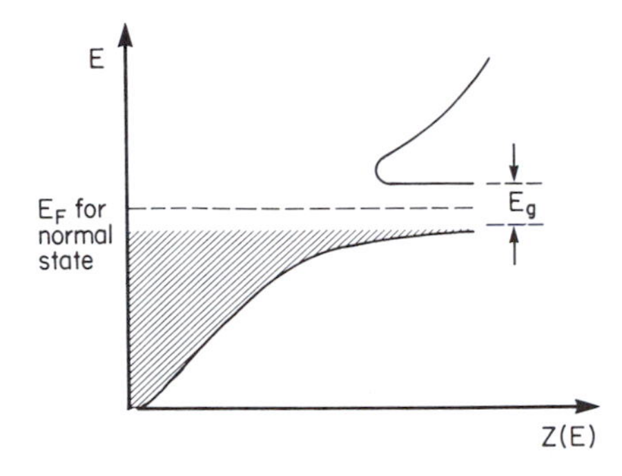

Figure 7.17. Density of states, $Z(E)$, versus electron energy in the superconducting state.

One further aspect has to be considered. We just mentioned that the electrons of a Cooper pair have a lower energy than two unpaired electrons. Thus, the Fermi energy in the superconducting state may be considered to be lower than that for the nonsuperconducting state. This lower state is separated from the normal state by an energy gap E_g (Fig. 7.17). The energy gap stabilizes the Cooper pairs against small changes of net momentum, i.e., prevents them from breaking apart. Such an energy gap of about 10^{-4} eV has indeed been observed by impinging IR radiation on a superconductor at temperatures below T_c and observing an onset of absorption of the IR radiation.

An alternate method for measuring this gap energy is by utilizing the Josephson effect. The experiment involves two pieces of metal, one in the superconducting state and the other in the normal state. They are separated by a thin insulating film of about 1 nm thickness (Fig. 7.18(a)). A small voltage of proper polarity in the millivolt range applied to this device raises the energy bands in the superconductor. Increasing this voltage eventually leads to a configuration where some filled electron states in the superconductor are opposite to empty states in the normal conductor (Fig. 7.18(b)). Then the Cooper pairs are capable of tunneling across the junction similarly as described in Section 4.3. The gap energy is calculated from the threshold voltage at which the tunneling current starts to flow.

In closing, we would like to revisit the electron–phonon coupling mechanism which is believed to be the essential concept for the interpretation of superconduction, at least for metals and alloys. It has been explained above that in the normal state of conduction (above T_c) strong interactions between electrons and phonons would lead to collisions (or scattering of the electron waves), and thus to electrical resistance, whereas at low temperatures the same interactions would cause Cooper pairs to form and thus promote superconduction. This would explain why the noble metals (which have small electron–phonon interactions) are not superconducting. In other words, poor conductors in the normal state of conduction are potential candidates for high T_c superconductors (and vice versa). Ceramic and organic superconduc-

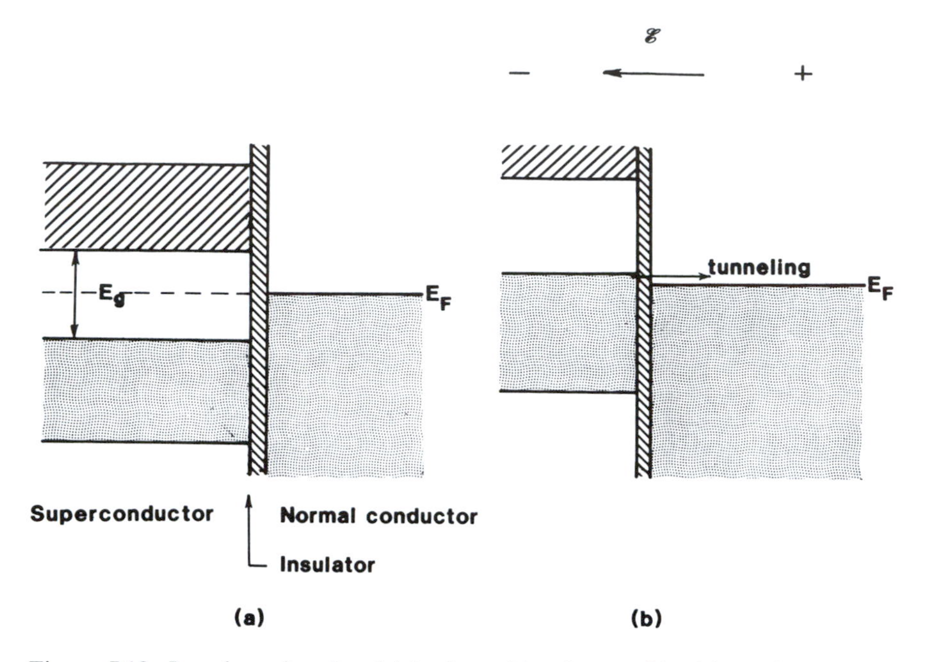

Figure 7.18. Josephson junction (a) in the unbiased state (b) with applied voltage across the junction which facilitates tunneling in the indicated direction.

tors fit into this scheme. Still, some scientists believe that phonons are involved in the coupling process only at very low temperatures (e.g., below 40 K). At somewhat higher temperatures, when phonons cause substantial scattering of the electrons, excitons (i.e., electron-hole pairs) may link electrons to form Cooper pairs, as suggested by A. Little for organic superconductors. Still other scientists propose resonating valence bonds as a coupling mechanism for high T_c superconductors.

Problems

1. Calculate the number of free electrons for gold using its density and its atomic mass.
2. Does the conductivity of an alloy change when long-range ordering takes place? Explain.
3. Calculate the time between two collisions and the mean free path for pure copper at room temperature. Discuss whether or not this result makes sense. *Hint*: Take the velocity to be the Fermi velocity, v_F, which can be calculated from the Fermi energy of copper $E_F = 7$ eV. Use otherwise classical considerations.
4. Electron waves are "coherently scattered" in ideal crystals at $T = 0$. What does this mean? Explain why in an ideal crystal at $T = 0$ the resistivity is small.

5. Calculate the number of free electrons per cubic centimeter (and per atom) for sodium from resistance data (relaxation time 3.1×10^{-14} s).

6. Give examples for coherent and incoherent scattering.

7. When calculating the population density of electrons for a metal by using (7.26), a value much larger than immediately expected results. Why does the result, after all, make sense? (Take $\sigma = 5 \times 10^5$ 1/Ω cm; $v_F = 10^8$ cm/s and $\tau = 3 \times 10^{-14}$ s.)

8. Solve the differential equation

$$m\frac{dv}{dt} + \frac{e\mathscr{E}}{v_F}v = e\mathscr{E} \qquad (7.10)$$

   and compare your result with (7.11).

9. Consider the conductivity equation obtained by the classical electron theory. According to this equation, a bivalent metal, such as zinc, should have a larger conductivity than a monovalent metal, such as copper, because zinc has about twice as many free electrons than copper. Resolve this discrepancy by considering the quantum mechanical equation for conductivity.

CHAPTER 8

Semiconductors

8.1. Band Structure

We have seen in Chapter 7 that *metals* are characterized by *partially* filled *valence* bands and that the electrons in these bands give rise to electrical conduction. On the other hand, the valence bands of *insulators* are *completely* filled with electrons. Semiconductors, finally, represent in some respect a position in between metals and insulators. We mentioned in Chapter 6 that semiconductors have, at low temperatures, a completely filled valence band and a narrow gap between this and the next higher, unfilled band. The latter one is called the *conduction band*. We discuss this now in more detail.

Because of band overlapping, the valence as well as the conduction bands of semiconductors consist of mixed (hybrid) *s*- and *p*-states. The eight highest $s + p$ states (two *s*- and six *p*-states)[6] split into two separate $(s + p)$ bands,[6] each of which consists of one *s*- and three *p*-states (see Fig. 8.1). The lower *s*-state can accommodate one electron per atom, whereas the three lower *p*-states can accommodate three electrons per atom. The valence band can, therefore, accommodate $4N_a$ electrons. (The same is true for the conduction band.) Because germanium and silicon possess four valence electrons per atom (group IV of the periodic table), the valence band is completely filled with electrons and the conduction band remains empty.

A deeper understanding of this can be gained from Fig. 8.2 which depicts part of a calculated band structure for silicon. Consider at first that electrons are "filled" into these bands like water being poured into a vessel. Then, of course, the lowest *s*-state will be occupied first. Since no energy gap exists

[6] See Appendix 3.

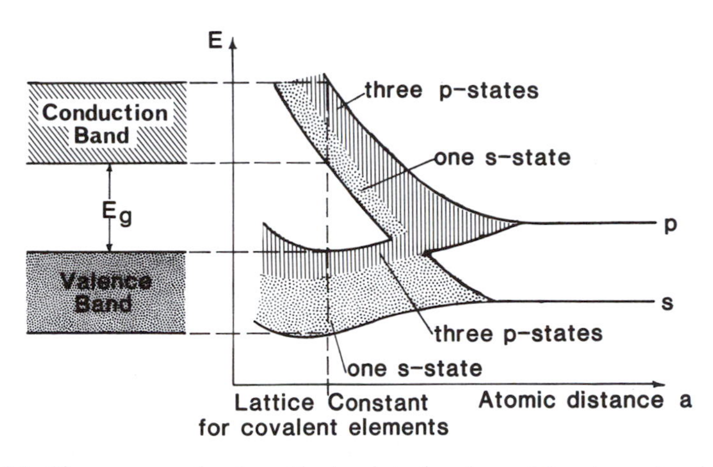

Figure 8.1. Sharp energy levels, widening into bands, and band overlapping with decreasing atomic distance for covalent elements. (Compare with Fig. 4.14.)

between the top of the *s*-state and the next higher *p*-state, additional electrons will immediately start to occupy the *p*-states. This process proceeds until all three lower *p*-states are filled. All of the $4N_a$ electrons of the semiconductor are accommodated now. Note that no higher energy band touches the *p*-states of the valence band. Thus, an energy gap exists between the filled valence and the empty conduction band. (As was shown in Fig. 5.23 the bands in different directions in **k**-space usually have a different shape so that a complete assessment can only be made by inspecting the entire structure.)

All materials which have bonds characterized by electron sharing (covalent bonds) have in common the above-mentioned hybrid bands (Fig. 8.1). An

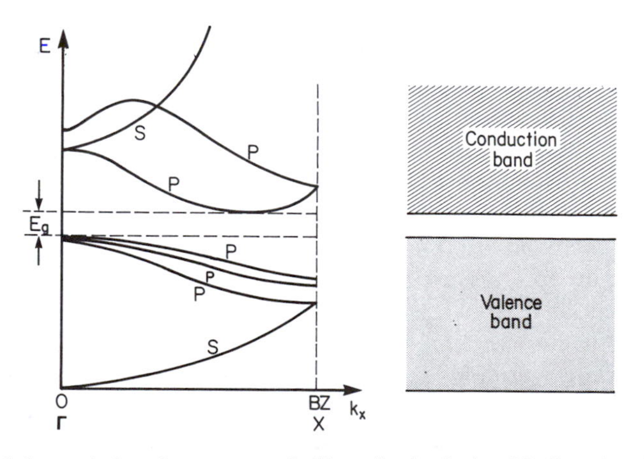

Figure 8.2. Schematic band structure of silicon in the k_x (or X) direction (plotted in the reduced zone scheme). The separation of the two highest *p*-states in the valence band is strongly exaggerated. Compare with the complete band structure of Fig. 5.23.

Table 8.1. Gap Energies for Some Covalent Elements at 0 K (see also Appendix 4).

Element	E_g [eV]
C (diamond)	5.48
Si	1.1
Ge	0.7
Sn (gray)	0.08

important difference is the magnitude of the gap energy E_g between the conduction band and the valence band. As can be seen from Table 8.1, the gap energies for group IV elements decrease with increasing atomic number. Diamond, for example, has a gap energy of 5.48 eV and is, therefore, an insulator (at least at room temperature) whereas the E_g for silicon and germanium is around 1 eV. Gray tin, finally, has an energy gap of only 0.08 eV. (It should be noted in passing that the utilization of diamond as an extrinsic semiconductor has been recently contemplated, see Section 8.3.)

8.2. Intrinsic Semiconductors

Semiconductors become conducting at elevated temperatures. In an intrinsic semiconductor, the conduction mechanism is predominated by the properties of the *pure* crystal. In order for a semiconductor to become conducting, electrons have to be excited from the valence band into the conduction band where they can be accelerated by an external electric field. Likewise, the electron holes which are left behind in the valence band contribute to the conduction. They migrate in the opposite direction to the electrons. The energy for the excitation of the electrons from the valence band into the conduction band stems usually from thermal energy. The electrons are transferred from one band into the next by *interband transitions.*

We turn now to a discussion of the *Fermi energy* in semiconductors. We learned in Section 6.2 that the Fermi energy is that energy for which the Fermi distribution function equals $\frac{1}{2}$. (It is advisable to keep only this "definition" of the Fermi energy in mind. Any other definition which might give a correct understanding for metals could cause confusion for semiconductors!) The probability that any state in the valence band of an intrinsic semiconductor at $T = 0$ K is occupied by electrons is 100%, i.e., $F(E) = 1$ for $E < E_V$ (Fig. 8.3). At higher temperatures, however, some of the electrons close to the top

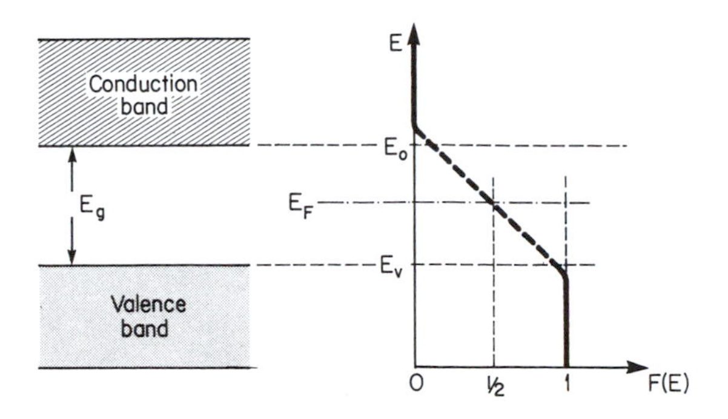

Figure 8.3. Schematic Fermi distribution function and Fermi energy for an intrinsic semiconductor for $T > 0$ K. The "smearing out" of the Fermi distribution function at E_0 and E_V is exaggerated. For reasons of convenience, the zero-point of the energy scale is placed at the bottom of the conduction band.

of the valence band have been excited into the conduction band. As a consequence, the probability function $F(E)$ is slightly reduced at the top of the valence band for $T > 0$ K.

On the other hand, no electrons are found at $T = 0$ K in the conduction band. Thus, the Fermi distribution function for $E > E_0$ must be zero. Again, for higher temperatures, a small deviation from $F(E) = 0$ near the bottom of the conduction band is expected (Fig. 8.3). The connection between the two branches of the $F(E)$ curve just discussed is marked with a dashed line in Fig. 8.3. This connecting line does *not* imply that electrons can be found in the forbidden band since $F(E)$ is merely the probability of occupancy of an *available* energy state. (A detailed calculation provides a slightly modified $F(E)$ curve whose vertical branches extend further into the forbidden band.)

Our discussion leads to the conclusion that the Fermi energy, E_F (i.e., that energy where $F(E) = \frac{1}{2}$), is located in the center of the forbidden band. In other words, for intrinsic semiconductors we find $E_F = -E_g/2$.

We may also argue somewhat differently: For $T > 0$ K the same amount of current carriers can be found in the valence as well as in the conduction band. Thus, the *average* Fermi energy has to be halfway between these bands. A simple calculation confirms this statement. (Problem 3 in this chapter should be worked at this point to deepen the understanding.) We implied in our consideration that the effective masses of electrons and holes are alike (which is not the case; see Appendix 4).

Of special interest to us is the number of electrons in the conduction band. From the discussion carried out above, we immediately suspect that a large number of electrons can be found in the conduction band if E_g is small and, in addition, if the temperature is high. In other words, we suspect that the number of electrons in the conduction band is a function of E_g and T. A detailed calculation, which we will carry out now, verifies this suspicion.

In Section 6.4 we defined N^* to be the number of electrons which have an energy equal to or smaller than a given energy E_n. For an energy interval between E and $E + dE$, we obtained, (6.9),

$$dN^* = N(E)\,dE, \tag{8.1}$$

where

$$N(E) = 2 \cdot Z(E) \cdot \mathrm{F}(E) \tag{8.2}$$

was called the population density (6.7) and

$$Z(E) = \frac{V}{4\pi^2}\left(\frac{2m}{\hbar^2}\right)^{3/2} E^{1/2} \tag{8.3}$$

is the density of states (6.5). In our particular case, the Fermi distribution function $\mathrm{F}(E)$ can be approximated by

$$\mathrm{F}(E) = \frac{1}{\exp\left(\frac{E - E_\mathrm{F}}{k_\mathrm{B} T}\right) + 1} \simeq \exp\left[-\left(\frac{E - E_\mathrm{F}}{k_\mathrm{B} T}\right)\right] \tag{8.4}$$

because $E - E_\mathrm{F}$ is about 0.5 eV and $k_\mathrm{B} T$ at room temperature is of the order of 10^{-2} eV. Therefore, the exponential factor is large compared to 1 (Boltzmann tail). We integrate over all available electrons which have energies larger than the energy at the bottom of the conduction band ($E = 0$), and obtain, with (8.1), (8.3), and (8.4),[7]

$$N^* = \frac{V}{2\pi^2} \cdot \left(\frac{2m}{\hbar^2}\right)^{3/2} \int_0^\infty E^{1/2} \cdot \exp\left[-\left(\frac{E - E_\mathrm{F}}{k_\mathrm{B} T}\right)\right] dE \tag{8.5}$$

or

$$N^* = \frac{V}{2\pi^2} \cdot \left(\frac{2m}{\hbar^2}\right)^{3/2} \exp\left(\frac{E_\mathrm{F}}{k_\mathrm{B} T}\right) \int_0^\infty \cdot E^{1/2} \cdot \exp\left[-\left(\frac{E}{k_\mathrm{B} T}\right)\right] dE. \tag{8.6}$$

Integration[8] yields

$$\begin{aligned} N^* &= \frac{V}{2\pi^2}\left(\frac{2m}{\hbar^2}\right)^{3/2} \exp\left(\frac{E_\mathrm{F}}{k_\mathrm{B} T}\right) \frac{k_\mathrm{B} T}{2} (\pi k_\mathrm{B} T)^{1/2} \\ &= \frac{V}{4}\left(\frac{2m k_\mathrm{B} T}{\pi \hbar^2}\right)^{3/2} \exp\left(\frac{E_\mathrm{F}}{k_\mathrm{B} T}\right). \end{aligned} \tag{8.7}$$

Introducing $E_\mathrm{F} = -E_\mathrm{g}/2$ (see above) and the effective mass m^* (Section 6.7), and equating $m_\mathrm{e}^* \equiv m_\mathrm{h}^*$, then we obtain, for the number of conduction-band

---

[7] The integration should actually be done over the states in the conduction band only. However, since the probability factor $\mathrm{F}(E)$ is rapidly approaching zero for energies $E > E_\mathrm{F}$, the substitution of infinity for the upper limit does not change the result appreciably. This substitution brings the integral into a standard form.

[8] $\int_0^\infty X^{1/2} e^{-nx}\,dx = 1/2n\sqrt{\pi/n}$.

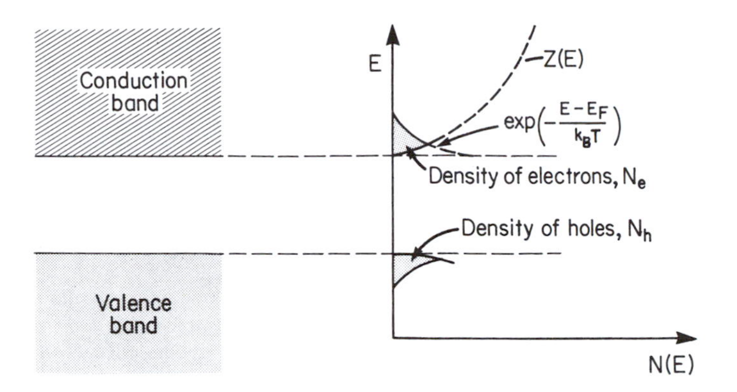

Figure 8.4. Density of electrons (N_e) and holes (N_h) for an intrinsic semiconductor.

electrons per unit volume $N_e = N^*/V$,

$$N_e = \frac{1}{4}\left(\frac{2mk_B}{\pi\hbar^2}\right)^{3/2}\left(\frac{m_e^*}{m_0}\right)^{3/2} T^{3/2} \exp\left[-\left(\frac{E_g}{2k_B T}\right)\right]. \tag{8.8}$$

The constant factor $\frac{1}{4}\left(\frac{2mk_B}{\pi\hbar^2}\right)^{3/2}$ has the value 4.84×10^{15} ($\mathrm{cm}^{-3}\ \mathrm{K}^{-3/2}$) so that we can write for (8.8)

$$N_e = 4.84 \times 10^{15}\left(\frac{m_e^*}{m_0}\right)^{3/2} T^{3/2} \exp\left[-\left(\frac{E_g}{2k_B T}\right)\right]. \tag{8.9}$$

We see from (8.9) that the number of electrons in the conduction band is a function of the energy gap and the temperature, as expected. We further notice that the contribution of a temperature increase to N_e is mostly due to the exponential term and only, to a lesser extent, due to the term $T^{3/2}$. A numerical evaluation of (8.9) tells us that the number of electrons per cubic centimeter in silicon at room temperature is about 10^9 (see Problem 1). This means that, at room temperature, only one in every 10^{13} atoms contributes an electron to the conduction. We shall see in the next section that in impurity semiconductors many more electrons can be found in the conduction band.

The electron and hole density is shown in Fig. 8.4 for an intrinsic semiconductor. The number of electrons is given by the area enclosed by the $Z(E)$ curve and $F(E) = \exp[-(E - E_F)/k_B T]$ (8.5).

As implied before, the number of electrons in the conduction band must equal the number of holes in the valence band. This means that an identical equation to (8.8) can be written for the holes if we assume $m_e^* \equiv m_h^*$, which is not strictly true.[9] (An additional term, which is usually neglected, modifies E_F slightly.)

[9] For numerical values, see the tables in Appendix 4.

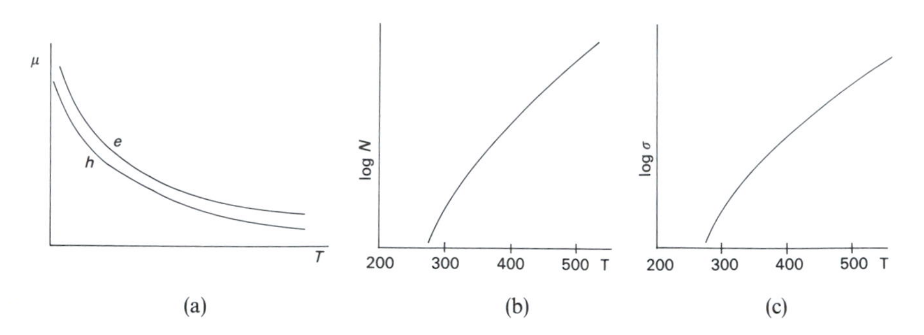

Figure 8.5. Schematic representation of the temperature dependence of (a) electron and hole mobilities, (b) number of carriers in an intrinsic semiconductor, and (c) conductivity for an intrinsic semiconductor. (T is given in degrees Kelvin.)

The conductivity[9] of an intrinsic semiconductor is not determined by the number of electrons and holes alone. The mobility[9] (μ) of the current carriers,

$$\mu = \frac{v}{\mathscr{E}}, \tag{8.10}$$

i.e., their (drift) velocity per unit electric field, also contributes its share to the conductivity, σ. An expression for the conductivity is found by combining (7.2),

$$j = \sigma\mathscr{E}, \tag{8.11}$$

and (7.4),

$$j = Nve, \tag{8.12}$$

with (8.10), which yields

$$\sigma = N\frac{v}{\mathscr{E}}e = N\mu e. \tag{8.13}$$

Taking both electrons and holes into consideration we can write

$$\begin{aligned} \sigma &= N_e e\mu_e + N_h e\mu_h, \\ \sigma &= 4.84 \times 10^{15}\left(\frac{m^*}{m_0}\right)^{3/2} T^{3/2} e(\mu_e + \mu_h)\exp\left[-\left(\frac{E_g}{2k_B T}\right)\right], \end{aligned} \tag{8.14}$$

where the subscripts e and h stand for electrons and holes, respectively. With increasing temperatures, the mobility of the current carriers is reduced by lattice vibrations (Fig. 8.5(a)). On the other hand, around room temperature, an increasing amount of electrons is excited from the valence band into the conduction band, thus increasing the number of current carriers, N_e and N_h (Fig. 8.5(b)). The conductivity is, according to (8.14), a function of these two factors (Fig. 8.5(c)).

At low temperatures the electrons are incoherently scattered by impurity atoms and lattice defects. It is therefore imperative that semiconductor materials are of extreme purity. Methods to achieve this high purity will be discussed in Section 8.7.11.

8.3. Extrinsic Semiconductors

8.3.1. Donors and Acceptors

We learned in the previous section that in intrinsic semiconductors only a very small number of electrons (about 10^9 electrons per cubic centimeter) contribute to the conduction of the electric current. In most semiconductor devices, a considerably higher number of charge carriers are, however, present. They are introduced by *doping*, i.e., by adding small amounts of impurities to the semiconductor material. In most cases, elements of group III or V of the periodic table are used as dopants. They replace some regular lattice atoms in a substitutional manner. Let us start our discussion by considering the case where a small amount of phosphorus (e.g., 0.0001%) is added to silicon. Phosphorus has five valence electrons, i.e., one valence electron more than silicon. Four of these valence electrons form regular electron-pair bonds with their neighboring silicon atoms (Fig. 8.6). The fifth electron, however, is only loosely bound to silicon, i.e., the binding energy

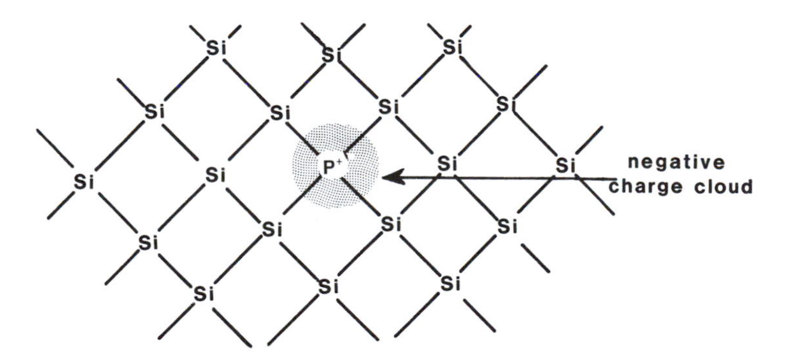

Figure 8.6. Two-dimensional representation of the silicon lattice. An impurity atom of group V of the periodic table (P) is shown to replace a silicon atom. The charge cloud around the phosphorus atom stems from the extra phosphorus electron. Each electron pair between two silicon atoms constitutes a covalent bond (electron sharing). The two electrons of such a pair are indistinguishable, but must have opposite spin to satisfy the Pauli principle.

is about 0.045 eV (see Appendix 4 and Problem 10.) At slightly elevated temperatures this extra electron becomes disassociated from its atom and drifts through the crystal as a conduction electron when a voltage is applied to the crystal. Extra electrons of this type are called "*donor electrons.*" They populate the conduction band of a semiconductor, thus providing a contribution to the conduction process.

It has to be noted that at sufficiently high temperatures, in addition to these donor electrons, some electrons from the valence band are also excited into the conduction band in an intrinsic manner. The conduction band contains, therefore, electrons from two sources, the amount of which depends on the device temperature (see Section 8.3.3). Since the conduction mechanism in semiconductors with donor impurities (P, As, Sb) is predominated by negative charge carriers (electrons) these materials are called *n-type semiconductors*. The electrons are the *majority carriers*.

A similar consideration may be done with impurities which stem from the third group of the periodic chart (B, Al, Ga, In). They possess one electron less than silicon and, therefore, introduce a *positive* charge cloud into the crystal around the impurity atom. The conduction mechanism in these semiconductors with *acceptor* impurities is predominated by positive carriers (holes) which are introduced into the valence band. They are therefore called *p-type semiconductors*.

8.3.2. Band Structure

The band structure of impurity or *extrinsic* semiconductors is essentially the same as for intrinsic semiconductors. It is desirable, however, to represent in some way the presence of the impurity atoms by *impurity states*. It is common to introduce into the forbidden band so-called *donor* or *acceptor* levels (Fig. 8.7). The distance between the donor level and the conduction band repre-

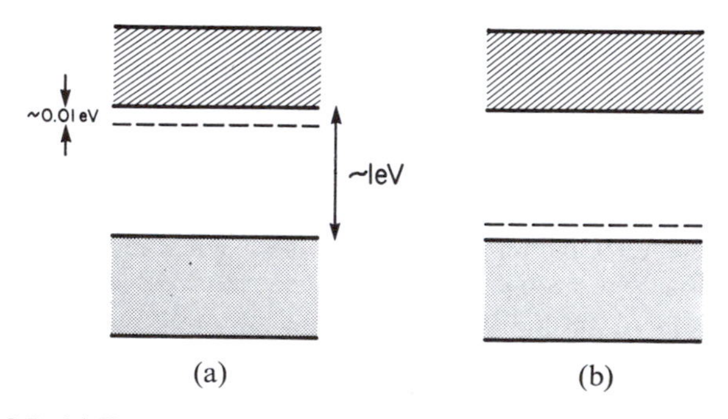

Figure 8.7. (a) Donor and (b) acceptor levels in extrinsic semiconductors.

sents the energy which is needed to transfer the extra electrons into the conduction band. (The same is true for the acceptor level and valence band.) It has to be emphasized, however, that the introduction of these impurity levels does not mean that mobile electrons or holes are found in the forbidden band of, say, silicon. The impurity states are only used as a convenient means to remind the reader of the presence of extra electrons or holes in the crystal.

8.3.3. Temperature Dependence of the Number of Carriers

At 0 K the excess electrons of the donor impurities remain in close proximity to the impurity atom and do not contribute to the electric conduction. We express this fact by stating that all donor levels are filled. With increasing temperature, the donor electrons overcome the small potential barrier (Fig. 8.7(a)) and are excited into the conduction band. Thus, the donor levels are increasingly emptied and the number of negative charge carriers in the conduction band increases exponentially, obeying an equation similar to (8.9). Once all electrons have been excited from the donor levels into the conduction band, any further temperature increase does not create additional electrons and the N_e versus T curve levels off (Fig. 8.8). As mentioned before, at still higher temperatures intrinsic effects create additional electrons which, depending on the amount of doping, can outnumber the electrons supplied by the impurity atoms.

Similarly, the acceptor levels do not contain any electrons at 0 K. At increasing temperatures, electrons are excited from the valence band into the acceptor levels, leaving behind positive charge carriers. Once all acceptor

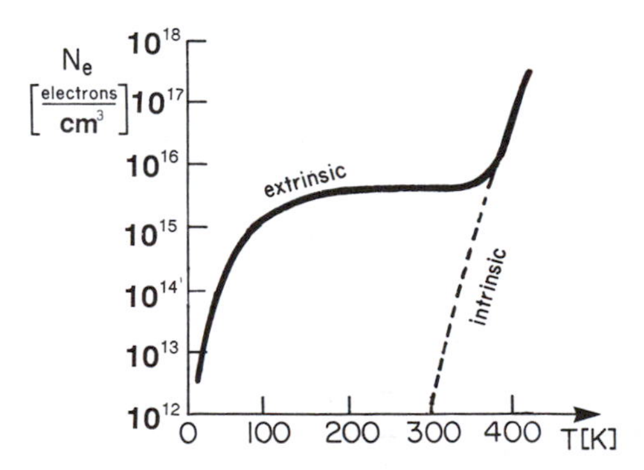

Figure 8.8. Schematic representation of the number of electrons per cubic centimeter in the conduction band versus temperature for an extrinsic semiconductor with low doping.

levels are filled, the number of holes in the valence band is not increased further until intrinsic effects set in.

8.3.4. Conductivity

The conductivity of extrinsic semiconductors can be calculated, similarly as in the previous section (8.13), by multiplying the number of carriers with the mobility μ and electron charge e. Around room temperature, however, only the majority carriers need to be considered. For electron conduction, for example, one obtains

$$\sigma = N_{de} e \mu_e, \tag{8.15}$$

where N_{de} is the number of donor electrons and μ_e is the mobility of the donor electrons in the conduction band. As mentioned above, it is reasonable to assume that, at room temperature, essentially all donor electrons have been excited from the donor levels into the conduction band (Fig. 8.8). Thus, for pure n-type semiconductors, N_{de} is essentially identical to the number of impurities (i.e., donor atoms) N_d. At substantially lower temperatures, i.e., at around 100 K, the number of conduction electrons needs to be calculated using an equation similar to (8.8).

Figure 8.9 shows the temperature dependence of the conductivity. We notice that the magnitude of the conductivity, as well as the temperature dependence of σ, is different for various doping levels. For low doping rates and low temperatures, for example, the conductivity decreases with increasing temperature (Fig. 8.9(b)). This is similar to the case of metals where the lattice vibrations present an obstacle to the drifting electrons (or, expressed differently, where the mobility of the carriers is decreased by incoherent scattering of the electrons). However, at room temperature intrinsic effects set in, which increases the number of carriers and therefore enhance the conductivity. As a

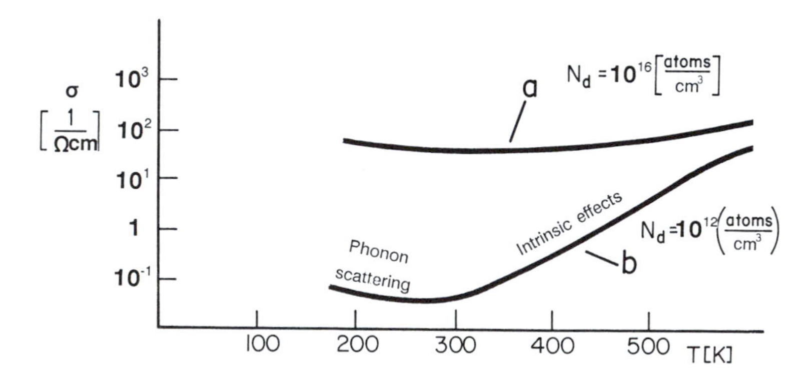

Figure 8.9. Conductivity of two extrinsic semiconductors, (a) high doping and (b) low doping. N_d = number of donor atoms per cubic centimeter.

consequence, two competing effects determine the conductivity above room temperature: an increase of σ due to an increase in the number of electrons, and a decrease of σ due to a decrease in mobility. (It should be mentioned that the mobility of electrons or holes also decreases slightly when impurity atoms are added to a semiconductor.) For high doping levels, the temperature dependence of σ is less pronounced due to the already higher number of carriers (Fig. 8.9(a)).

8.3.5. Fermi Energy

In an n-type semiconductor, more electrons can be found in the conduction band than holes in the valence band. This is particularly true at low temperatures. The Fermi energy must therefore be between the donor level and the conduction band (Fig. 8.10). With increasing temperatures, an extrinsic semiconductor becomes progressively intrinsic and the Fermi energy approaches the value for an intrinsic semiconductor, i.e., $-(E_g/2)$. [Similarly, the Fermi energy for a p-type semiconductor rises with increasing temperature from below the acceptor level to $-(E_g/2)$.]

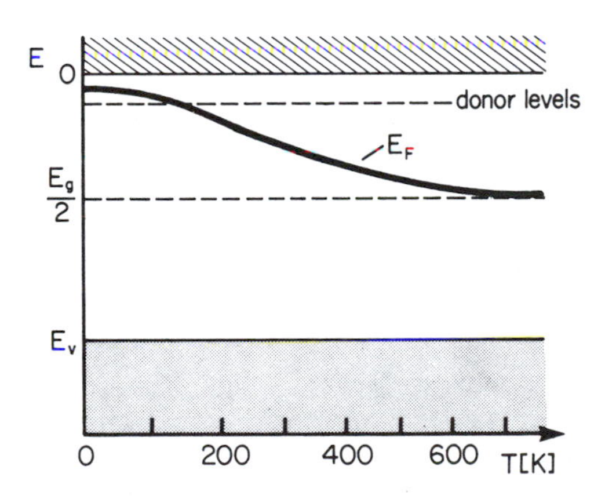

Figure 8.10. Fermi level of an n-type semiconductor as a function of temperature. $N_d \approx 10^{16}$ (atoms per cubic centimeter).

*8.4. Effective Mass

Some semiconductor properties can be better understood and calculated by evaluating the effective mass of the charge carriers. We mentioned in Section 6.7 that m^* is inversely proportional to the curvature of an electron band. We now make use of this finding.

Let us first inspect the upper portion of the valence bands for silicon near Γ (Fig. 8.2). We notice that the curvatures of these bands are convex downward. It is known from Fig. 6.8 that in this case the charge carriers have a negative effective mass, i.e., these bands can be considered to be populated by electron holes. Further, we observe that the curvatures of the individual bands are slightly different. Thus, the effective masses of the holes in these bands must likewise be different. One distinguishes appropriately between *light holes* and *heavy holes.* Since two of the bands, namely those having the smaller curvature, are almost identical, we conclude that two out of the three types of holes are heavy holes.

We turn now to the *conduction* band of silicon and focus our attention on the lowest band (Fig. 8.2). We notice a minimum (or valley) at about 85% between the Γ and X points. Since the curvature at that location is convex upward, we expect this band to be populated by electrons. (The energy surface near the minimum is actually a spheroid. This leads to *longitudinal* and *transverse* masses m_l^* and m_t^*.) Values for the effective masses are given in Appendix 4. Occasionally, *average effective masses* are listed in the literature. They may be utilized for estimates.

8.5. Hall Effect

The number and type of charge carriers (electrons or holes) which were calculated in the preceding sections can be elegantly measured by making use of the Hall effect. Actually, it is quite possible to measure concentrations of less than 10^{12} electrons per cubic centimeter in doped silicon, i.e., one can measure one donor electron (and therefore one donor atom) per 10^{10} silicon atoms. This sensitivity is several orders of magnitude better than in any chemical analysis.

We assume for our discussion an n-type semiconductor in which the conduction is predominated by electrons. Suppose an electric current j flows in the positive x-direction and a magnetic field (of magnetic induction B) is applied normal to this electric field in the z-direction (Fig. 8.11). Each electron is then subjected to a force, called the Lorentz force, which causes the electron paths to bend, as shown in Fig. 8.11. As a consequence, the electrons accumulate on one side of the slab (in Fig. 8.11 on the right side) and are deficient on the other side. Thus, an electric field is created in the (negative) y-direction which is called the Hall field. In equilibrium, the Hall force

$$F_{\mathrm{H}} = -e\mathscr{E}_y \tag{8.16}$$

balances the above mentioned Lorentz force

$$F_{\mathrm{L}} = v_x B_z e, \tag{8.17}$$

where v_x is the velocity of the electrons, and e the electron charge. $F_{\mathrm{H}} + F_{\mathrm{L}} =$

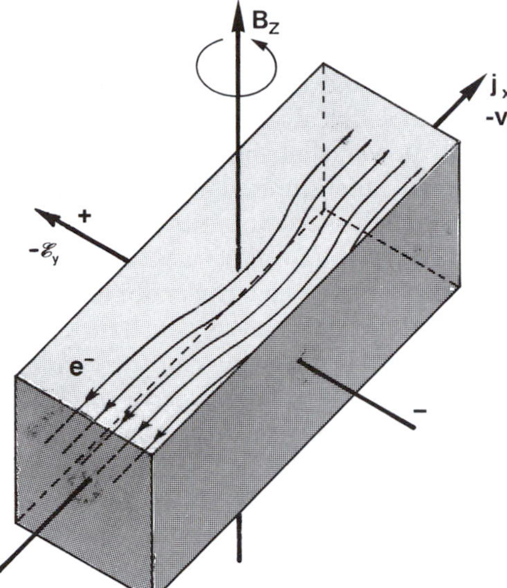

Figure 8.11. Schematic representation of the Hall effect in an *n*-type semiconductor (or a metal in which electrons are the predominant current carriers).

0 yields, for the Hall field,

$$\mathscr{E}_y = v_x B_z. \tag{8.18}$$

Combining (8.18) with (7.4) (and knowing that the current is defined to be directed in the opposite direction to the electron flow)

$$j_x = -N v_x e \tag{8.19}$$

yields for the number of conduction electrons (per unit volume)

$$N = -\frac{j_x B_z}{e\mathscr{E}_y}. \tag{8.20}$$

The variables on the right side of (8.20) can all be easily measured and the number of conduction electrons can then be calculated. Quite often, a *Hall constant*

$$R_{\mathrm{H}} = -\frac{1}{Ne} \tag{8.21}$$

is defined which is inversely proportional to the density of charge carriers, N. The sign of the Hall constant indicates whether electrons or holes predominate in the conduction process. R_{H} is negative when electrons are the predominant charge carriers. (The electron holes are deflected in the same direction as the electrons but travel in the opposite direction.)

8.6. Compound Semiconductors

Gallium arsenide (a compound of group III and group V elements of the periodic chart) is of great technical interest, partially because of its large band gap[10] which essentially prevents intrinsic contributions in impurity semiconductors even at elevated temperatures, partially because of its larger electron mobility[10] which aids in high-speed applications, and particularly because of its optical properties which result from the fact that GaAs is a direct band-gap material (see Fig. 5.24). The large electron mobility in GaAs is caused by a small value for the electron effective mass which in turn results from a comparatively large convex upward curvature of the conduction electron band near Γ. (See in this context the band structure of GaAs in Fig. 5.24.) The electrons which have been excited into the conduction band (mostly from donor levels) are most likely populating this high curvature region near Γ.

The atomic bonding in III–V and II–VI semiconductors resembles that of the group IV elements (covalent) with the additional feature that the bonding is partially ionic because of the different valences of the participating elements. The ionization energies[10] of donor and acceptor impurities in GaAs are as a rule one order of magnitude smaller than in germanium or silicon, which ensures complete ionization even at relatively low temperatures. The crystal structure of GaAs is similar to that of silicon. The gallium atoms substitute the corner and the face atoms, whereas arsenic takes the places of the four interior sites (zinc-blende structure).

The high expectations which have been set into GaAs as the semiconductor material of the future have not yet materialized to date. It is true that GaAs devices are two and a half times faster than silicon-based devices, and that the "noise" and the vulnerability to cosmic radiation is considerably reduced in GaAs because of its larger band gap. On the other hand, its ten times higher price and its much greater weight ($\delta_{Si} = 2.3$ g/cm^3 compared to $\delta_{GaAs} = 5.3$ g/cm^3) are serious obstacles to broad computer-chip usage or for solar panels. Thus, GaAs is predominantly utilized for special applications such as high-frequency devices (e.g., 10 GHz), certain military projects, or satellite preamplifiers. One of the few places, however, where GaAs seems to be, so far, without serious competition is in optoelectronics (though even this domain appears to be challenged according to the most recent research results).

We will learn in Part III that only direct band-gap materials such as GaAs are useful for lasers and light-emitting diodes (LED). Indirect band-gap materials, such as silicon, possess instead the property that part of the energy of an excited electron is removed by lattice vibrations (phonons). Thus, this energy is not available for light emission. We shall return to GaAs devices in Section 8.7.9.

GaAs is, of course, not the only compound semiconductor material which

---

[10] See the tables in Appendix 4.

has been heavily researched or is being used. Indeed, most compounds consisting of elements of groups III and V of the periodic chart are of some interest. Among them are GaP, InP, InAs, InSb, and AlSb to mention a few.[10] But also group II–VI compounds, such as ZnO, ZnS, ZnSe, CdS, CdTe, or HgS are considered for applications. These compounds have in common that the combination of the individual elements possesses an average of four valence electrons per atom because they are located at equal distances from either side of the fourth column. Another class of compound semiconductors are the group IV–VI materials[10] which include PbS, PbSe, or PbTe. Finally, ternary alloys such as $Al_xGa_{1-x}As$, or quaternary alloys such as $Al_xGa_{1-x}As_ySb_{1-y}$ are used. Most of the compounds and alloys are utilized in optoelectronic devices, e.g., $GaAs_{1-x}P_x$ for LEDs which emit light in the visible spectrum (see Part III). $AlGa_{1-x}As$ is also used in modulation-doped field-effect transistors (MODFET).

Finally, silicon carbide is the most important representative of the group IV–IV compounds. Since its band-gap is around 3 eV, α-SiC can be used for very high-temperature (700° C) device applications and for LEDs which emit light in the blue end of the visible spectrum. SiC is, however, expensive and cannot yet be manufactured with reproducible properties.

8.7. Semiconductor Devices

8.7.1. Metal–Semiconductor Contacts

If a semiconductor is coated on one side with a metal, a *rectifying* contact or an *ohmic* contact is formed, depending on the type of metal used. Both cases are equally important. Rectifiers are widely utilized in electronic devices, e.g., to convert alternating current into direct current. However, the type discussed here has been mostly replaced by $p - n$ rectifiers. On the other hand, all semiconductor devices need contacts in which the electrons can easily flow in both directions. They are called ohmic contacts because their current–voltage characteristic obeys Ohm's law (7.1).

At the beginning of our discussion let us assume that the surface of an n-type semiconductor has somehow been negatively charged. The negative charge repels the free electrons that had been near the surface and leaves positively charged donor ions behind (e.g., As^+). Any electron which drifts toward the surface (negative x-direction in Fig. 8.12(a)) "feels" this repelling force. As a consequence, the region near the surface has less free electrons than the interior of the solid. This region is called the *depletion layer* (or sometimes *space charge region*).

In order to illustrate the repelling force of an external negative charge it is customary to curve the electron bands upward near the surface. The depletion can then be understood by stating that the electrons assume the lowest

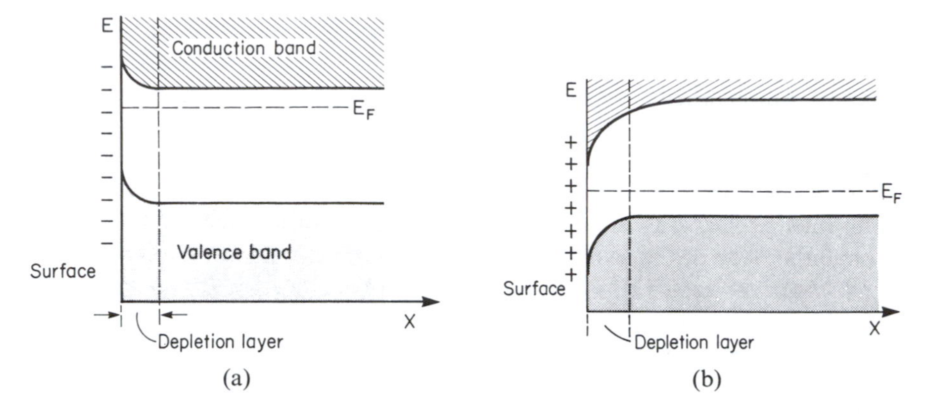

Figure 8.12. (a) Band diagram for an *n*-type semiconductor whose surface has been negatively charged. (b) Band diagram for a *p*-type semiconductor, the surface of which is positively charged. *X* is the distance from the surface.

possible energy state (or colloquially expressed: "the electrons like to roll downhill").

Similarly, if a *p*-type semiconductor is positively charged at the surface, the positive carriers (holes) are repelled toward the inner part of the crystal and the band edges are bent downward (Fig. 8.12(b)). This represents a potential barrier for holes (because holes "want to drift upward" like a hydrogen-filled balloon).

8.7.2. Rectifying Contacts (Schottky Barrier Contacts)

It is essential for further discussion to introduce the *work function* ϕ which is the energy difference between the Fermi energy and the ionization energy. In other words, ϕ is the energy which is necessary to transport an electron to infinity. (Values for ϕ are given in Appendix 4.)

Let us consider a metal and an *n*-type semiconductor before they are brought into contact. In Fig. 8.13(a) the Fermi energy of a metal is shown to be lower than the Fermi energy of the semiconductor, i.e., $\phi_M > \phi_S$. Immediately after the metal and semiconductor have been brought into contact, electrons start to flow from the semiconductor "down" into the metal until the Fermi energies of both solids are equal (Fig. 8.13(b)). As a consequence, the metal will be charged negatively and a potential barrier is formed just as shown in Fig. 8.12. This means that the energy bands in the bulk semiconductor are lowered by the amount $\phi_M - \phi_S$ with respect to a point A.

In the equilibrium state electrons from both sides cross the potential barrier. This electron flow constitutes the so-called *diffusion current*. The number of electrons diffusing in both directions must be identical for the following reason: the metal contains more free electrons, but these electrons have to

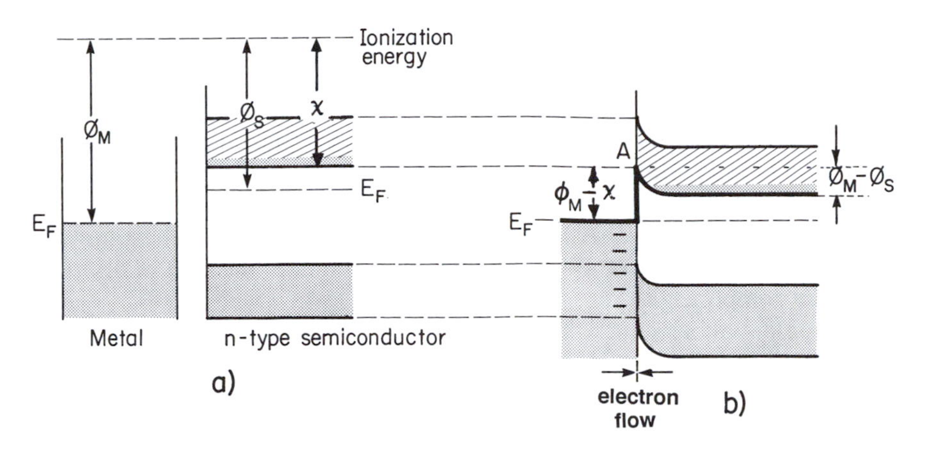

Figure 8.13. Energy bands for a metal and an *n*-type semiconductor (a) before and (b) after contact. $\phi_M > \phi_S$. The potential barrier is marked with heavy lines. χ is the *electron affinity.*

climb a higher potential barrier than the electrons in the semiconductor whose conduction band contains less free electrons.

Similarly, if a *p*-type semiconductor is brought into contact with a metal and $\phi_M < \phi_S$, electrons diffuse from the metal into the semiconductor and charge the metal and, therefore, the surface of the semiconductor positively. Consequently, a "downward" potential barrier (for the holes) is formed (Fig. 8.14).

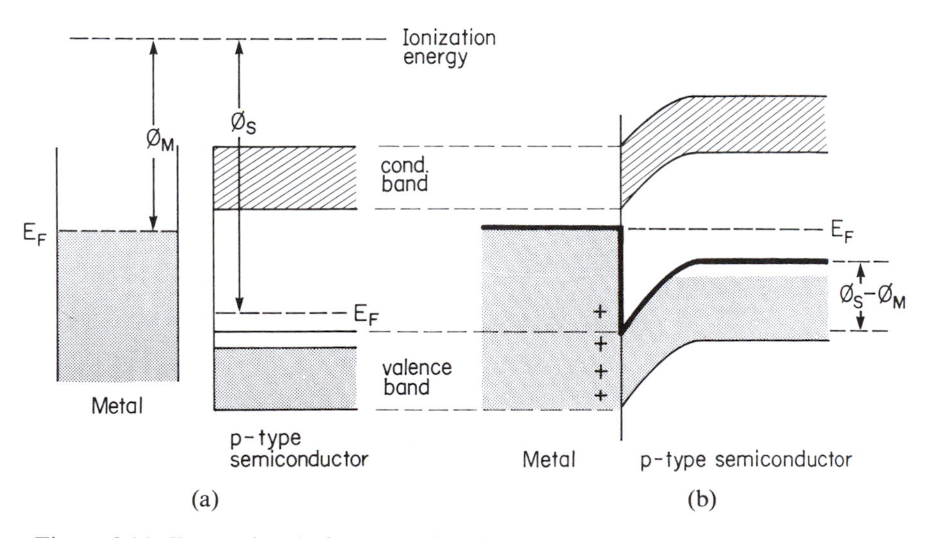

Figure 8.14. Energy bands for a metal and a *p*-type semiconductor (a) before and (b) after contact. $\phi_M < \phi_S$.

In addition to the diffusion current just mentioned, a "*drift current*" needs to be taken into consideration. Let us assume that an electron–hole pair was thermally created in or near the depletion layer of a *p*-type semiconductor (Fig. 8.14(b)). Then the thermally created electron in the conduction band is immediately swept down the barrier, and the hole in the valence band is swept up the barrier. This drift current is usually very small (particularly if the band gap is large such as in GaAs) and is relatively insensitive to the height of the potential barrier. The total current across a junction is the sum of drift and diffusion components.

The potential barrier height for an electron diffusing from the semiconductor into the metal is $\phi_M - \phi_S$ (see Fig. 8.13(b)). This potential difference is called the *contact potential.* The height of the potential barrier from the metal side is $\phi_M - \chi$, where χ is the *electron affinity* measured from the bottom of the conduction band to the ionization energy (vacuum level) (Fig. 8.13).

We shall now estimate the net current which flows across the potential barrier when a metal and an *n*-type semiconductor are connected to a d.c. source (*biasing*). At first, the metal is assumed to be connected to the negative terminal of a battery. As a result, the metal is charged even more negatively than without bias. Thus, the electrons in the semiconductor are repelled even more, and the potential barrier is increased (Fig. 8.15(a)). Further, the depletion layer has become wider. Because both barriers are now relatively high, the diffusion currents in both directions are negligible. However, the small and essentially voltage-independent drift current still exists which results in a very small and constant net electron current from the metal into the semiconductor (*reverse bias*, Fig. 8.15(a)).

If the polarity of the battery is reversed, the potential barrier in the semiconductor is reduced, i.e., the electrons are "driven" across the barrier so that

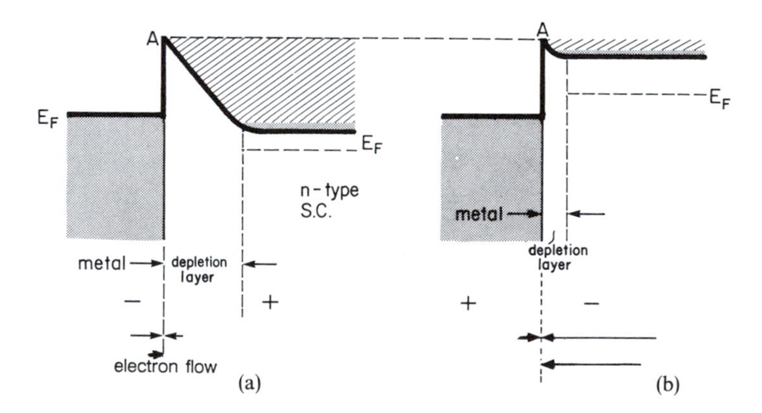

Figure 8.15. Metal–semiconductor contact with two polarities: (a) reverse bias and (b) forward bias. The amount of electrons which flow in both directions and the net current is indicated by the length of the arrows. The potential barriers are marked with heavy lines.

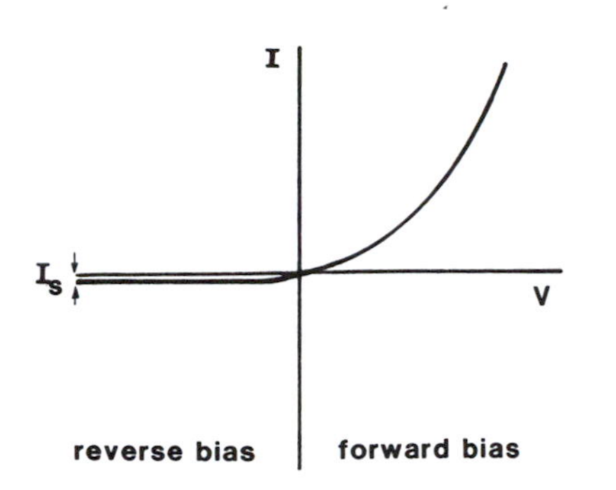

Figure 8.16. Characteristic of a rectifier. The reverse current is grossly exaggerated!

a large net current from the semiconductor into the metal results (*forward bias*). The depletion layer is narrow (Fig. 8.15(b)). The voltage–current characteristic of a rectifier is shown in Fig. 8.16. Rectifiers of this type are used to convert alternating current into direct current.

The current which flows from the metal into the semiconductor is

$$I_{MS} = ACT^2 \exp\left[-\left(\frac{\phi_M - \chi}{k_B T}\right)\right], \tag{8.22}$$

(see Fig. 8.13b) where A is the area of the contact and C is a constant. The current flowing from the semiconductor into the metal is

$$I_{SM} = ACT^2 \exp\left[-\left(\frac{\phi_M - \phi_S - eV}{k_B T}\right)\right], \tag{8.23}$$

where V is the bias voltage (which has the sign of the polarity of the metal) and e is the electronic charge. The net current $I_{net} = I_{SM} - I_{MS}$ consists of two parts, namely, the saturation current (occasionally called the generation current)[11]

$$I_S = ACT^2 \exp\left[-\left(\frac{\phi_M - \phi_S}{k_B T}\right)\right] \tag{8.24}$$

and a voltage-dependent term. The net current is then obtained by combining (8.22), (8.23), and (8.24)

$$I_{net} = I_S\left[\exp\left(\frac{eV}{k_B T}\right) - 1\right]. \tag{8.25}$$

We see from (8.25) that for forward bias (positive V) the net current increases exponentially with voltage. Figure 8.16 reflects this behavior. On the other hand, for reverse bias (negative V) the current is essentially constant and equal to $-I_S$. The saturation current is about three orders of magnitude smaller than the forward current. (It is shown exaggerated in Fig. 8.16.)

[11] For low enough temperatures, one can assume $\phi_S \approx \chi$; see Figs. 8.10 and 8.13.

We shall learn in Section 8.7.4 that the same rectifying effect as discussed above can also be achieved by using a p–n diode. There are, however, a few advantages in using the metal/semiconductor rectifier. First, the conduction in a metal/semiconductor device involves naturally one type of conduction carrier (e.g., electrons) only. Thus, no mutual annihilation of electrons and holes can occur. As a consequence of this lack of "carrier recombination," the device may be switched more quickly from forward to reverse bias and is therefore better suited for microwave frequency detectors. Second, the metal base provides better heat removal than a mere semiconductor chip, which is helpful in high power devices.

8.7.3. Ohmic Contacts (Metallizations)

In Fig. 8.17 band diagrams are shown for the case where a metal is brought into contact with an n-type semiconductor. It is assumed that $\phi_M < \phi_S$. Thus, electrons flow from the metal into the semiconductor, charging the metal positively. The bands of the semiconductor bend "downward" and no barrier exists for the flow of electrons in either direction. In other words, this configuration allows the injection of a current into and out of the semiconductor without suffering a sizable power loss. The current increases, in essence, linearly with increasing voltage and is symmetric about the origin as Ohm's law requires. Accordingly, this junction is called an *ohmic contact*. A similar situation exists for a p-type semiconductor and $\phi_M > \phi_S$.

Aluminum is frequently used for making the contact between a device (e.g., the p-region of a rectifier) and the external leads. Aluminum bonds readily to Si or SiO_2 if the device is briefly heated to about 550° C after Al deposition. Since aluminum has a larger work function than silicon (see Appendix 4) the contact to a p-region is ohmic. Additionally, the diffusion of aluminum into silicon yields a shallow and highly conductive p^+-region.[12]

[12] The superscript plus means *heavily doped region.*

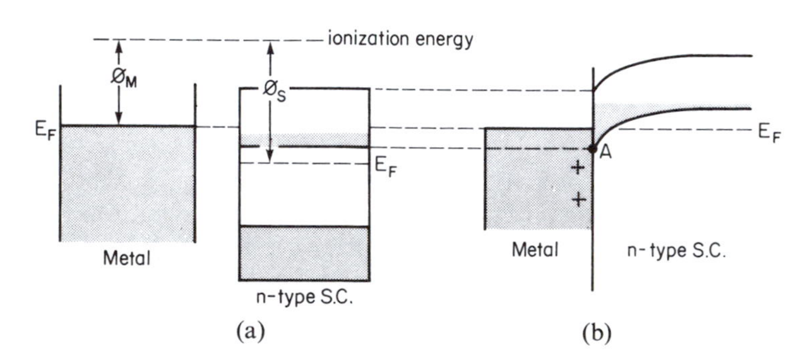

Figure 8.17. Ohmic contact between metal and n-type semiconductor ($\phi_M < \phi_S$).

Now, aluminum is likewise used as a contact material for n-type silicon. To prevent a rectifying contact in this case, one usually lays down a heavily doped and shallow n^+-layer[12] on top of the n-region. Since this n^+-layer is highly conductive and is made to be very thin, tunneling through the barrier accomplishes the unhindered electron flow (see Sections 4.3 and 8.7.8).

We note in passing that thin (less than 1 μm thick) aluminum films are commonly used to interconnect many individual electronic devices which are manufactured on one *chip* in order to achieve large-scale integration. These thin-film metallizations have to sustain a high current density and are thus potentially vulnerable for circuit failure by a mechanism called *electrotransport* or *electromigration*. Here, a momentum exchange between the accelerated electrons and the metal ions pushes some metal ions from the negative side to the positive side of the thin-film stripe, eventually causing voids to form near the cathode. Interconnection of these voids makes the metal films discontinuous. Small additions of copper to aluminum or alternatively, gold metallizations, or metal silicides (for high-temperature applications and GaAs contacts) are used to diminish this problem.

8.7.4. *p–n* Rectifier (Diode)

We learned in Section 8.7.2 that when a metal is brought into contact with an extrinsic semiconductor, a potential barrier may be formed which gives rise to the rectifier action. A similar potential barrier is created when a p-type and an n-type semiconductor are joined.

As before, electrons flow from the higher level (n-type) "down" into the p-type semiconductor so that the p-side is negatively charged. This proceeds until equilibrium is reached and both Fermi energies are at the same level. The resulting band diagram is shown in Fig. 8.18.

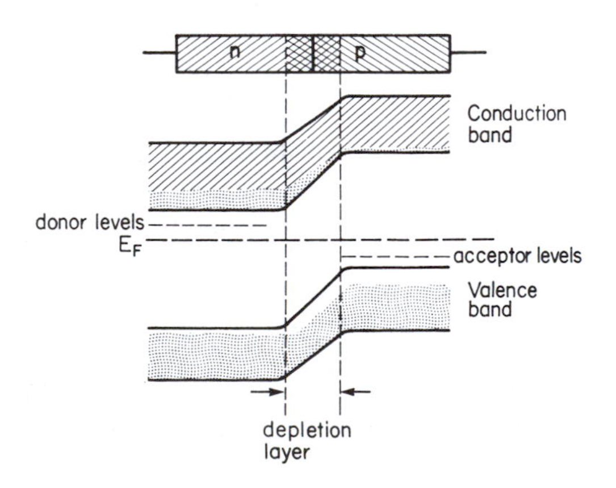

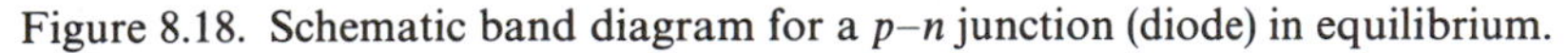
Figure 8.18. Schematic band diagram for a p–n junction (diode) in equilibrium.

Consider first the conduction band only. The electrons which want to diffuse from the n-region into the p-region encounter a potential barrier near the junction. For statistical reasons, only a few of them have enough energy to climb the barrier and diffuse into the p-region. The electrons in the p-region, on the other hand, can easily diffuse "down" the potential barrier into the n-region. Note that only a few electrons exist in the conduction band of the p-region. (They have been thermally excited into this band by intrinsic effects.) In the equilibrium state the number of electrons crossing the junction in both directions is therefore identical. (The same is true for the holes in the valence band.)

When an external potential is applied to this device, effects similar to the ones described in Section 8.7.2 occur: connecting the positive terminal of a d.c. source to the n-side withdraws electrons from the depletion area which becomes wider and the potential barrier grows higher (Fig. 8.19(a)). As a consequence, only a small drift current (from intrinsic effects) exists (reverse bias). On the other hand, if the n-side is charged negatively, the barrier decreases in height and the space charge region narrows. A large net electron flow occurs from the n-type region to the p-type region (forward bias, Fig. 8.19(b) and (c)).

In Fig. 8.19(a) and (b) "quasi-Fermi levels" for electrons and holes are shown. They are caused by the fact that the electron density varies in the junction from the n-side to the p-side by many orders of magnitude, while the electron current is almost constant. Consequently, the Fermi level must also be almost constant over the depletion layer.

It has to be emphasized that the current in a p–n rectifier is the sum of both electron and hole currents. The net current may be calculated by using an equation similar to (8.25) whereby the saturation current, I_S, in the present case is a function of the equilibrium concentration of the holes in the n-region (C_{hn}), the concentration of electrons in the p-region (C_{ep}), and other device parameters. The saturation current in the case of reverse bias is given by the *Shockley equation* which is also called the *ideal diode law*:

$$I_S = Ae\left(\frac{C_{ep}D_{ep}}{L_{ep}} + \frac{C_{hn}D_{hn}}{L_{hn}}\right), \tag{8.26}$$

where the D's and L's are diffusion constants and diffusion lengths, respectively (e.g., D_{ep} = diffusion constant for electrons in the p-region, etc.). The diffusion constant is connected with the mobility μ through the Einstein relation

$$D_{ep} = \frac{\mu_{ep} k_B T}{e} \tag{8.27}$$

(see textbooks on thermodynamics). The minority carrier diffusion length is given by a reinterpretation of a well-known equation of thermodynamics

$$L_{ep} = \sqrt{D_{ep} \cdot \tau_{ep}}, \tag{8.28}$$

where τ_{ep} is the lifetime of the electrons in the p-type region before these

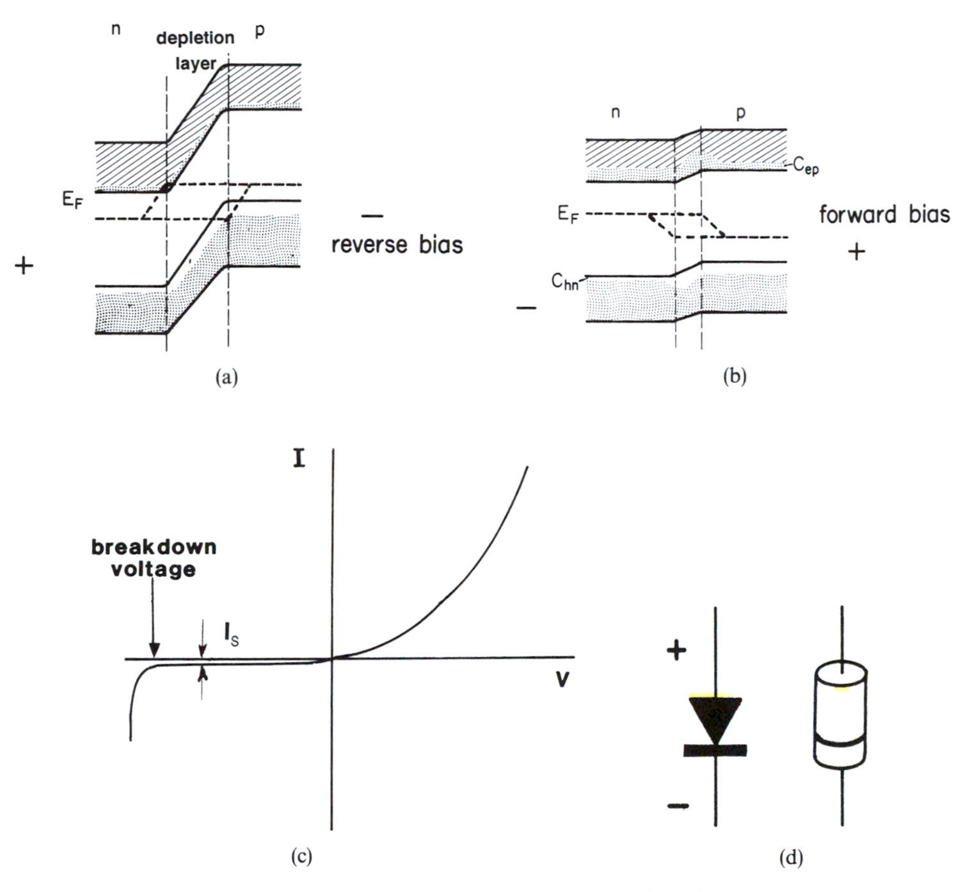

Figure 8.19. (a) Reverse and (b) forward bias of a p–n junction (diode). (c) Voltage–current characteristic of a p–n junction (diode). As in Fig. 8.16, I_s is shown grossly exaggerated. (d) Symbol of a p–n rectifier in a circuit and designation of polarity in an actual rectifier.

electrons are annihilated by recombination with holes. In order to keep the reverse current small, both C_{hn} and C_{ep} (minority carriers) have to be kept at low levels (compared to electrons and holes introduced by doping). This can be accomplished by selecting semiconductors having a large energy gap (see tables in Appendix 4) and by high doping.

8.7.5. Zener Diode

When the reverse voltage of a p–n diode is increased above a critical value, the high electric field strength causes some electrons to become accelerated to a velocity at which impact ionization occurs [Fig. 8.20(a)]. In other words, some electrons are excited by the electric field from the valence band into the

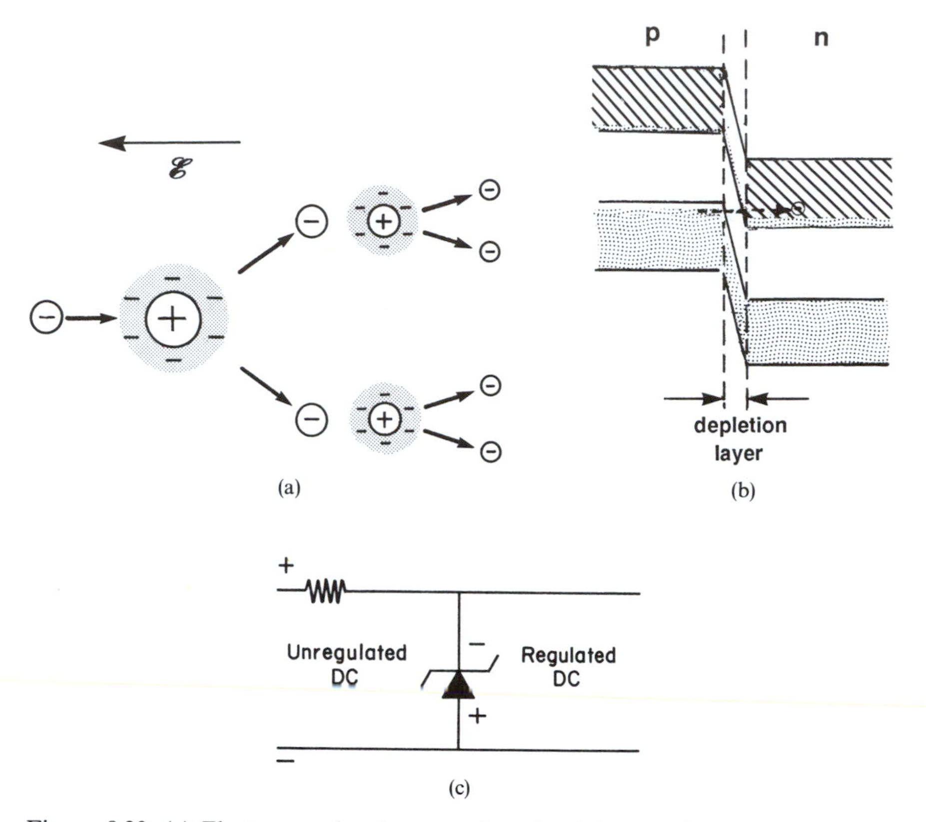

Figure 8.20. (a) Electron avalanche created at breakdown voltage. (b) Tunneling (Zener breakdown). (c) Zener diode in a circuit for voltage regulation.

conduction band, leaving behind an equal number of holes. The free electrons (and holes) thus created are likewise accelerated and create new electron–hole pairs, etc., until eventually a *breakdown* occurs, i.e., the reverse current increases quite rapidly (Fig. 8.19(c)). The breakdown voltage which is the result of this *avalanching* process depends on the degree of doping: the higher the doping, the lower the breakdown voltage. Alternatively to this avalanche mechanism, a different breakdown process may take place under certain conditions. It occurs when the doping is heavy and thus the barrier width becomes very thin (i.e., < 10 nm). Applying a high enough reverse voltage causes the bands to shift to the degree that some electrons in the valence band of the *p*-side are opposite to empty states in the conduction band of the *n*-material. These electrons can then tunnel through the depletion layer as described in Sections 4.3 and 8.7.8 and are depicted in Fig. 8.20(b). *Tunneling* (or *Zener breakdown*) takes place usually at low reverse voltages (e.g., below about 4 volts for silicon-based diodes) whereas avalanching is the mechanism which occurs when the reverse voltage is large.

The breakdown effect just described is used in a circuit to hold a given voltage constant at a desired level (Fig. 8.20(c)). The Zener diode is therefore utilized as a circuit protection device. The Zener diode is generally not destroyed by the breakdown, unless excessive heat generation causes it to melt.

8.7.6. Solar Cell (Photodiode)

A photodiode consists of a p–n junction (Fig. 8.21). If light of sufficiently high energy falls on such a device, electrons are lifted from the valence into the conduction band, leaving holes in the valence band. The electrons in the depleted area immediately "roll down" into the n-region, whereas the holes are swept into the p-region. These additional carriers can be measured in an external circuit (photographic exposure meter) or used to generate electrical energy. In order to increase the effective area of the junction, the p-type region is made extremely thin (1 μm) and light is radiated through the p-layer (Fig. 8.21).

The electron–hole pairs which are created some distance away from the depleted region are generally not separated by the junction field and eventually recombine; they do not contribute to the electric current. However, some electrons or holes which are within a diffusion length from the depleted region drift into this area and thus contribute to the current. In semiconducting materials which contain only a few defects (such as grain boundaries, dislocations, and impurities) the electrons or holes may diffuse up to 200 μm before they get trapped, whereas in semiconducting materials containing a large number of defects the diffusion length reduces to 10 μm. The closer a carrier

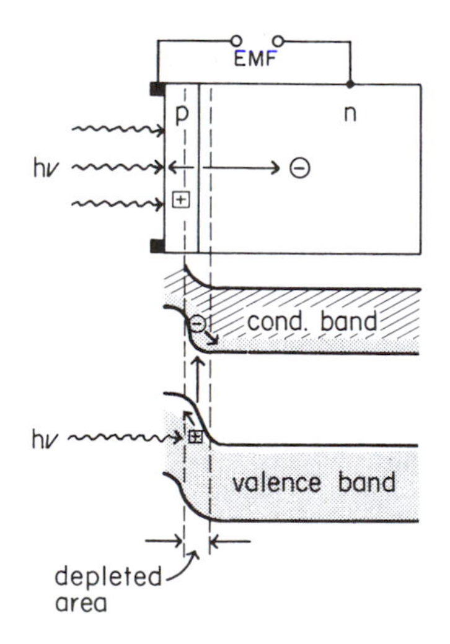

Figure 8.21. Solar cell; the p-region is only about 1 μm thick.

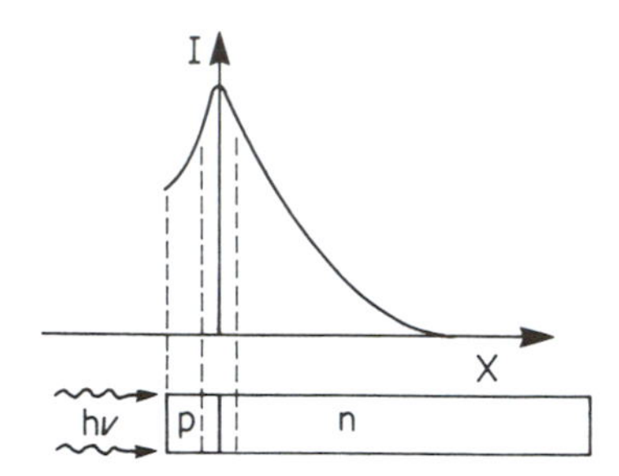

Figure 8.22. Schematic representation of the contribution of electrons and holes to the photocurrent (I) with respect to the distance x from the p–n junction.

was created to the p–n boundary, the larger is its chance of contributing to the current (Fig. 8.22).

The thin p-type layer introduces an internal resistance to the collection current which reduces the efficiency of the energy conversion. At present the maximal efficiency of a photovoltaic device, made of crystalline silicon and involving a three-layer technology (see Part III), is about 20–24%. The energy which is needed to produce such a device (including mounting and installation) is recovered in about 6 years when the collector is located in North Africa or Central America. (Installation in central Europe or the northern states of the USA and Canada may double the energy recovery time.) The cost of photovoltaic devices (presently \$8–\$10 per installed watt) can be reduced by utilizing polycrystalline, less purified, or amorphous silicon, but at the expense of efficiency. As an example, photovoltaics made of commerical, hydrogen-doped amorphous silicon (see Section 9.4) have an efficiency of only 6–8%, but its invested energy for production and mounting is recovered in just 1 year. The efficiency of this device has been enhanced to 12% in laboratory experiments. The goal is to produce for terrestrial applications inexpensive solar cells, having 20% efficiency or better and a lifetime of about 20 years. The lifetime is reduced when the metal contacts (grids) to the semiconductor corrode. The most recent development employs dye-coated titanium dioxide and an electrochemical cell which mimics the role of chlorophyll in photosynthesis.

The photovoltaic cell depicted in Fig. 8.21 has one inherent disadvantage: the impinging light has to travel first through the p-type layer (however thin it may be) before it eventually reaches the depleted (active) area. This attenuates its intensity to a certain degree. In addition, the incoming light is somewhat blocked by the metal electrodes which cover part of the face of the cell. The resulting loss in efficiency is a trade-off for a large surface area (which is often desirable to increase power). For telecommunication applications however, for which high *efficiency* is more important, a rather ingenious alternative design can be used. Imagine that the light impinges *transversely* on (or better, *along*) the depletion layer. For this the beam is channeled-in from the *side* by a light-conducting device such as an optical fiber or a wave guide (Fig. 8.23). In order to increase the effective area, i.e., the width, W,

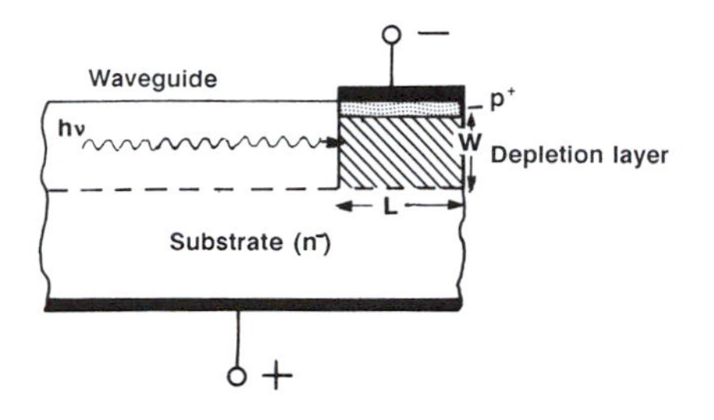

Figure 8.23. Schematic of a transverse-type photodiode which is connected to a light carrying medium such as an optical fiber or a waveguide.

of the depletion region, the photodiode is strongly reverse-biased and the doping of one of the semiconductors is comparatively light. (For details refer to Fig. 8.19(a).) The efficiency is further maximized by increasing the length of the depletion layer, L. This device yields almost 100% quantum efficiency.

The quantum efficiency can be calculated by the equation

$$\eta = 1 - \frac{\exp(-\alpha W)}{1 + \alpha L}, \tag{8.29}$$

where α is a parameter which determines the degree of photon absorption by the electrons (α is defined in (10.21a)). As an example, for a GaAs photodiode the n-region is lightly doped because the *electron* mobility in GaAs is much larger than the hole mobility, see Appendix 4. This shifts the depleted region towards the n-side. On the other hand, the p-region is *heavily* doped (and thin) in order to minimize its resistance.

The incoming light which is modulated by information (such as the spoken word in telecommunications) modulates in turn the electrical current in the photodiode. This transforms a signal which is transmitted by light into an electrical signal. We shall return to this topic and to other *optoelectronic devices* in Part III.

*8.7.7. Avalanche Photodiode

This device is a p–n photodiode which is operated in a high reverse bias mode, i.e., at near-breakdown voltage. The electrons and holes which were created by transitions from the valence into the conduction band by the incident light are accelerated through the depleted area with a high velocity. As a consequence, they ionize the lattice atoms and generate secondary hole–electron pairs which in turn are accelerated, thus generating even more hole–electron pairs. The result is a photocurrent gain, which may be between 10 and 1000. The avalanche photodiode is ideally suited for low light level applications because of its high signal-to-noise ratio, and for very high frequencies (GHz). It is also used for detectors in long-distance, fiber-optics telecommunication systems.

*8.7.8. Tunnel Diode

So far, we have restricted our discussion mostly to the case for which the electrons drift from the *n*-type to the *p*-type semiconductor by way of "climbing" a potential barrier. Another electron transfer mechanism is possible, however. If the depleted area is very narrow (approximately 10 nm) and if certain other requirements (see below) are fulfilled, electrons may tunnel *through* the potential barrier. (See in this context Fig. 4.7, Fig. 8.20(b), and equation (4.39).) Heavy doping (e.g., 10^{20} impurity atoms per cubic centimeter) yields this condition.

The situation can best be understood by inspecting Fig. 8.24(a) in which a schematic band diagram of a tunnel diode is shown. Because of the high doping level, the Fermi energy extends into the valence band of the *p*-type semiconductor and into the conduction band of the *n*-type semiconductor. In the equilibrium state, the same amount of electrons is tunneling through the potential barrier in both directions, i.e., no net current flows.

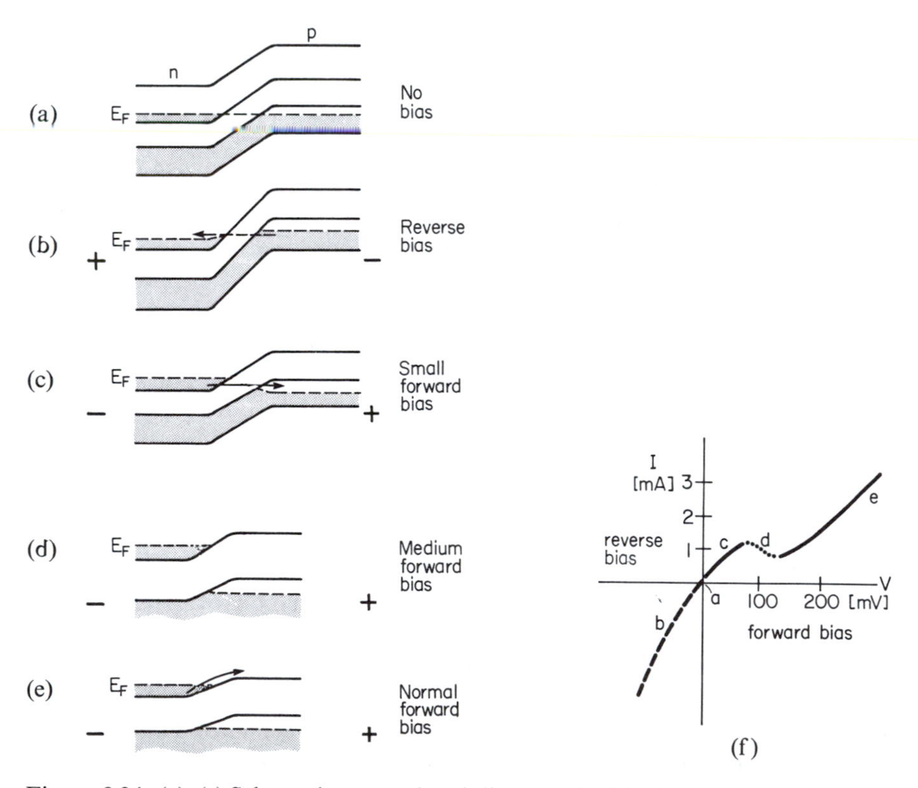

Figure 8.24. (a)–(e) Schematic energy band diagrams for highly doped *n*- and *p*-type semiconductors (tunnel diode). (a) No bias. (b) Reverse bias. (c) Small forward bias. (d) Medium forward bias. (e) "Normal" forward bias. (f) Voltage–current characteristic for a tunnel diode.

If a small *reverse* bias is applied to this device (Fig. 8.24(b)), the potential barrier is increased as usual and the Fermi energy, along with the top and bottom of the bands in the *p*-area, is raised. This creates empty electron states in the conduction band of the *n*-type semiconductor, opposite from filled states in the valence band of the *p*-type semiconductor. As a consequence, some electrons tunnel from the *p*-type to the *n*-type semiconductor, as indicated by an arrow. An increase in the *reverse* voltage yields an increase in the electron current through the device (see Fig. 8.24(f)).

Let us now consider several forward voltages. A small forward bias (Fig. 8.24(c)) creates just the opposite of that seen in Fig. 8.24(b). Electrons are tunneling through the potential barrier from the conduction band of the *n*-type semiconductor into empty states of the valence band of the *p*-type semiconductor. The applied voltage needs to be only several millivolts and it produces a forward current of about one milliamp.

If, however, the voltage is increased to, say, 100 mV, the potential barrier might be decreased so much that, opposite to the filled *n*-conduction states, no allowed empty states in the *p*-area are present [Fig. 8.24(d)]. (The area opposite to the filled *n*-conduction states may be the forbidden band.) In this case, no tunneling takes place. As a consequence of this, the current decreases with increasing forward voltage, as shown in Fig. 8.24(f). We experience a *negative current–voltage characteristic.*

Finally, if the forward voltage is increased even more, the electrons in the conduction band of the *n*-type semiconductor obtain enough energy to climb the potential barrier to the *p*-side just as in a regular *p–n* junction. As a consequence, the current increases with voltage, just as in Fig. 8.19(c).

Of particular interest is the range in which a negative voltage–current characteristic is experienced. One has to bear in mind that all other electrical devices have a positive voltage–current characteristic, i.e., they dissipate energy. Therefore, if a tunnel diode is connected to properly dimensioned resistors and capacitors, a simple oscillator can be built which does not lose energy because the net resistance is zero. Those devices can oscillate at frequencies up to 10^{11} cycles per second.

8.7.9. Transistors

Bipolar Junction Transistor. An *n–p–n* transistor may be considered to be an *n–p* diode back-to-back with a *p–n* diode. A schematic band diagram for an unbiased *n–p–n* transistor is shown in Fig. 8.25. The three connections of the transistor are called emitter (E), base (B), and collector (C).

If the transistor is used for the amplification of a signal, the "diode" consisting of emitter and base is forward biased, whereas the base–collector "diode" is strongly reverse biased (Fig. 8.26(a)). The electrons ejected into the emitter, therefore, need to have enough energy to be able to "climb" the potential barrier into the base region. Once there, the electrons diffuse through the base

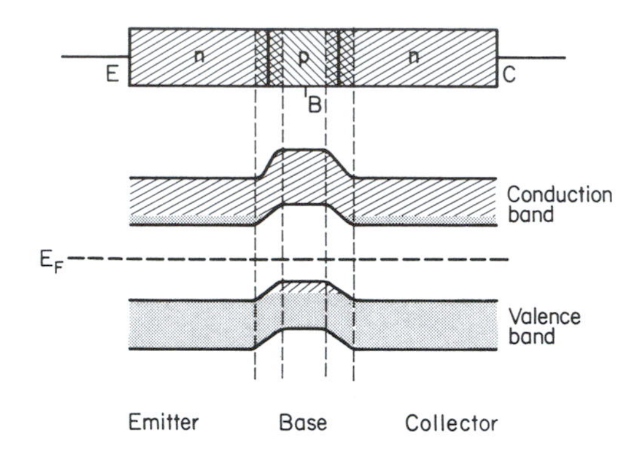

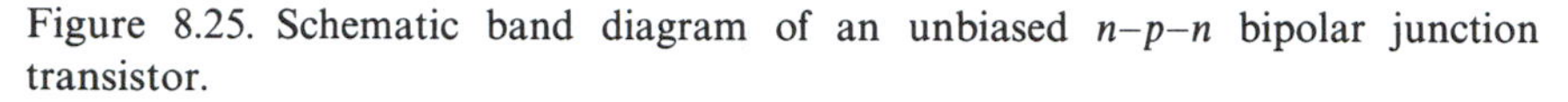

Figure 8.25. Schematic band diagram of an unbiased *n–p–n* bipolar junction transistor.

area until they have reached the depletion region between base and collector. Here, the electrons are accelerated in the strong electric field produced by the collector voltage (Fig. 8.26(b)). This acceleration causes amplification of the input a.c. signal.

One may consider this amplification from a more quantitative point of view. The forward biased emitter–base diode is made to have a small resistance (approximately 10^{-3} Ω cm), whereas the reverse biased base–collector diode has a much larger resistance (about 10 Ω cm). Since the current flowing through the device is practically identical in both parts, the power ($N = I^2R$) is larger in the collector circuit. This results in a power gain.

The electron flow from emitter to collector can be controlled by the bias voltage on the base: a *large* positive (forward) bias decreases the potential barrier and the width of the depleted region between emitter and base (Fig.

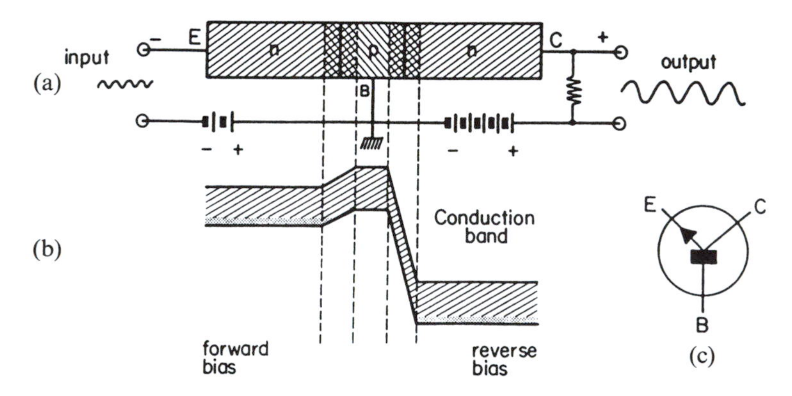

Figure 8.26. (a) Biasing of an *n–p–n* bipolar transistor. (b) Schematic band diagram (partial) of a biased *n–p–n* bipolar transistor. (c) Symbol used for a bipolar *n–p–n* transistor.

8.19). As a consequence, the electron injection into the *p*-area is relatively high. In contrast, a *small*, but still positive base voltage, results in a comparatively larger barrier height and in a wider depletion area, which causes a smaller electron injection from the emitter into the base area. In short, the voltage applied to the base modulates the transfer of the electrons from the emitter into the base area. As a consequence, the strong collector signal mimics the waveform of the input signal. This feature is utilized for the amplification of music or voice, etc.

In another application, a transistor may be used as an electronic switch. The electron flow from emitter to collector can be stopped completely (or turned on) by an appropriate base voltage. This virtue is used for logic and memory functions in computers (see Section 8.7.12).

The device shown in Fig. 8.26 is called a "bipolar transistor"; the current passes in series through *n*-type as well as through *p*-type semiconductor materials.

Some details need to be added about technical features of the bipolar transistor. In order to obtain a large electron density in the emitter, this area is heavily doped. In the *p*-doped base area, the drifting electrons are subject to possible recombination with holes. Therefore, the number of holes there has to be kept to a minimum, which is accomplished by light doping. (Light doping also reduces the unwanted injection of hole current into the base.) Recombination is further decreased by making the base region extremely thin, i.e., 10^{-5}–10^{-7} m. A narrow base region has a beneficial side effect: it increases the frequency response. (The reciprocal of the electron transit time equals the highest possible frequency at which amplification can be achieved.) The doping rate of the collector area is in general not critical. Usually, the doping is light for high gain and low capacitance of the device. The voltage–current characteristics for a transistor are shown in Fig. 8.27.

In *p–n–p* transistors, the majority carriers are holes. The function and features of a *p–n–p* transistor are similar to an *n–p–n* transistor.

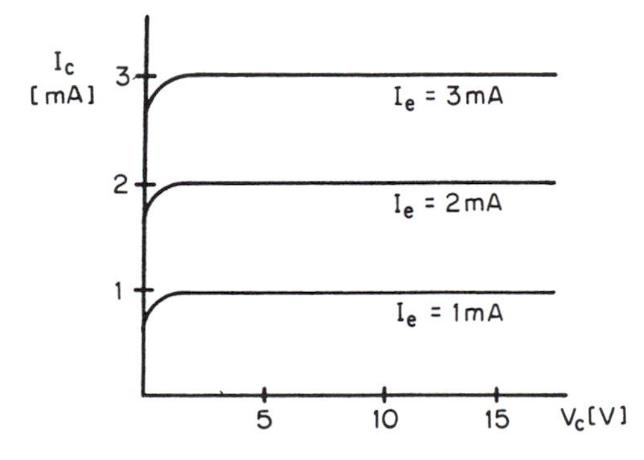

Figure 8.27. Schematic collector voltage–current characteristics of a transistor for various emitter currents. I_c = collector current, I_e = emitter current, and V_c = collector voltage.

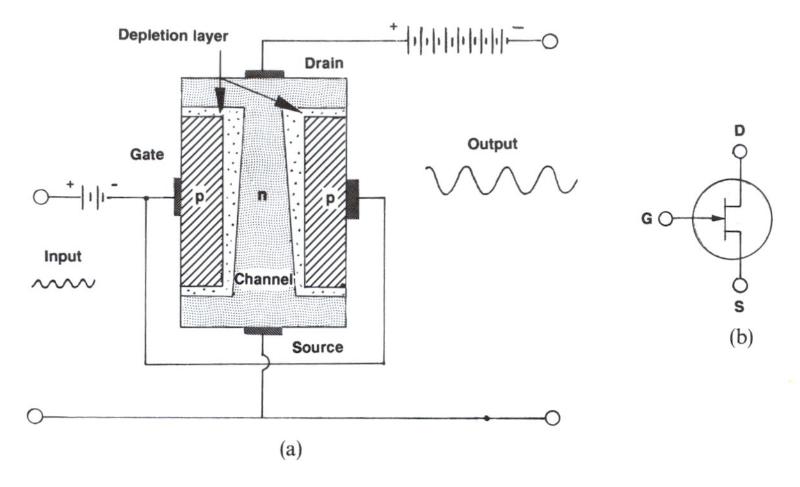

Figure 8.28. (a) Schematic representation of an *n*-channel junction field-effect transistor. The dark areas symbolize the metal contacts (e.g., aluminum). (b) Circuit symbol for an *n*-channel JFET. *Note:* In a *p*-channel JFET the arrow points away from the channel.

Junction Field-Effect Transistor (JFET). The junction field-effect transistor consists of a *channel* through which the charge carriers (e.g., electrons in Fig. 8.28)) need to pass on their way from a *source* (S) to the *drain* (D). The conducting path (source, channel, and drain) is made of *one* semiconducting material only, e.g., *n*-type. (This is in contrast to the *bipolar* transistor shown in Fig. 8.26, in which the current passes in series through *n*-type as well as through *p*-type semiconductor materials.) Field-effect transistors are therefore designated as *unipolar*. The electrons that flow from the source to the drain can be controlled by an electric field which is established by applying a negative voltage to the *p*-doped *gate* (G) (to stay within the example of Fig. 8.28). In other words, the *p–n* gate-to-channel diode is reverse biased. This reverse biasing increases the width of the depletion layer (see Fig. 8.19) thus causing the conducting channel to become narrower. (Close to the drain terminal, the *p–n* junction is more reverse biased which results in a wider depletion layer near the drain.) A zero bias voltage on the gate results in a maximal source-to-drain current. A reverse voltage on the gate *depletes* the source-to-drain electron flow. A very large reverse current eventually *pinches* the current *off*. Junction field-effect transistors are therefore said to be of the *depletion* or "*normally-on*" type. A periodic variation of the gate voltage naturally varies the source to drain current in the same manner (quite similar to the way the electron flow between emitter and collector in a bipolar transistor is modulated by the base voltage).

Junction field-effect transistors can be used as amplifiers, exploiting the effect that a small change in the gate voltage causes a large change in the channel current. Since the gate-to-channel p–n junction is reverse biased only a minute current flows in the gate/source circuit (as known from Fig. 8.16); the input impedance[13] is therefore high.

JFETs which use n-type semiconductors for the channel material, as depicted in Fig. 8.28, are appropriately called *n-channel field-effect transistors.* The reader may correctly suspect that a *p-channel field-effect transistor* uses holes as charge carriers, n-type semiconductors as gate materials, and a reversal of the polarities of all voltages for its operation. The arrow in the circuit symbol (Fig. 8.28(b)) for p-channel transistors points away from the gate.

Metal–Oxide–Semiconductor Field-Effect Transistor (MOSFET). The depletion type MOSFET consists of high-doped source and drain regions and a low-doped channel, all of the same polarity (n- or p-type). (The high doping facilitates low-resistance connections.) The n-channel MOSFET is laid down on a p-type substrate (called the body, see Fig. 8.29). It differs from the JFET in that a thin oxide layer, on which a metal film has been deposited, electrically insulates the gate metal from the channel. Thus, the input impedance is even larger and the gate current is substantially smaller than for JFET devices. Again, the channel width is controlled by the voltage between gate and body. Specifically, a negative charge on the gate drives the channel electrons away from the gate and towards the substrate, similarly as is illustrated in Fig. 8.12. In short, the channel can be made to be partially depleted of electrons, i.e., the conductive region of the channel becomes narrowed by a negative gate voltage. The more negative the gate voltage (V_G), the smaller the current through the channel from source to drain until eventually the current is *pinched off* (see Fig. 8.29(c).) For the above reasons, this device is called a *depletion-type metal-oxide semiconductor field-effect transistor* or "*normally on*" MOSFET.

An alternative to the depletion-type MOSFET which we just discussed is the *enhancement-type* MOSFET. Figure 8.30 shows that this device does not possess a built-in channel for electron conduction, i.e., at least as long as no gate voltage is applied. In essence, there is no electron flow from source to drain for a zero gate voltage. The device is therefore called a "*normally-off*" MOSFET. If, however, a large enough positive voltage is applied to the gate, most of the holes immediately below the gate oxide are repelled, i.e., they are driven into the substrate, thus removing possible recombination sites. Concomitantly, negative charge carriers are attracted into this channel (called the *inversion layer*). In short, a path (or a bridge) for the electrons between source and drain can be created by a positive gate voltage. The metal-oxide semiconductor technology, in particular, the enhancement-type MOSFETs, dominate the integrated circuit industry at present. They are utilized in memories,

[13] The term *impedance* is used to describe the a.c. resistance which may consist of ohmic, capacitive, and inductive parts.

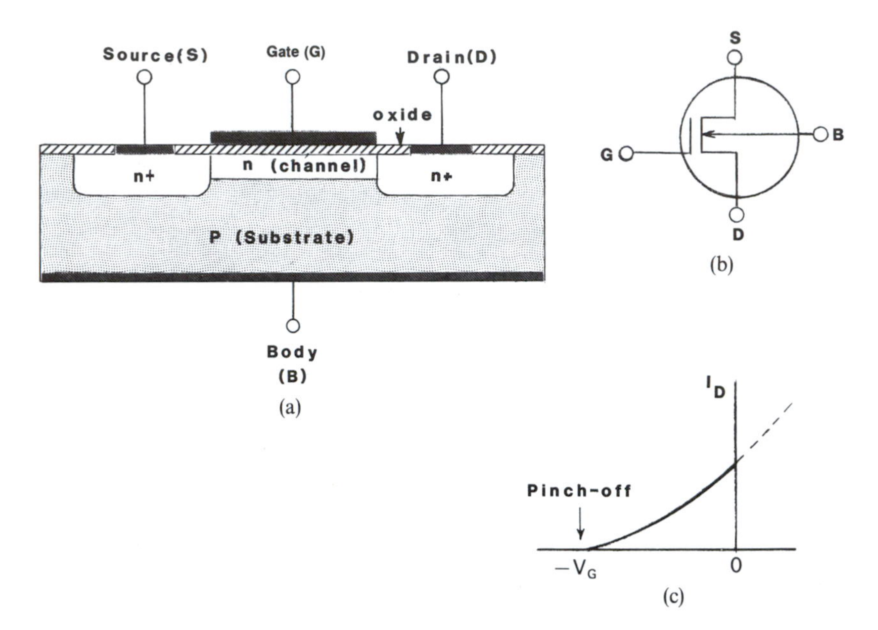

Figure 8.29. (a) Schematic representation of an n-channel depletion- (normally on) type MOSFET. The dark areas symbolize the (aluminum) metallizations. The "oxide" layer may consist of SiO_2, nitrides (Si_3N_4), oxinitrides (Si_3N_4–SiO_2), or multilayers of these substances. This layer is about 10 nm thick. The gate voltage is applied between terminals G and B. Quite often the B and S terminals are interconnected. (b) Circuit symbol for n-channel depletion-type MOSFET. (c) Gate voltage/Drain current characteristics ("Transfer" characteristics). For positive gate voltages (dashed portion of the curve) the device can operate in the "enhancement mode" (see Fig. 8.30(c)).

microcomputers, logic circuits, amplifiers, analog switches, and operational amplifiers with very high input impedances.

Depletion-type and enhancement-type MOSFET technologies which utilize n-channels (as depicted in Figs. 8.29 and 8.30) are summarized by the name "NMOSFET" (in contrast to "PMOSFET" which employs devices with p-channels). If both an n-channel and a p-channel device are integrated on one chip and wired in series, the technology is labeled "CMOSFET" which stands for *complementary MOSFET*. This tandem device has become the dominant technology for information processing, because of its low operating voltage (0.1 V), low power consumption (heat!), and short channel length with accompanying high speed. Alternative names for MOSFET are MOST (metal–oxide–semiconductor transistor) or MISFET (metal–insulator–semiconductor field-effect transistor). Bipolar transistors in combination with JFETs are called "BIFETs." They are used in high-performance linear circuits. If a JFET structure employs a metal–semiconductor junction often in combination with n-type GaAs a "MESFET" device is created which is used

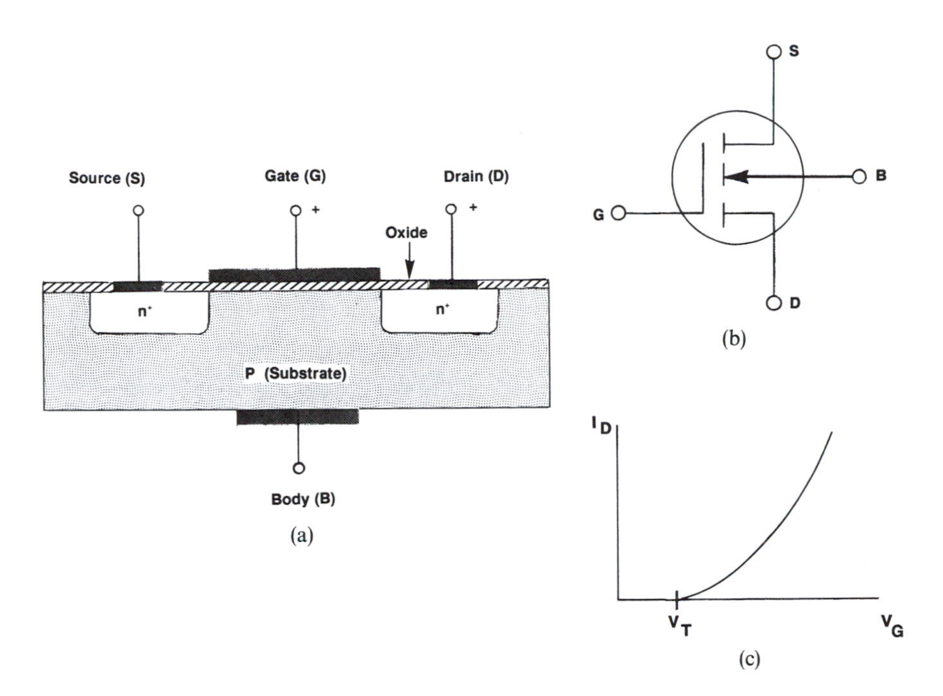

Figure 8.30. (a) Enhancement (normally-off)-type *n*-channel MOSFET. For details, see the caption of Fig. 8.29. (b) Circuit symbol. (The broken line indicates that the path between S and D is normally interrupted.) (c) Gate Voltage (V_G)/Drain Current (I_D) characteristics. V_T is the threshold gate voltage above which a drain current sets in.

for amplifiers and logic circuits in the gigahertz range (see next section). A MODFET (modulation doped field-effect transistor) consists of a thin layer of aluminum–gallium–arsenide which is deposited on an undoped GaAs substrate. This device is even faster than a MESFET because the absence of impurity atoms increases the distance which an electron or a hole can travel before a collision with a foreign atom occurs.

Finally, a few words on device geometry, etc., of a MOSFET, as shown in Fig. 8.29, may be useful. In order to obtain a short switching time and a high-frequency response, the channel length has to be short. The highest possible frequency at which amplification can be achieved equals the inverse of the electron source-to-drain transit time. The width of the device has to be kept small in order to reduce the cross-sectional area and, thus, the power density. (This reduces the heat which needs to be removed.) As an example, the channel length may be about 1 μm, the device width may be a few micrometers, and the field oxide thickness may be near 0.05 μm. The doping of the *p*-area needs to be small to sustain a high resistance and thus, a high electric field ($\sim 10^6$ V/cm) across the junction without current breakdown. The metal layer is generally made of aluminum. Alternate materials are highly doped silicon, refractory metals such as tungsten, or silicides of refractory metals such as TiSi or MoSi.

**Gallium–Arsenide Metal–Semiconductor Field-Effect Transistor (MESFET).* Users of computers demand still higher switching speeds than the present 10^{-9} s cut-off or cut-on times achieved with silicon technology. Gallium–arsenide with its almost sixfold larger electron mobility compared to silicon (see Appendix 4) seems to be the answer. A quick inspection of the relevant band diagrams (Figs. 5.23 and 5.24) indeed confirms that the curvature of the conduction band near Γ is larger for GaAs than the comparable band for silicon (close to the X symmetry point) which translates into a smaller effective mass and, thus, into the just-mentioned larger electron mobility for GaAs. However, the *upper valence* bands for both materials are almost identical and fairly flat. Thus, the effective masses of the *holes* for GaAs and silicon are rather large and their hole mobilities are consequently small (see also Appendix 4). A transistor which aims to exploit the higher electron mobility in GaAs should therefore utilize *n*-type GaAs only.

Figure 8.31 depicts a metal–semiconductor field-effect transistor (MESFET) which consists of an *n*-doped, thin GaAs *active layer* situated over a semi-insulating (Cr-doped) GaAs slab. Three metal contacts provide the source, the gate, and the drain areas. The gate metal forms, together with the underlying semiconductor, a Schottky barrier (see Section 8.7.2). If ϕ_M is larger than ϕ_S and the gate metal is negatively charged, a reverse bias results (Fig. 8.15(a)). The larger the reverse bias, the wider the depletion region. If the depletion region is caused to fill essentially the entire active layer, any attempted electron flow from source to drain is stopped (or *pinched off*). A small negative gate voltage (or no gate voltage at all) allows an almost unhindered source-to-drain electron flow. The device shown in Fig. 8.31 is therefore a *depletion-* (or *normally-on*) *type* FET (see also Fig. 8.29(c)).

For high-speed, low-power applications, however, the *normally-off* GaAs MESFET is even better suited. For this device, the active layer is made so

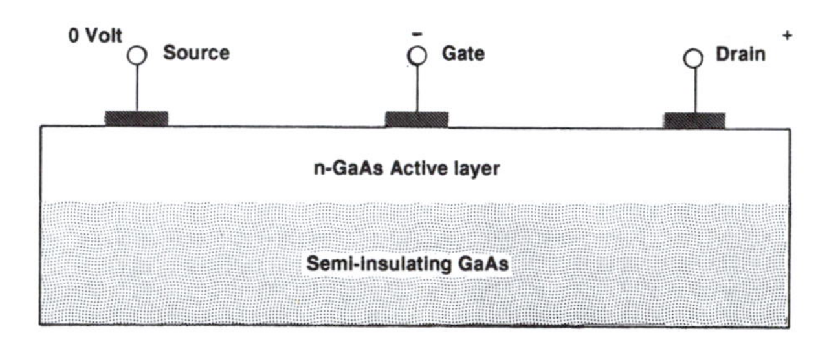

Figure 8.31. Schematic representation of a GaAs MESFET (Metal–semiconductor field-effect transistor). Source and drain metallizations (dark areas) are selected to form ohmic contacts with the *n*-doped GaAs. The gate metal forms, with the *n*-doped GaAs, a Schottky-barrier contact.

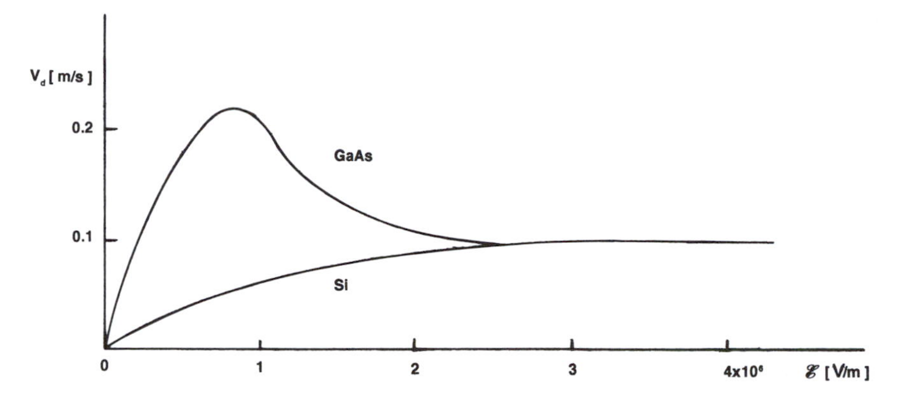

Figure 8.32. Average electron drift velocity as a function of electric field strength for GaAs and silicon.

thin that the depletion area between the metal and the GaAs (Fig. 8.15) fills the entire active layer.[14] As a consequence, the active layer below the gate metal electrode is depleted of electrons without necessitating an applied voltage. A *positive* gate voltage is then required to attract electrons into the depletion area, thus making it conductive. Given the above-described GaAs device, the speed, i.e., the response time of the source-to-drain current to a change in the gate voltage, can be further increased by decreasing the length of the gate, which is presently about 1 μm.

Several effects may, however, offset the superior electron mobility in GaAs. First, the time required to reach the breakdown voltage under the influence of a reverse voltage (see Fig. 8.19(c)) is only two and a half times faster than in silicon. As we know from Fig. 8.20(a), this breakdown electric field triggers a helpful self-ionizing avalanche which multiplies the number of electrons. Second, a transistor of any type can be made to switch faster by applying more power to it. This, in turn, increases the heat which needs to be dissipated. Now, silicon has a three times larger thermal conductivity than GaAs (see Table 19.3). Thus, silicon switches can be made much smaller than those made of GaAs. Since the speed of a device also depends on the length the electrons have to travel, a very small silicon device may well switch as fast as a large device made of GaAs. Third, the electron drift velocity depends upon the electric field strength. At low field strengths, the GaAs drift velocity is indeed substantially larger than for silicon (Fig. 8.32). However, as the field strength increases, the drift velocity for silicon and GaAs becomes nearly identical. This has its reason in the extra and slightly higher energy states which silicon possesses near the X-symmetry point (Fig. 5.23), in which elec-

[14] The depletion layer width in GaAs varies with impurity concentration between 3 μm for 10^{14} cm^{-3} and 0.05 μm for 10^{18} impurity atoms per cubic centimeter.

trons can be scattered after they have collided with structural imperfections of the crystal lattice.

Knowing the facts presented above, it seems understandable why some leading semiconductor manufacturers have left the GaAs field. However, the pendulum may soon swing in the other direction, as suggested in the next section.

*8.7.10. Quantum Semiconductor Devices

It is the ultimate goal of industry to make semiconductor switches for computer applications as small, as fast, as inexpensive, and as efficient as possible. Conventional field-effect transistors pose, ultimately, certain limitations towards progressive miniaturization: the smaller they become, the less effective they switch, owing to current leakage, and particularly because of impurities or lattice defects which scatter the moving electrons in ultrasmall devices to an intolerable degree. There are also processing limitations caused by the presently used photolithography techniques. Quantum structures are said to be the devices of the future which may overcome these shortcomings.

In order to explain the nature of a quantum device we need first to recall that the electron states for bulk crystalline solids consist of continuous energy bands, such as the valence band or the conduction band (Fig. 8.2). We also recall that the density-of-states curve has a parabolic shape in this case (Fig. 6.4). If, however, the dimensions of a crystalline solid are reduced to the size of the wavelength of electrons (e.g., 20 nm for GaAs), the formerly continuous energy bands split into discrete energy levels, similarly as is known from Section 4.2, where we treated the behavior of *one* electron in a potential well. In essence, the same type of calculation as in Section 4.2 is carried out for quantum devices. Thus, results equivalent to (4.18) are obtained. Further, when the dimensions are reduced to the degree as outlined above, and under certain other conditions (see below), the density of states becomes discontinuous, i.e., $Z(E)$ also becomes quantized (see Fig. 8.33(c)). The mechanism associated with these effects is, therefore, quite appropriately called *size quantization.*

Let us demonstrate size quantization for a particular case in which a small band-gap material is sandwiched between two layers of a "wide" band-gap material. Specifically, a cube-shaped piece of GaAs whose lateral dimensions are made to be about 20 nm is layered between two similarly shaped cubes made of aluminum–gallium–arsenide, which in turn are sandwiched between two longer slabs consisting of *n*-doped GaAs (Fig. 8.33(a)). This configuration, for which all three dimensions of the center materials have values which are near the electron wavelength, is called a *quantum dot* (in contrast to a two-dimensional "confinement," which is termed *quantum wire* or a one-dimensional confinement named *quantum well*).

Figure 8.33(b) depicts simplified electron bands for the quantum dot struc-

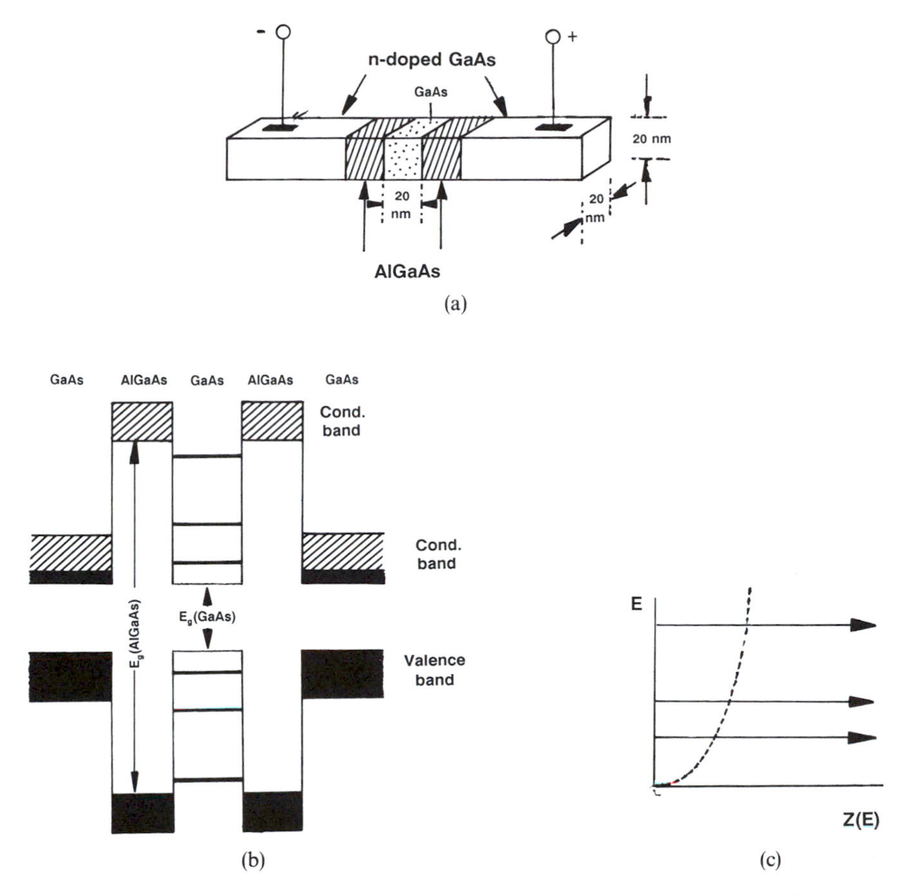

Figure 8.33. (a) Schematic representation of a quantum dot structure. (b) Energy levels for GaAs for the quantum dot structure depicted in (a). (*Note*: The gap energy difference between GaAs ($E_g = 1.42$ eV) and AlGaAs is greatly exaggerated. This difference may be as small as 0.2 eV.) (c) Discontinuous density of energy states for a quantum dot structure. The dashed parabola indicates the density of states for a bulk crystal, as is known from Fig. 6.4.

ture shown in Fig. 8.33(a). AlGaAs is a "wide" band-gap material whose electron affinity (Fig. 8.13) is smaller than that of GaAs. Thus, its conduction band is at a higher energy compared to the conduction band of GaAs. This results in a *potential barrier* between the two GaAs regions. In general, an electron in the *n*-doped GaAs regime does not possess enough energy for climbing this potential barrier, or otherwise diffusing into the adjacent regions (Fig. 8.34(a)). If, however, a sufficiently large voltage is applied to this device the conduction band of the *n*-doped GaAs is raised to a level at which its conduction electrons are at the same height as an empty energy state of the center GaAs region (Fig. 8.34(b)). At this point the electrons are capable of

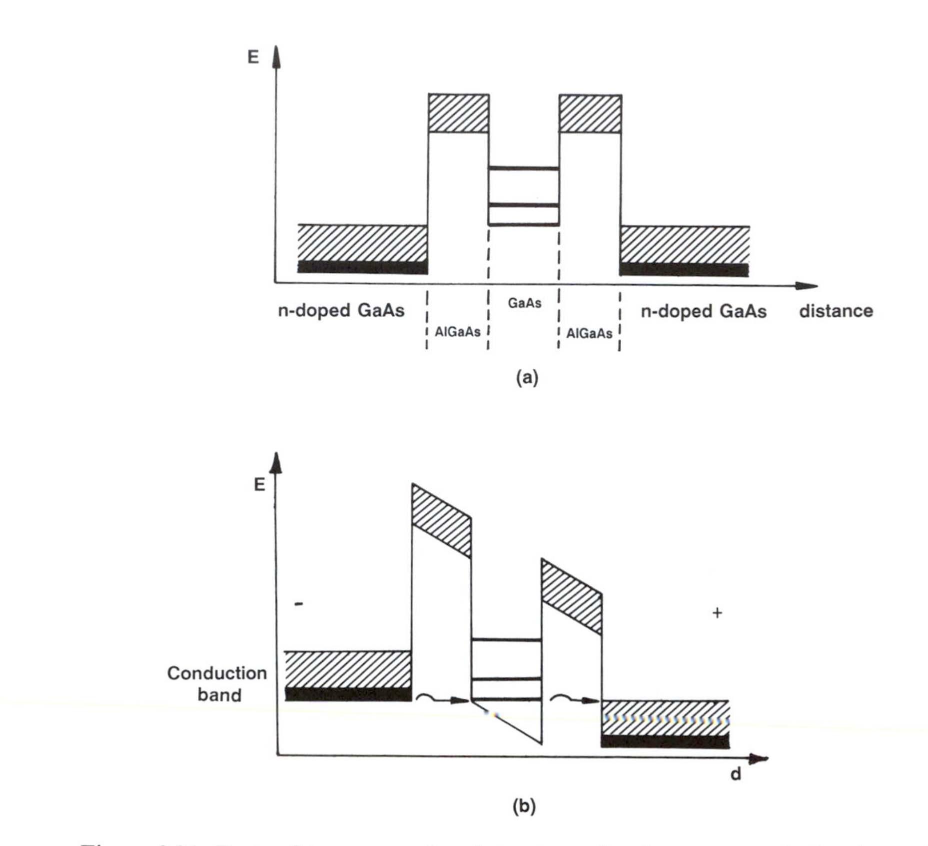

Figure 8.34. Parts of two energy band structures for the quantum device shown in Fig. 8.33. For simplicity, only the conduction bands are shown. (a) No applied voltage. (b) With applied voltage which facilitates electron tunneling from the conduction band of the *n*-doped GaAs into an empty energy level of the center GaAs region.

tunneling through the potential barrier formed by the AlGaAs region and thus reach one of these discrete energy levels. The tunneling is quite effective because of the large density of states which is associated with these quantum states (Fig. 8.33(c)).

If a slightly higher (or somewhat smaller) voltage is applied, the electrons of the *n*-doped GaAs are no longer at par with an empty energy level and the tunneling comes to a near standstill. This causes a current–voltage characteristic with negative differential resistance, i.e., a region in which the current decreases as the applied voltage increases (see Fig. 8.35).

An interrelated effect to size quantization is *resonance* which enhances the tunneling current. Once a specific voltage, the resonating voltage, has been reached, the electron waves inside the center region are reflected back and forth between the walls. In essence, constructive interference occurs between the waves traveling in opposite directions.

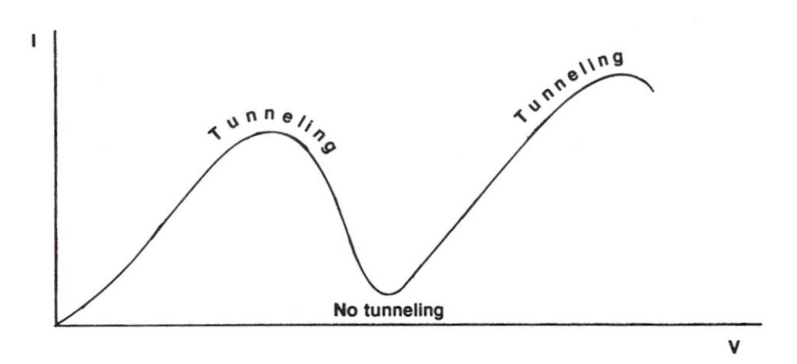

Figure 8.35. Current–voltage characteristic of a quantum dot device as depicted in Figs. 8.33 and 8.34.

A further advancement of the quantum device introduced so far consists of an *array* of a multitude of quantum wells stacked on top of each other. This periodic arrangement of wide band-gap and narrow band-gap materials is called a *superlattice*. It introduces an artificial periodicity into the solid, caused by the multiple atomic layers of one type of material in sequence with multiple atomic layers of another type. By this mode of varying the structural parameters of a solid, new electronic properties can be engineered.

Quantum devices are about one-hundredth of the size of presently known FETs. Thus, major problems have still to be overcome concerning interconnections, device architecture, and with respect to creating three-terminal devices. It has been speculated, however, that once these problems have been solved the reduction in cost per function might be as large as ten-thousand fold.

8.7.11. Semiconductor Device Fabrication

The evolution of solid state microelectronic technology started in 1947 with the invention of the point contact transistor by Bardeen, Brattain, and Shockley at Bell Laboratories. (Until then, electronic devices used vacuum tubes invented in 1906 by Lee deForest, and silicon rectifiers which were known since the beginning of this century.) The development went via the *germanium junction transistor* (Shockley, 1950), the *silicon transistor* (Shockley, 1954), the first *integrated circuit* (Kilby, Texas Instruments, 1959), the *planar transistor* (Noyce and Fairchild, 1962), and the *planar epitaxial transistor* (Texas Instruments, 1963) to the *ultra large-scale integration* (ULSI) of today with several millions of transistors on one chip. Attempts are now made to reach one billion transistors per chip, called *gigascale integration* (GSI). We have discussed in the previous sections some limits towards this goal which are imposed to a large degree by the "*materials barrier*." (However, device limits, circuit limits, and system limits likewise play a role.) Silicon has been the principal semiconductor material used in the past 40 years. No other electronic material has a combination of so many favorable properties. Most of all, silicon is abundant; 28% of the earth's crust consists of silicon in one way or the other. (Silicon is behind oxygen, the second most abundant chemical element.) The raw material (sand, i.e., quartzite) is inexpensive. The native oxide, silicon dioxide (SiO_2), is an excellent insulator. The band gap is large enough to guarantee stable electrical properties at moderate temperatures. The heat conductivity is relatively large. Further, silicon forms almost perfect (dislocation-free) single crystals. And finally, silicon is nontoxic, i.e., environmentally safe. Still, for special applications and possibly gigascale integration, compound semiconductors need to be considered, as discussed in the previous chapters.

The starting material for silicon wafer fabrication is sand (SiO_2) which is electromet reduced (in an arc furnace) with coal, etc., to 98% silicon. This powdered raw silicon is reacted with hydrogen chloride to form trichlorosilane gas ($Si + 3HCl \rightarrow SiHCl_3 + H_2$) which is fractionally distilled for purification and subsequently reduced with hydrogen to polycrystalline silicon ($SiHCl_3 + H_2 \rightarrow Si + 3HCl$). From here on several methods for single crystal growth are used. In the predominantly utilized crucible pulling process, introduced by *Czochralski*, the high-purity silicon is melted in a fused-silica (SiO_2) crucible which is, in turn, supported by a carbon crucible (Fig. 8.36(a)). A seed crystal (mainly (100) or (111) orientation), held on a rod, initially touches the melt and is then slowly lifted, employing a withdrawal speed of about 1 mm per minute. Concomitantly, the crucible as well as the pulling rod are rotated in opposite directions with about 50 revolutions per minute. The entire system is enclosed in a chamber which is either slightly evacuated (a few Torrs) or backfilled with argon or helium. Proper cooling and pulling speeds allow one to control the diameter of the evolving single crystal rod. The starting crystal must initially have a thin neck to produce a dislocation-free crystal.

Since the crucible consists of SiO_2 and of carbon, some oxygen and carbon are introduced into the silicon during melting (about 5×10^{17} oxygen atoms and about 2×10^{16} carbon atoms per cubic centimeter). Other foreign elements of high-purity silicon are generally in the 10^{10}–10^{13} per cubic centimeter range. Oxygen and carbon impurities are electrically inactive because they form inert compounds with silicon (e.g., SiO_2 or SiC). However, their presence in high concentrations leads to the premature breakdown of p–n junctions. Harmful impurities and tiny defects can be trapped (*gettered*) either at a specially prepared back side of the wafer (e.g., by mechanically introduced dislocations) or inside the crystal on very small SiO_2 precipitates. On the other hand, impurities such as phosphorus or boron may be intentionally added to the melt for doping purposes to yield n-type or p-type materials.

A lower oxygen concentration (10^{16} atoms/cm^3) can be achieved involving the crucibleless *float-zone technique* (Fig. 8.36(b)). At first, a pure, polycrystalline silicon rod is manufactured by resistive heating a silicon filament (8 mm wide, 2 m long) in a trichlorosilane and hydrogen atmosphere. As men-

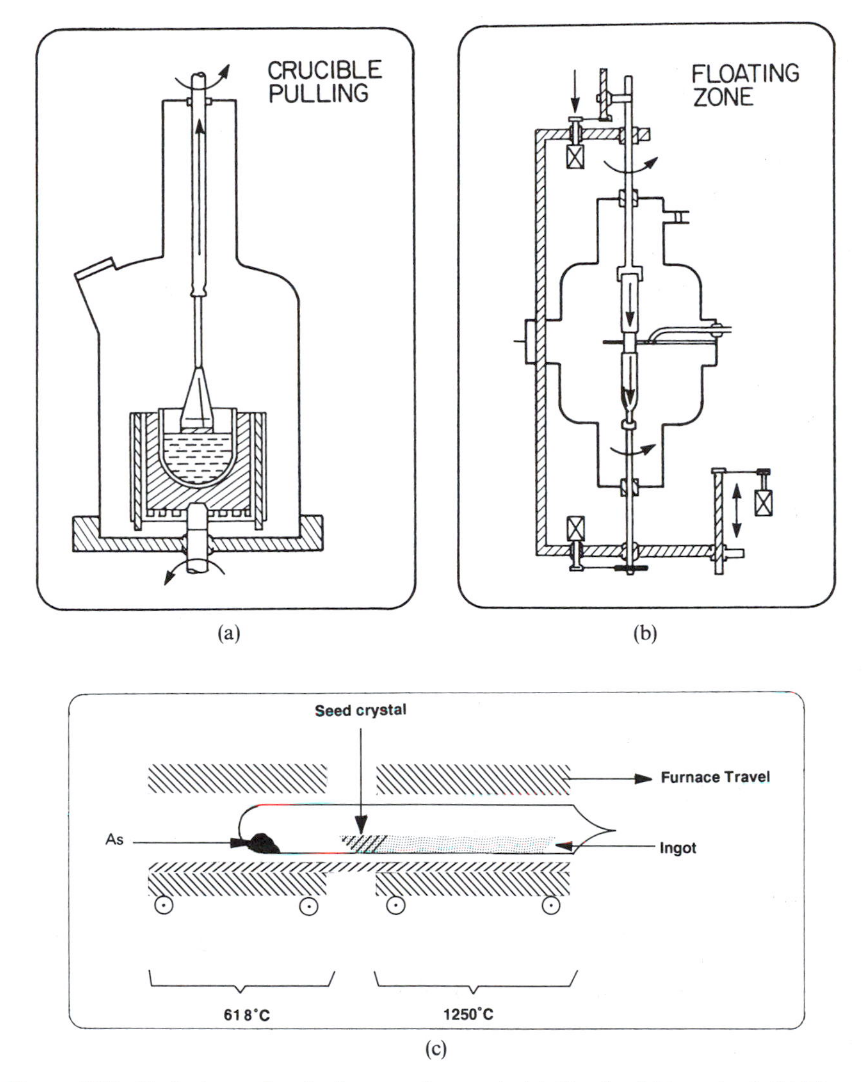

Figure 8.36. Techniques for single-crystal growth. (a) Czochralski method. (b) Floating zone method. (c) Bridgman method (demonstrated for GaAs).

tioned above, this reduces the $SiHCl_3$ to silicon which is slowly deposited on the 1000° C hot silicon filament. The polysilicon rod thus grown is vertically inserted into a vacuum chamber with a single crystalline seed crystal at its bottom, and then rotated. An induction-heated ring-shaped furnace is slowly moved along the rod which melts, at first, a part of the seed crystal and then consecutively small zones (a few cm long) from the bottom up, thus eventually forming a large single crystal as an extension of the seed crystal. The float-zone technique is also used for purification purposes (zone refining).

The *Bridgman technique* is rarely used for silicon production. It is, however, frequently applied to grow single crystalline GaAs. The Bridgman method involves the melting of polycrystalline material in a long (silicon–nitride coated) carbon crucible which, in turn, is placed into a horizontally arranged, sealed quartz tube. A traveling furnace with two different heating zones melts the ingot as well as part of a single crystalline seed which is placed next to it. In the case of GaAs (Fig. 8.36(c)) some extra arsenic, located in the low temperature (618° C) part of the tube, provides an overpressure of arsenic to maintain stoichiometry. Moving the hot zone of the furnace slowly away from the melt causes gradual solidification and eventually an extension of the single crystal into the entire rod.

The two-zone furnace is also used to melt separately arsenic and gallium in individual boats, which facilitates the synthesis of GaAs over a period of many hours.

Once the rods have been obtained, they are sliced, lapped, etched, and polished to obtain the 0.3–0.4 mm thick wafers. At present up to 8 in. diameter silicon wafers are commercially available; the trend goes, however, towards the 12 in. disc.

Next, the devices are fabricated on (or in) these wafers in extremely clean rooms, applying surface oxidation, photolithography, etching, and (most of all) by introducing various dopants involving successive and often quite elaborate manufacturing steps. The most important of these production steps are illustrated in Fig. 8.37 and described in detail below. A simplified example of the final product is depicted in Fig. 8.38, which contains some basic electronic components on one common substrate. Millions of these elements are squeezed on a, say, $6 \times 6\ \mathrm{mm}^2$ area, and hundreds of these complete circuits (properly interconnected) are fabricated together on one wafer. After electrical testing, the individual chips are cut apart with a diamond-tipped saw. The functional chips are finally packaged in hermetically sealed containers and sold.

The individual manufacturing steps shown in Fig. 8.37 are as follows:

Oxidation. Silicon dioxide forms readily on silicon by placing the wafer into a tube furnace, heated between 900° and 1200° C, and exposing it to water vapor and possibly oxygen. This *wet* or *steam oxidation* is much faster than dry oxidation for which oxygen without steam is reacted with the silicon slice. Occasionally, silicon nitride replaces SiO_2.

Photolithography. In order to be able to etch small openings through an SiO_2 layer at a desired place, the silicon dioxide needs first to be coated with a protective layer called the *photoresist* which, after exposure to UV light and subsequent developing, remains on the substrate. Thus, a mask (comparable to a photographic negative) has to be produced which contains a pattern of nontransparent areas. The mask/photoresist/wafer sandwich is then exposed to UV light. During the subsequent developing process, the unexposed photo-

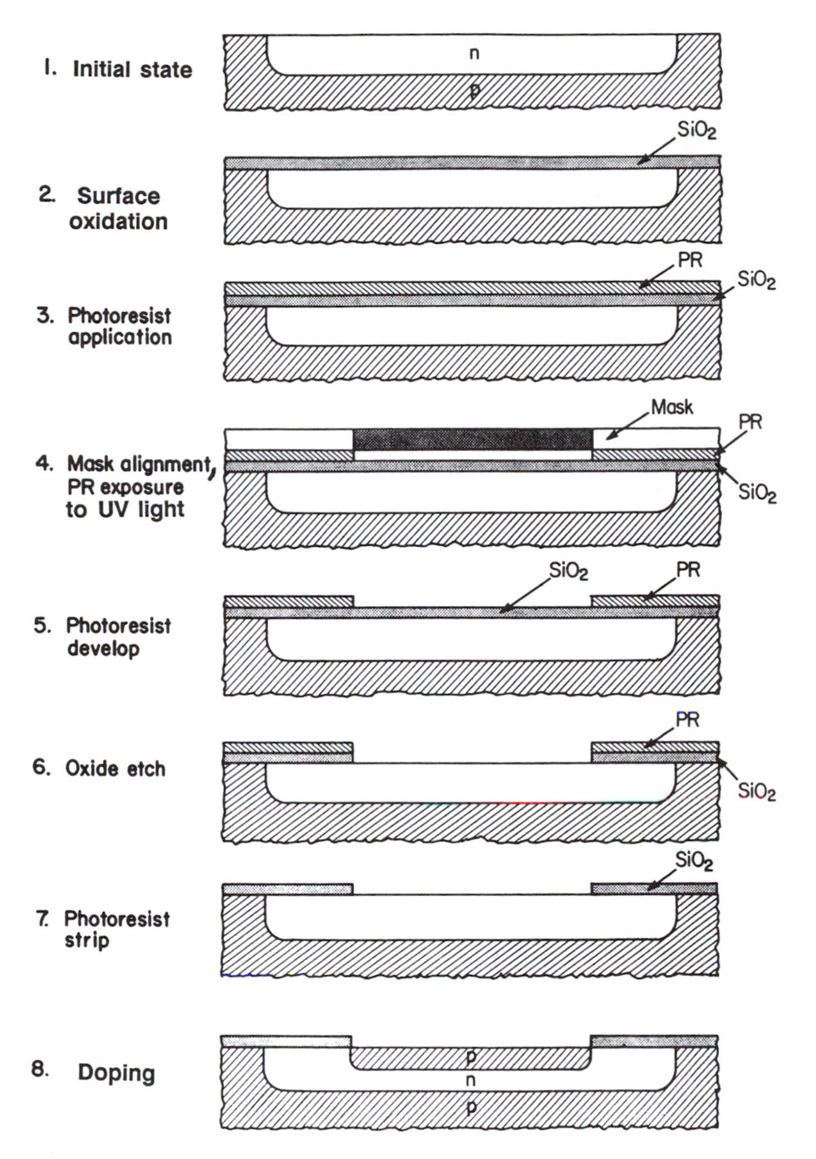

Figure 8.37. Photoresist (PR) masking sequence to obtain a *p–n–p* transistor.

resist is dissolved at the places where the SiO_2 needs to be removed. The remaining photoresist protects the SiO_2 from the etching solution.

Oxide Etch. Wet chemical removal of the SiO_2 layer is accomplished by applying hydrofluoric acid (HF) at room temperature. The underlying silicon is not attacked by this etchant (this would require a HNO_3/HF solution). *Wet chemical etching* poses, however, some problems if submicron geometries

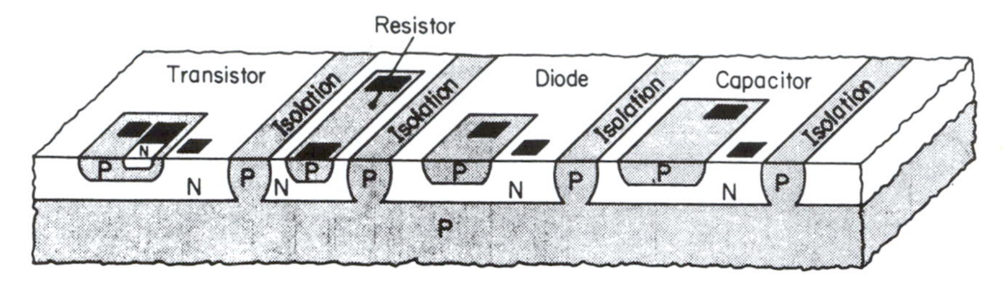

Figure 8.38. Basic components of integrated circuits (bipolar). The dark areas are the contact pads.

need to be produced, since the etchant attacks not only vertically but also laterally, causing line broadening by undercutting. Thus, *dry etching* techniques, such as ion etching or reactive plasma etching, are increasingly utilized. Ion etching involves the removal of the exposed material by bombardment with energetic noble gas ions (such as argon). Since ion etching removes SiO_2 as well as the underlying silicon, the bombarding time has to be carefully controlled. Plasma etching, on the other hand, uses a chemical reaction which converts the substance to be etched into a volatile compound by utilizing a chemically active gas such as halocarbons in a plasma chamber.

Photoresist Strip. This process is accomplished by a simple chemical dissolution reaction.

Doping. In the pioneering times of semiconductor fabrication, the infusion of donor or acceptor elements into silicon was mainly done out of the gas phase with subsequent drive-in diffusion at temperatures near or above 1000° C. Though this process worked quite well initially, increasing miniaturization demanded a more precise doping technique. Thus, ion implantation has mainly been utilized since the early 1970s. This technique involves the ionization of the species to be implanted and their subsequent acceleration towards the substrate in an electric field. (The silicon needs to be shielded where implantation is unwanted.) The range where the heavy dopants come to rest in the silicon substrate obeys a Gaussian distribution. The collision-induced lattice damage, however, needs to be removed in a subsequent processing step. Annealing between 700° and 1000° C for a short time restores the original lattice symmetry and also causes the dopants to become electrically active.

Packaging. The individual chips (after having been cut from the wafer) are bonded to "headers" consisting of ceramics, such as aluminum nitride or silicon carbide for efficient heat removal, or of plastics for low cost and automation. The contact pads are then connected to the "pins" using extremely fine gold or aluminum wires. The bottleneck for accomplishing ever-increasing speeds in computers is the transfer time of the electrons from one chip to the next. Thus, multichip modules (MCM), i.e., the mounting of many chips into one package, will be increasingly used in the future to eliminate interchip delays. The drawback of this technique is that only one defective chip could make all the remaining chips worthless. In addition, it requires intercompany cooperation since one manufacturer does not usually produce all types of chips.

At the end of this chapter, a few remarks on economics may be of interest. The fabrication of solid state microelectronic devices (chips) is a \$50 billion per year industry (worldwide). This figure does *not* include the factory sales of complete electronic systems in which these chips are incorporated (which is a factor of 10 higher). The increase in value of the packaged circuit compared to the raw material is roughly 10 millionfold if the starting material (sand) is valued at a transportation cost of 3 cents per kilogram. The raw silicon can be purchased for a few dollars per kg. The polysilicon already represents a value of \$60 per kilogram and the silicon wafer sells for \$2,000 per kilogram (about \$120 for an 8 in. wafer). The next steps are big jumps: the processed wafer is valued at \$25,000 per kilogram, the raw chip costs \$110,000, and the packaged chip is finally sold to the computer manufacturers for \$300,000 per kilogram.

Let us look at the economic picture from another point of view. As one might expect, it takes a substantial amount of energy to produce a wafer. The largest part (about 400 kWh/kg) is already consumed to produce the polysilicon. Or put differently: the energy consumption for melting and purification alone is 1000 kWh for 1 m^2 of wafer surface. Production of single crystals by the Czochralski method requires another 150 kWh for 1 m^2 of silicon surface. Doping, etc., consumes 25–50 kWh/m^2 (depending on the complexity of the device). All taken, including packaging, etc., roughly 1400 kWh are expended until microelectronic devices have been fabricated on a 1 m^2 silicon surface. (See, in this context, also Sections 8.7.6 and 9.4.)

*8.7.12. Digital Circuits and Memory Devices

The reader might legitimately wonder at this point how transistors are used in computers and similar devices. Even though this topic sidetracks the flow of our presentation somewhat, a few introductory remarks on switching devices, information processing, and information storage may nevertheless be of interest. We need to start with the recognition that electronic data-processing systems use binary digits, i.e., *zeros* and *ones* as carriers for information. As an example, the numeral sequence "0010" means in the binary system the decimal number "two," whereas 0101 represents the decimal number 5. The first digit at the right of a binary number represents 2^0, the next ones represent, consecutively, 2^1, 2^2, 2^3, etc. A **bi***nary digi***t**, or a *bit* is the smallest possible piece of information. (A group of related bits, e.g., 8 bits for word processing, is called a byte.) A "zero" in the present context means that the

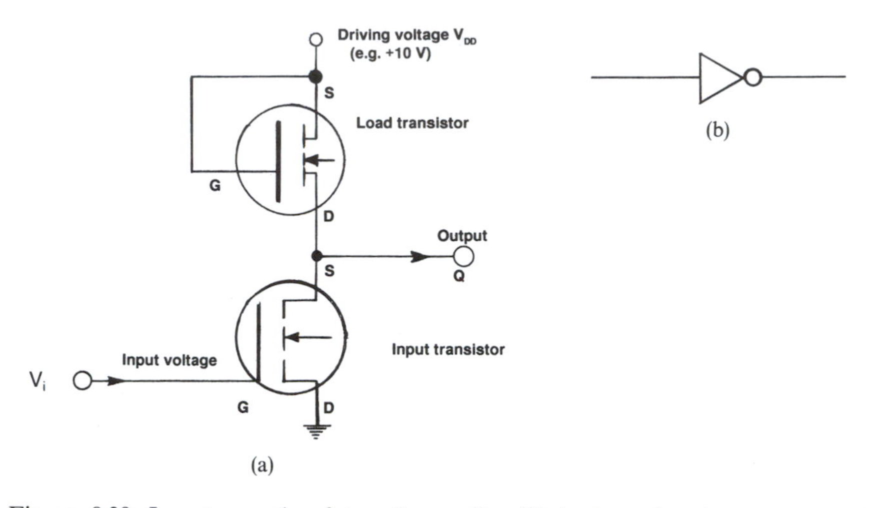

Figure 8.39. Inverter made of two "normally-off" (*n*-channel, enhancement-type) MOSFETs (NOT gate). (a) circuit; (b) symbol in wiring diagram. (V_{DD} means "Drain power supply voltage".) Compare with Fig. 8.30. The load transistor may be replaced by a (polysilicon) resistor.

electrical current is off, whereas a "one" means that the current is on. So much about preliminaries.

Let us begin with an "AND" device. We inspect the normally-off MOSFET in Fig. 8.30 and see that a voltage on the drain terminal is *only* obtained if we apply voltages *simultaneously* to the source *and* the gate terminals. In other words, a source voltage AND a gate voltage cause a voltage on the drain terminal. (The circuit symbol of an AND gate is explained in Fig. 8.40).

Next, we discuss the *inverter* circuit. It consists of two *normally-off* MOS transistors which are wired in series (Fig. 8.39). The upper or *load transistor* (whose channel is made long and narrow to restrict the current flow) is always kept "on" by connecting the driving voltage to its gate. If a high enough voltage is simultaneously applied to the gate of the *lower* or *input* transistor, then this lower MOSFET likewise becomes conducting and the driving voltage drains through both transistors into the ground. As a consequence, the output voltage at terminal Q is nearly zero. Thus, the inverter circuit inverts a "one" signal on the input terminal into a "zero" signal on the output terminal (and vice versa). The circuit symbol for an inverter (or "NOT gate") is shown in Fig. 8.39(b).

We have just mentioned that the current through the load transistor is relatively small due to its special design. Still, an inverter which essentially does not consume *any* power (except during switching) would be even more desirable. This is accomplished by CMOS technology, i.e., by using an enhancement-type *p*-channel MOSFET as a load transistor, and an enhancement-type *n*-channel MOSFET as an input transistor. Unless the circuit is

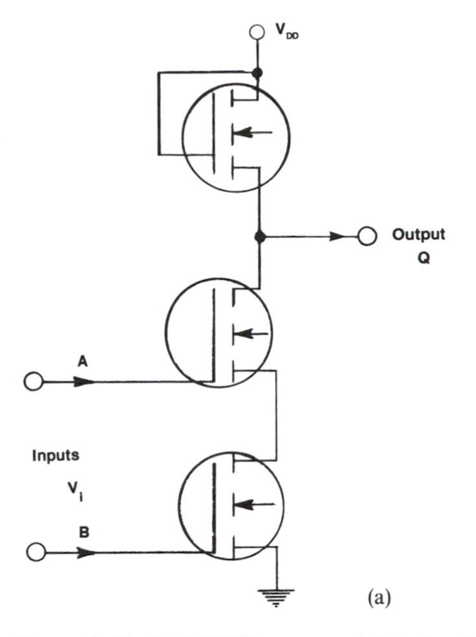

Figure 8.40. (a) NAND gate and (b) circuit symbol for a NAND digital function. (Note that the circuit symbol for an AND gate is the same as above without the open circle at the right.)

switching, one MOSFET is always off (not conducting current) whereas the other is on. Since the two MOSFETs are connected in series similarly as in Fig. 8.39, little power (except due to leakage current) is consumed. It is left to the reader to draw up and discuss the appropriate circuit diagram. (See Problem 18.)

The next logic device which we discuss is a "NAND" circuit. It consists of a load MOSFET, wired in series with two (or more) input transistors. All MOSFETs shown in Fig. 8.40 are of the "normally-off" type. Assume that high enough voltages are applied to the gates of *both* input transistors to make them conducting. Thus, the output terminal Q is connected to ground, i.e., the output voltage is almost zero. Since input voltages on the gates of the A *and* the B transistors inverts the input signal from "one" to "zero," we call the present logic building block an "AND" gate combined with a "NOT circuit" and term the entire digital functional "NOT–AND" or a "NAND" gate for short. The reader may convince himself that the output is always "one" when at least one of the input voltages is "zero." On the other hand, if both inputs are "one," the output is "zero."

Finally, in a "NOR" circuit, the input transistors are wired in parallel (Fig. 8.41). Applying to one *or* all of them high enough gate voltages causes the driving voltage to be "zero." The circuit is appropriately called "NOT–OR" or "NOR." Evidently, the output voltage is only "one" if all input voltages are "zero."

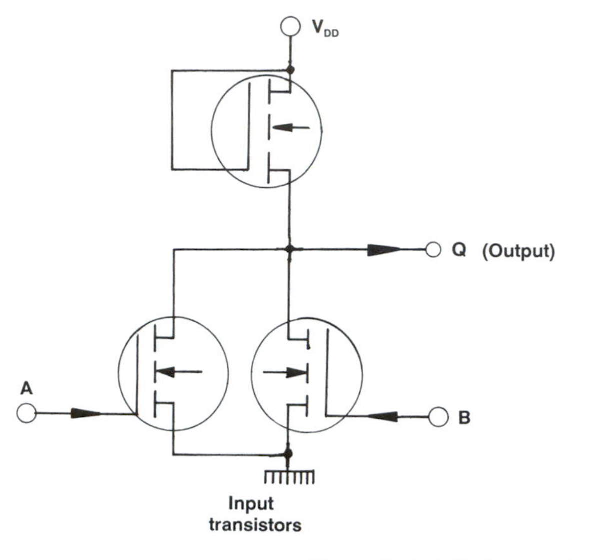

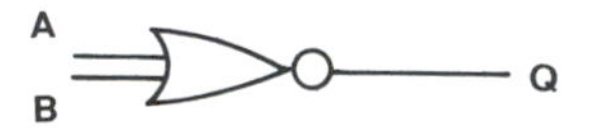

Figure 8.41. NOR logic circuit.

In short, the five basic *building blocks* which are obtained by properly circuiting one or more transistors are AND, NAND, OR, NOR, and NOT. (See also Problem 21.)

We are ultimately interested in knowing how a memory device works, i.e., we are interested in a unit made of transistors which can store information. For comparison, a toggle switch which turns the light on or off can be considered to be a digital information storage device. It can be flipped on and then flipped off to change the content of its information. Appropriately, the device which we are going to discuss is called a flip-flop. Let us assume that a flip-flop has a built-in latch to prevent the accidental change of information. This is done electronically by combining a NOT gate with two AND gates, as shown in the left part of Fig. 8.42. (The output of an AND gate is zero as long as *one* of the inputs is zero!) It is left to the reader to figure out the various combinations. As an example, if the gate is unlatched (1) and the data input is 1, we obtain "zero" on the R terminal and "one" on the S terminal.

The flip-flop on the right part of Fig. 8.42 consists of two NOR gates which are cross coupled. (Remember that the *output* of a NOR is always *zero* when *at least one input is one*, and it is *one* when *both inputs are zero*.) The above example with R = 0 and S = 1 yields a "one" at the output terminal Q (and a "zero" at the complement output $\overline{Q}$), i.e., D and Q are identical. The information which is momentarily fed to the D terminal and into the system is *permanently stored* in the flip-flop even if the wires R and S are cut off. The cross coupling keeps the two NOR gates mutually in the same state at least as long as a driving voltage remains on the devices. In short, one bit of information has been stored.

Let us now latch the gating network, i.e., let G be "zero." Whatever option

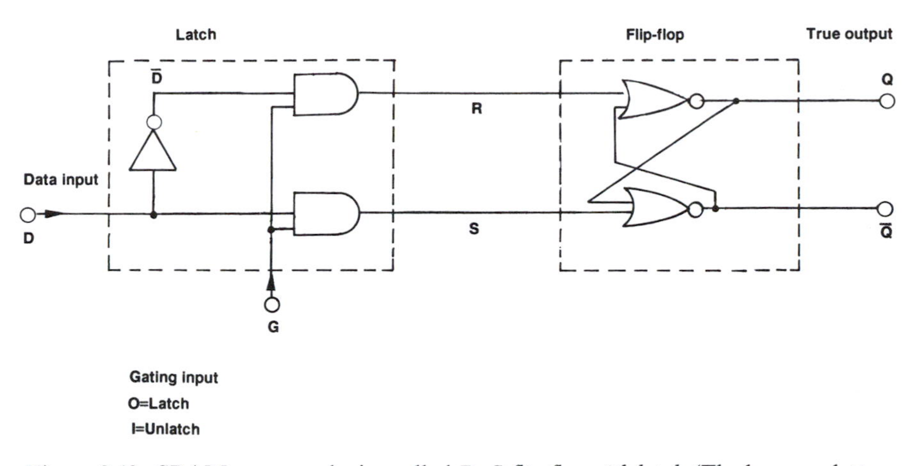

Figure 8.42. SRAM memory device called *R–S flip-flop with latch.* (The bar on a letter signifies the complement information.)

the data input D will assume in this case, the output Q will always be "one," as the reader should verify. In other words, the latching prevents an accidental change of the stored information. On the other hand, latching and unlatching by itself does not change the information content of the flip-flop either!

The device described above is called a *static random-access memory* or, in short, an "S-RAM" because the information remains permanently in the storage unit. The memory cell shown in Fig. 8.42 can evidently store only *one* bit of information. Let us now imagine a two-dimensional array of these storage elements, connected in a number of horizontal and vertical lines (Fig. 8.43). A designated memory element, being located at the crosspoint of a specific row and a specific column wire, can then be exclusively addressed by sending an electrical impulse through *both* wires simultaneously. One of these

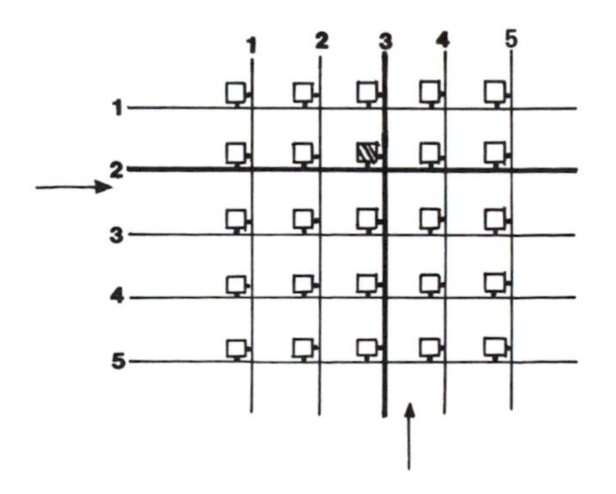

Figure 8.43. Schematic representation of a two-dimensional memory addressing system. By activating the #2 row wire and the #3 column wire, the content of the cross-hatched memory element (situated at their intersection) can be changed.

wires operates on the gating input (Fig. 8.42), the other one activates the data input. As we have discussed above, only a simultaneous activation of both input wires can change the information content of a flip-flop. An array of 32 columns and 32 rows of memory elements constitutes 1024 bits of information storage or one *kilobit*. (Yes, a K-bit is *not* 1,000 bits.)

In order to reduce the area on a chip and the power consumption of a storage device, a different memory cell than the above-introduced flip-flop is frequently used. It is called the *one-transistor dynamic random-access memory* (DRAM, pronounced D-RAM). The information is stored in a capacitor which can be accessed through an enhancement-type transistor (Fig. 8.44). Only concomitant voltages on gate and source allow access to the capacitor. Since the stored charge in a capacitor leaks out in a few milliseconds, the information has to be periodically "refreshed" every 2 milliseconds by means of refresh circuits. No voltage on the capacitor is used as a "zero," whereas a certain voltage on the capacitor represents the "one" logic.

The *16 Megabit chip* combines a multitude of these or similar building blocks through ultralarge-scale integration (ULSI) on one piece of silicon having the size of a finger nail.

The memory devices discussed so far are of the "*volatile*" type, i.e., they lose their stored information once the electric power of the computer is interrupted. In *nonvolatile memories*, such as the *read-only memory* (ROM), information is permanently stored in the device. Let us consider, for example, the MOSFET depicted in Fig. 8.30(b). Assume that the connection to the gate has been permanently interrupted during fabrication. Then the transistor will never transmit current from source to drain. Thus, a "zero" is permanently stored without the necessity to maintain a driving voltage. If, on the other hand, the path to the gate is left intact, the MOSFET can be addressed and current between source and drain may flow, which constitutes a "one." The stored information can be read but it cannot be altered.

In the *programmable read-only memory* (PROM) the information may be written by the user, for example, by blowing selected fuse links to the gate. As above, the alteration is permanent and the information can be read only.

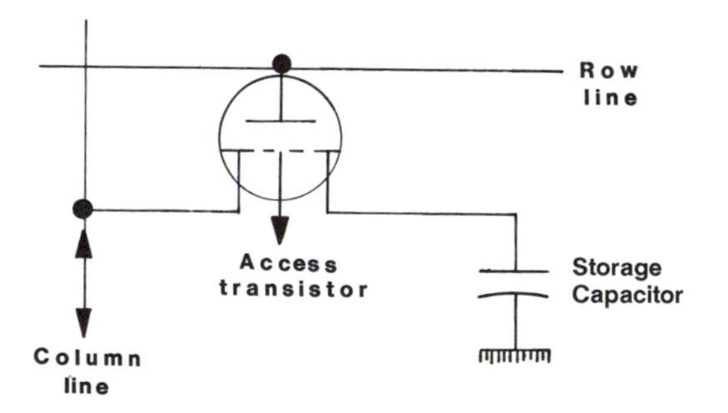

Figure 8.44. One-transistor dynamic random-access memory (DRAM). The information flows in and out through the column line.

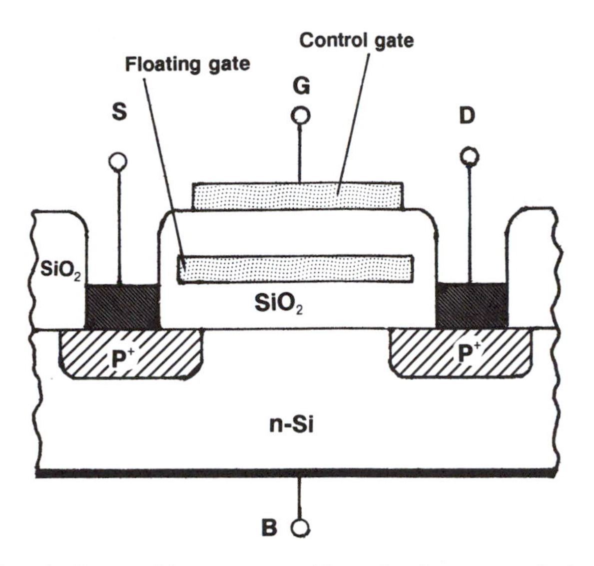

Figure 8.45. Electrically erasable-programmable read-only memory device (EEPROM), also called stacked-gate avalanche-injected MOS (SAMOS), or, with some modifications, flash memory device.

The *erasable-programmable read-only memory* (EPROM) allows the user to program the device as well as erase the stored information. An EPROM contains a "*floating gate*," i.e., a gate (consisting of heavily doped polysilicon) which is completely imbedded in SiO_2, see Fig. 8.45. For programming, the drain-substrate junction is strongly reverse biased until avalanche-breakdown sets in (see Fig. 8.20), and electrons are injected from the drain region into the SiO_2 layer. A large voltage ($\sim$25 V) between a second gate, the control gate, and the substrate allows some electrons to cross the insulator, thus negatively and permanently charging the floating gate. The oxide thickness is on the order of 100 nm which assures a charge retention time of about 100 years. A permanent charge on the floating gate constitutes a "1"; no charge represents the zero state. Exposure of the EPROM to ultraviolet light or X-rays through a window (not shown in Fig. 8.45) increases the conductivity of the insulator and allows the charge to leak out of the floating gate, thus erasing any stored information.

For electrical erasure, a large positive voltage can be applied to the control gate which removes the stored charge from the floating gate. This returns the "*electrically erasable-programmable ROM*" (EEPROM) to the zero state. Alternately, electrical erasure can be performed by applying a positive voltage to the source, which also pulls charge from the floating gate. The "*flash memory*" device uses this erasure method while employing a thinner (10 nm) and higher quality oxide below the floating gate. This improves efficiency and reliability. *Flash memory cards*, which are the size of a credit card, are predicted to cut substantially into the floppy (magnetic) disc and perhaps the hard disc memory market because of their small weight, their lower power consumption, and their fast access time (twenty times faster than a floppy disk). These virtues are important features for laptop computers. Sales for 1995 are projected to be about $1.5 billion.

It should be mentioned in closing that magnetic storage devices are discussed in Sections 17.4 and 17.5. Optical storage devices are explained in Section 13.10.

Problems

Intrinsic Semiconductors

1. Calculate the number of electrons in the conduction band for silicon at $T = 300$ K. (Assume $m_e^*/m_0 = 1$.)
2. Would germanium still be a semiconductor if the band gap was 4 eV wide? Explain! (*Hint*: Calculate N_e at various temperatures. Also discuss extrinsic effects.)
3. Calculate the Fermi energy of an intrinsic semiconductor at $T \neq 0$ K. (*Hint*: Give a mathematical expression for the fact that the probability of finding an electron at the top of the valence band plus the probability of finding an electron at the bottom of the conduction band must be 1.) Let $N_e \equiv N_p$ and $m_e^* \equiv m_h^*$.
4. At what (hypothetical) temperature would all 10^{22} (cm^{-3}) valence electrons be excited to the conduction band in a semiconductor with $E_g = 1$ eV?
5. The outer electron configuration of neutral germanium in its ground state is listed in a textbook as 4 $s^2 4p^2$. Is this information correct? Someone argues against this configuration that the p-states hold six electrons. Therefore, the p-states in germanium and therefore the valence band are only partially filled. Who is right?
6. In the figure below, σ is plotted as a function of the reciprocal temperature for an intrinsic semiconductor. Calculate the gap energy.

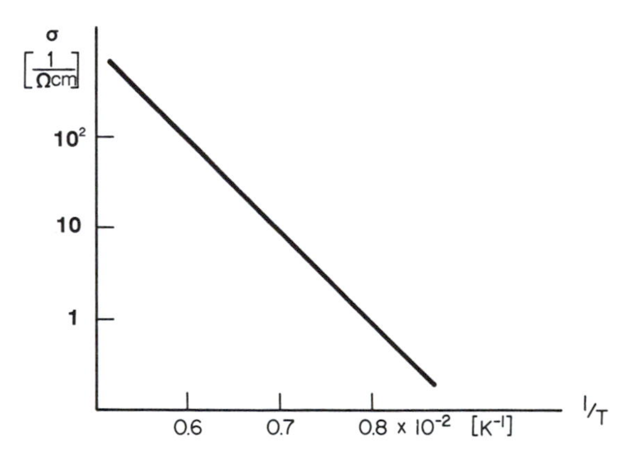

Extrinsic Semiconductors

7. Calculate the Fermi energy and the conductivity at room temperature for germanium containing 5×10^{16} arsenic atoms per cubic centimeter. (*Hint*: Use the mobility of the electrons in the host material.)

8. Consider a silicon crystal containing 10^{12} phosphorous atoms per cubic centimeter. Is the conductivity increasing or decreasing when the temperature is raised from 300° to 350° C? Explain by giving numerical values for the mechanisms involved.

9. Consider a semiconductor with 10^{13} donors/cm^3 which have an ionization energy of 10 meV.
   (a) What is the concentration of extrinsic conduction electrons at 300 K?
   (b) Assuming a gap energy of 1 eV (and $m^* \equiv m_0$), what is the concentration of intrinsic conduction electrons?
   (c) Which contribution is larger?

10. The binding energy of a donor electron can be calculated by assuming that the extra electron moves in a hydrogen-like orbit. Estimate the donor binding energy of an n-type impurity in a semiconductor by applying the modified equation (4.18a)

$$E = \frac{m^* e^4}{2\hbar^2 \varepsilon^2},$$

where $\varepsilon = 16$ is the dielectric constant of the semiconductor. Assume $m^* = 0.8\, m_0$. Compare your result with experimental values listed in Appendix 4.

11. What happens when a semiconductor contains both donor and acceptor impurities? What happens with the acceptor level in the case of a predominance of donor impurities?

Semiconductor Devices

12. You are given a p-type doped silicon crystal and are asked to make an ohmic contact. What material would you use?

13. Describe the band diagram and function of a p–n–p transistor.

14. Can you make a solar cell from metals only? Explain!

*15. A cadmium sulfide photodetector is irradiated over a receiving area of 4×10^{-2} cm^2 by light of wavelength 0.4×10^{-6} m and intensity of 20 W m^{-2}.
   (a) If the energy gap of cadmium sulfide is 2.4 eV, confirm that electron–hole pairs will be generated.
   (b) Assuming each quantum generates an electron–hole pair, calculate the number of pairs generated per second.

16. Calculate the room temperature saturation current and the forward current at 0.3 V for a silver/n-doped silicon Schottky-type diode. Take for the active area 10^{-8} m^2 and for $C = 10^{19}$ A/m^2 K^2.

17. Draw up a circuit diagram and discuss the function of an inverter made with CMOS technology. (*Hint*: An enhancement-type *p–n–p* MOSFET needs a negative gate voltage to become conducting; an enhancement type *n–p–n* MOSFET needs for this a positive gate voltage.)

18. Draw up a circuit diagram for an inverter which contains a *normally-on* and a *normally-off* MOSFET. Discuss its function.

19. Convince yourself that the unit in (8.26) is indeed the Ampere.

20. Calculate the thermal energy provided to the electrons at room temperature. Your will find that this energy is much smaller than the band gap of silicon. Thus, no intrinsic electrons should be in the conduction band of silicon at room temperature. Still, according to your calculations in Problem 1, there is a sizable amount of intrinsic electrons in the conduction band at $T = 300$ K. Why?

21. Propose an "OR" logic circuit.

22. Calculate the lateral dimensions of a quantum well structure made of GaAs. (*Hint*: Keep in mind that the lateral dimension has to equal the wavelength of the electrons in this material.) Refer to Section 4.2 and Fig. 4.4(a). Use the data contained in the tables of Appendix 4.

23. Calculate the number of electrons and holes per incident photon, i.e., the quantum efficiency in a transverse photodiode. Take $W = 8$ μm, $L = 8$ mm, and $\alpha = 40$ cm^{-1}.

CHAPTER 9

Electrical Conduction in Polymers, Ceramics, and Amorphous Materials

9.1. Conducting Polymers and Organic Metals

Materials which are electrical (and thermal) insulators are of great technical importance and are, therefore, used in large quantities in the electronics industry, e.g., as handles for a variety of tools, as coatings of wires, or for casings of electrical equipment. Most polymeric materials have the required insulating properties and have been used for decades for this purpose. It came, therefore, as a surprise when it was discovered that some polymers and organic substances may have electrical properties which resemble those of conventional semiconductors, metals, or even superconductors. We shall focus our attention mainly on these materials. This does not imply that conducting polymers are of technical importance at this time. Indeed, they are not yet. This is due to the fact that many presently known conducting polymers seem to be unstable in air or above room temperature. In addition, many dopants used to impart a greater conductivity are highly toxic, and doping often makes the polymers brittle. These problems might be overcome, however, as more conducting polymers are synthesized.

We now attempt to discuss conducting polymers in the light of solid state physics. Conventional solid state physics deals preferably with the properties of well-defined regular arrays of atoms. We have learned in Chapter 7 that a periodic array of lattice atoms is imperative for coherent scattering of electron waves and thus for a high conductivity. Further, the periodic arrangement of atoms in a crystal and the strong interactions between these atoms causes, as we know, a widening of energy levels into energy bands (see Section 4.4).

We know that highly conducting materials such as metals are characterized by partially filled bands, which allow a free motion of the conduction electrons in an electric field. Insulators and semiconductors, on the other hand, possess (at least at 0 K) completely filled valence bands, and empty conduction bands. The difference in band structure between crystalline insulators and semiconductors is a matter of degree rather than of kind: insulators have wide gaps between valence and conduction bands whereas the energy gaps for semiconductors are narrow. Thus, in the case of semiconductors, the thermal energy is large enough to excite some electrons across the gap into the conduction band. The conductivity in pure semiconductors is known to *increase* (exponentially) with increasing temperature and decreasing gap energy (8.14), whereas the conductivity in metals *decreases* with increasing temperature (Fig. 7.7). Interestingly enough, conducting polymers have a temperature dependence of the conductivity similar to that of semiconductors. This suggests that certain aspects of semiconductor theory may be applied to conducting polymers. The situation in polymers cannot be described, however, without certain modifications to the band model brought forward in the previous chapters. This is due to the fact that polymeric materials may exist in amorphous as well as in crystalline form or, more commonly, as a mixture of both. This needs to be discussed in some detail.

Polymers consist of molecules which are long and chainlike. The atoms which partake in such a chain (or macromolecule) are regularly arranged along the chain. Several atoms combine and form a specific building block, called a monomer, and thousands of monomers combine to a polymer. As an example, we depict polyethylene which consists of repeat units of one carbon atom and two hydrogen atoms, Fig. 9.1(a). If one out of four hydrogen atoms in polyethylene is replaced by a chlorine atom, polyvinylchloride (PVC) is formed upon polymerization (Fig. 9.1(b)). In polystyrene, one hydrogen atom is replaced by a benzene ring. More complicated macromolecules may contain side chains which are attached to the main link. They are appropriately named "branched polymers." Macromolecules whose backbones consist largely of carbon atoms, as in Fig. 9.1, are called "organic" polymers.

The binding forces which hold the individual atoms in polymers together are usually covalent and sometimes ionic in nature. Covalent forces are much stronger than the binding forces in metals. They are based on the same interactions which are responsible for forming a hydrogen molecule from two hydrogen atoms. Quantum mechanics explains covalent bonds by showing

Figure 9.1. (a) Polyethylene. (b) Polyvinylchloride. (The dashed enclosures mark the repeat unit. Polyethylene is frequently depicted as two CH_2 repeat units for historical reasons.)

that a lower energy state is achieved when two equal atomic systems are closely coupled and in this way exchange their energy (see Section 16.2). In organic polymers each carbon atom is often bound to four atoms (see Fig. 9.1) because carbon has four valencies.

In contrast to the strong binding forces between the atoms within a polymeric chain, the secondary interactions between the individual macromolecules are usually weak. The latter are of the Van der Waals type, i.e., they are based on forces which induce dipole moments in the molecules. (Similar weak interactions exist between the atoms in rare gases such as argon, neon, etc.)

In order to better understand the electronic properties of polymers by means of the electron theory and the band structure concept, one needs to know the degree of order or the degree of periodicity of the atoms because only ordered and strongly interacting atoms or molecules lead, as we know, to distinct and wide electron bands. Now, it has been observed that the degree of order in polymers depends on the length of the molecules and on the simplicity of the molecular structure. Certain heat treatments may influence some structural parameters. For example, if a simple polymer is slowly cooled below its melting point, one might observe that some macromolecules align parallel to each other. The individual chains are separated by regions of supercooled liquid, i.e., of amorphous material (Fig. 9.2). Actually, slow cooling yields, for certain polymers, a highly crystalline structure.

In other polymers, the cooling procedure might cause the entire material to go into a supercooled-liquid state. In this state the molecules can be considered to be randomly arranged. After further cooling, below a glass transition temperature, the polymer might transform itself into a glassy amorphous solid which is strong, brittle, and insulating. However, as stated before, we shall concern ourselves mainly with polymers which have a high degree of crystallinity. Amorphous materials will be discussed in Section 9.4.

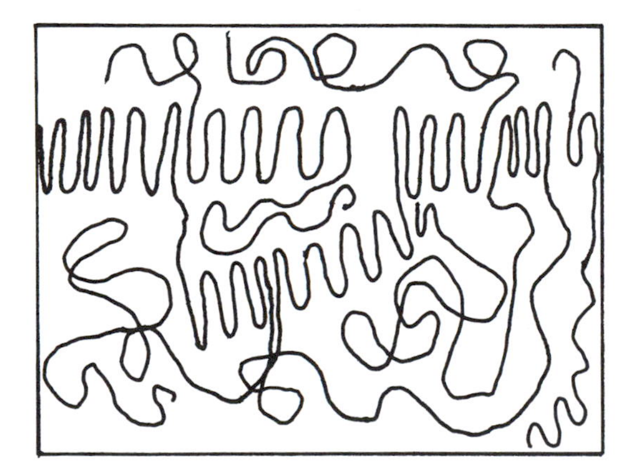

Figure 9.2. Simplified representation of a semicrystalline polymer (folded-chain model).

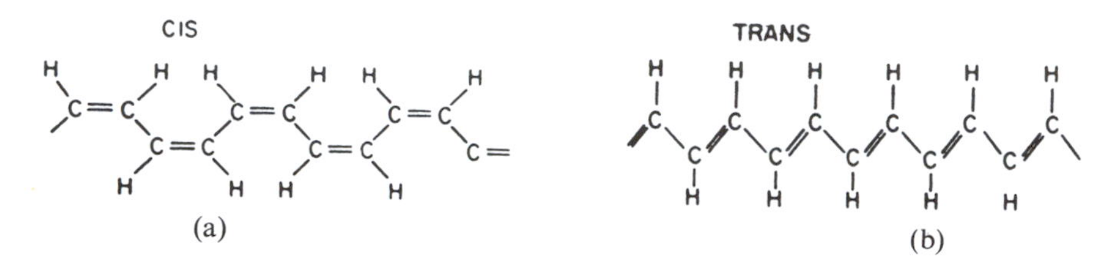

Figure 9.3. Theoretical isomers of polyacetylene (a) *cis*-transoidal isomer, (b) *trans*-transoidal isomer. Polyacetylene is synthesized as *cis*-$(CH)_x$ and is then isomerized into the *trans*-configuration by heating it at 150° C for a few minutes.

A high degree of crystallinity and a relatively high conductivity have been found in polyacetylene which is the simplest *conjugated* organic polymer. It is considered to be the prototype of a conducting polymer. A conjugated polymer has alternating single and double bonds between the carbons (see Fig. 9.3 which should be compared to Fig. 9.1(a)). Two principle isomers are important: in the *trans* form, the hydrogen atoms are alternately bound to opposite sides of the carbons (Fig. 9.3(b)), whereas in the *cis* form the hydrogen atoms are situated on the same side of the double-bond carbons (Fig. 9.3(a)). *Trans*-polyacetylene is obtained as a silvery, flexible film which has a conductivity comparable to that of silicon (Fig. 9.4).

Figure 9.5 shows three band structures for *trans*-$(CH)_x$ assuming different distances between the carbon atoms. In Fig. 9.5(a) all carbon bond lengths are taken to be equal. The resulting band structure is found to be characteristic for a metal, i.e., one obtains distinct bands, the highest of which is *partially* filled by electrons. Where are the free electrons in the conduction band coming from? We realize that one of the electrons in the double bond of a conjugated polymer can be considered to be only loosely bound to the neighboring carbon atoms. Thus, this electron is easily disassociated from its carbon atom by a relatively small energy, which may be provided by thermal energy. The delocalized electrons may be accelerated as usual in an electric field.

In reality, however, a uniform bond length between the carbon atoms does not exist in polyacetylene. Instead, the distances between the carbon atoms alternate because of the alternating single and double bonds. Band structure calculations for this case show, interestingly enough, some gaps between the individual energy bands. The resulting band structure is typical for a semiconductor (or an insulator)! The width of the band gap near the Fermi level depends mainly on the degree of alternating bond lengths (Fig. 9.5(b) and (c)).

It has been shown that the band structure in Fig. 9.5(b) best represents the experimental observations. Specifically, one finds a band gap of about 1.5 eV and a total width of the conduction band of 10–14 eV. The effective mass m^* is $0.6m_0$ at $k = 0$ and $0.1m_0$ at $k = \pi/a$. Assuming $\tau \rightarrow 10^{-14}$ s, the free carrier mobility, μ, along a chain is calculated to be about 200 cm^2/V s. The latter

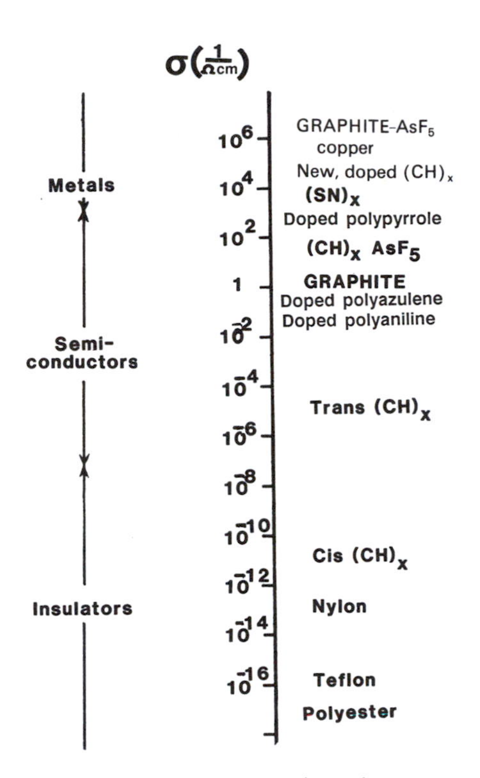

Figure 9.4. Conductivities of polymers in $\Omega^{-1}\ \mathrm{cm}^{-1}$. (Compare with Fig. 7.1.)

quantity is, however, hard to measure since the actual drift mobility in the entire solid is reduced by the trapping of the carriers which occurs during the "hopping" of the electrons between the individual macromolecules. In order to improve the conductivity of $(CH)_x$ one would attempt to decrease the differences of the carbon–carbon bond lengths, thus eventually approaching the uniform bond length as shown in Fig. 9.5(a). This has, indeed, been accomplished by synthesizing $(CH)_x$ via a *nonconjugated* precursor polymer which is subsequently heat treated. Conductivities as high as 10^2 $1/\Omega$ cm have been obtained this way.

Polyacetylene, as discussed so far, should be compared to conventional intrinsic semiconductors. Now we know from Section 8.3 that the conductivity of semiconductors can be substantially increased by doping. The same is true for polymer-based semiconductors. Indeed, arsenic–pentafluoride-doped polyacetylene has a conductivity which is about seven orders of magnitude larger than undoped $(CH)_x$. Thus, σ approaches the conductivity of metals, as can be seen in Figs. 9.4 and 9.6. Iodine or AsF_5 provide *p*-type semiconductors, whereas alkali metals are *n*-type dopants. The doping is achieved through the vapor phase or by electrochemical methods. The dopant molecules diffuse between the $(CH)_x$ chains and provide a charge transfer between

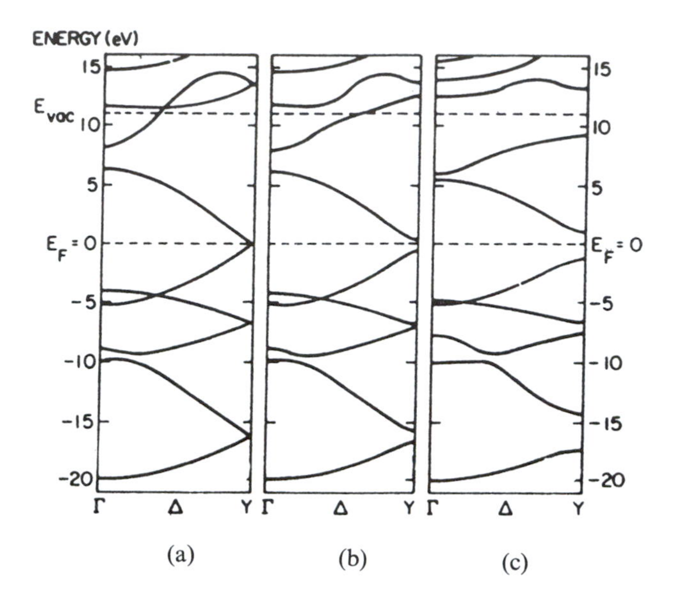

Figure 9.5. Calculated band structure of *trans*-$(CH)_x$ for different carbon–carbon bond lengths: (a) uniform (1.39 Å); (b) weakly alternating C=C, 1.36 Å; C—C, 1.43 Å); and (c) strongly alternating (C=C, 1.34 Å; C—C, 1.54 Å). Note the band gaps at Y as bond alternation occurs. Reprinted with permission from P.M. Grant and I.P. Batra, *Solid State Comm.* **29**, 225 (1979).

the polymer and the dopant. The dopant ends up as an anion in the case of acceptor and as a cation in the case of donor dopants. Another form of doping involves a series of oxidations (or reductions) of the polymer. For example, *polypyrrole*, when electrochemically oxidized, loses one electron which yields a "conjugated radical ion."

A refinement in the description of the conduction mechanism in polyacetylene can be provided by introducing the concept of *solitons*. A soliton is a structural distortion in a conjugated polymer and is generated when a single bond meets another single bond, as shown in Fig. 9.7. At the distortion point a localized nonbonding electron state is generated, similar to an n-type impurity state in a silicon semiconductor. The result is a localized level in the center of the forbidden band. It is believed that when an electron is excited from the valence band into the conduction band (leaving a hole in the valence band) this electron–hole pair decays in about 10^{-12} s into a more stable soliton–antisoliton pair.

Near the center of a soliton, the bond lengths are equal. We recall that uniform bond lengths constitute a metal. Thus, when many solitons have been formed and their spheres of influence overlap, a metal-like conductor would result.

It is also conceivable that one of the double bonds next to a soliton switches

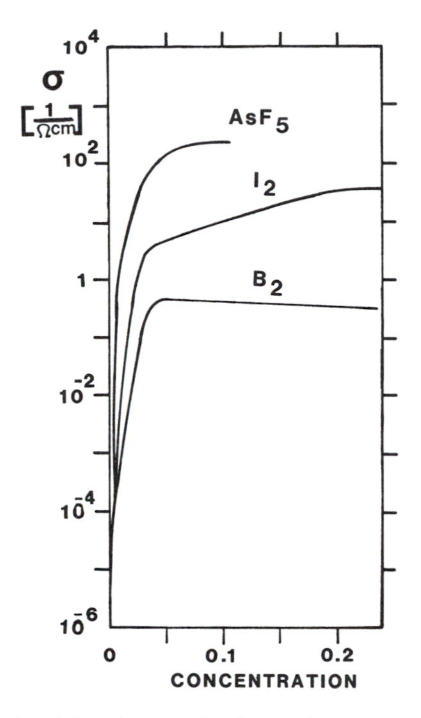

Figure 9.6. Conductivity change of polyacetylene as a result of doping.

over to a single bond. If this switching occurs consecutively in one direction, a soliton wave results. This can be compared to a moving electron.

Up to now, we discussed mainly the properties of polyacetylene. However, several additional conductive polymers have been discovered. The chains in inorganic poly(sulfur nitride) consist of alternating sulfur and nitrogen atoms. Because of the different valencies of the S^{2-} and N^{3-} ions, $(SN)_x$ is an electron-deficient material with an alternating bond structure. The bond length alternation is not severe, so that $(SN)_x$ has a room temperature conductivity of about 10^3 ohm^{-1} cm^{-1} along the chain direction. The conductivity increases with a reduction in temperature. At temperatures close to 0 K, poly(sulfur nitride) becomes superconducting. In brominated $(SN)_x$ the Br_3^- and Br_2^- ions are aligned along the chain axis, giving rise to a one-dimensional superlattice.

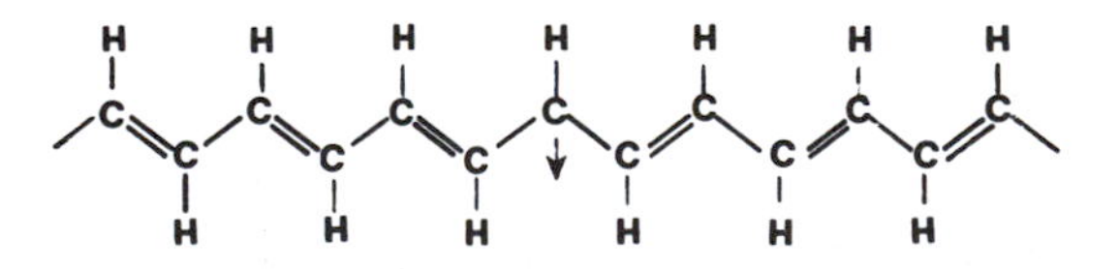

Figure 9.7. A broken symmetry in polyacetylene creates a domain wall or a *soliton.* (An *antisoliton* is the mirror image of a soliton.)

In graphite, an individual "molecule" consists of a "sheet" of carbon atoms. The conductivity is found to be nearly metallic, at least parallel to the layers (Fig. 9.4). AsF_5-doped graphite has an even higher conductivity. The conduction is increased by producing a mixture of easily ionized electron donors and electron acceptors. The charge is then shared between the donors and acceptors. These materials are called charge transfer complexes.

Polyaniline has a reasonably good conductivity and a high environmental stability. It can be used for electronic devices such as field-effect transistors, electrochromic displays, as well as for batteries. A rechargeable battery is on the market whose cathode consists of polyaniline (and whose anode is made of a lithium–aluminum alloy).

Another class of organic conductors are the *charge-transfer salts* in which a donor molecule, such as tetrathiafulvalene (TTF), transfers electrons to an acceptor molecule, such as tetracyanoquinodimethane (TCNQ). The planar molecules stack on top of each other in sheets, thus allowing an overlap of wave functions and a formation of conduction bands which are partially filled with electrons due to the charge transfer. It is assumed that, because of the sheetlike structure, the charge-transfer compounds are quasi-one-dimensional. Conductivities along the stacks, as high as $2 \times 10^3\ \Omega^{-1}\ \mathrm{cm}^{-1}$, have been observed at room temperature. Below room temperature the metallic conductors often transform into semiconductors or insulators. Even superconduction at very low temperatures (about 13 K) has been observed. In the presence of a magnetic field and at low temperatures these materials undergo, occasionally, a transition from a metallic, nonmagnetic state into a semimetallic, magnetic state. Organic metals are generally prepared by electrochemical growth in a solution. They are, as a rule, quite brittle, single crystalline, and relatively small.

Replacing metals and semiconductors by lightweight conducting polymers seems to be, at the present state of the art, nearly impossible, mainly because of their poor stability. However, this very drawback (i.e., the high reactivity of conducting polymers) concomitant with a change in conductivity can be profitably utilized in devices such as remote gas sensors, biosensors, or other remotely readable indicators which detect changes in humidity, radiation dosage, mechanical abuse, or chemical release. As an example, polypyrrole noticeably changes its conductivity when exposed to only 0.1% NH_3, NO_2, or H_2S. Further, experiments have been undertaken to utilize $(CH)_x$ for measuring the concentration of glucose in solutions.

9.2. Ionic Conduction

In ionic crystals (such as the alkali halides), the individual lattice atoms transfer electrons between each other to form positively charged cations and negatively charged anions. The binding forces between the ions are electrostatic in nature and are thus very strong. The room temperature conductivity of ionic crystals is about twenty-two orders of magnitude smaller than the conductivity of typical metallic conductors (Fig. 7.1). This large difference in σ can be understood by realizing that the wide bandgap in insulators allows only extremely few electrons to become excited from the valence into the conduction band.

The main contribution to the electrical conduction in ionic crystals (as little as it may be) is, however, due to a mechanism which we have not yet discussed, namely, ionic conduction. Ionic conduction is caused by the movement of some negatively (or positively) charged ions which *hop* from lattice site to lattice site under the influence of an electric field. (This type of conduction is similar to that which is known to occur in aqueous electrolytes.) The ionic conductivity

$$\sigma_{\text{ion}} = N_{\text{ion}} e \mu_{\text{ion}} \tag{9.1}$$

is, as outlined before (8.13), the product of three quantities. In the present case, N_{ion} is the number of ions per unit volume which can change their position under the influence of an electric field and μ_{ion} is the mobility of these ions.

In order for ions to move through a crystalline solid, they must have sufficient energy to pass over an *energy barrier* (Fig. 9.8). Further, an equivalent lattice site next to a given ion must be empty in order for an ion to be able to change its position. Thus, N_{ion} in (9.1) depends on the vacancy concentration in the crystal (i.e., on the number of *Schottky defects*). In short, the theory of ionic conduction contains essential elements of diffusion theory, with which the reader might be familiar.

Diffusion theory links the mobility of the ions, which is contained in (9.1), with the diffusion coefficient D, through the Einstein relation

$$\mu_{\text{ion}} = \frac{De}{k_B T}. \tag{9.2}$$

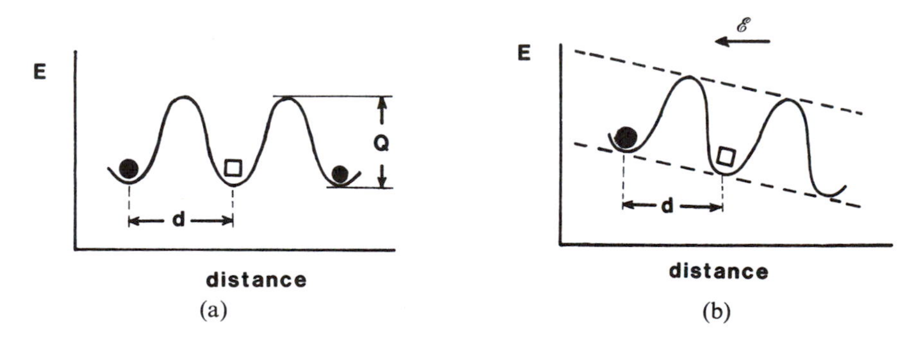

Figure 9.8. Schematic representation of a potential barrier which an ion (●) has to overcome to exchange its site with a vacancy (□). (a) Without an external electric field; (b) with an external electric field. d = distance between two adjacent, equivalent lattice sites; Q = activation energy.

The diffusion coefficient varies with temperature; this dependence is commonly expressed by an Arrhenius equation,

$$D = D_0 \exp\left[-\left(\frac{Q}{k_B T}\right)\right], \tag{9.3}$$

where Q is the activation energy for the process under consideration (Fig. 9.8), and D_0 is a pre-exponential factor which depends on the vibrational frequency of the atoms and some structural parameters. Combining (9.1) through (9.3) yields

$$\sigma_{\text{ion}} = \frac{N_{\text{ion}} e^2 D_0}{k_B T} \exp\left[-\left(\frac{Q}{k_B T}\right)\right]. \tag{9.4}$$

Equation (9.4) is shortened by combining the pre-exponential constants into σ_0:

$$\sigma_{\text{ion}} = \sigma_0 \exp\left[-\left(\frac{Q}{k_B T}\right)\right]. \tag{9.5}$$

Taking the natural logarithm yields

$$\ln \sigma_{\text{ion}} = \ln \sigma_0 - \left(\frac{Q}{k_B}\right)\frac{1}{T}. \tag{9.6}$$

Equation (9.6) suggests that if $\ln \sigma_{\text{ion}}$ is plotted verus $1/T$, a straight line with a negative slope would result. Figure 9.9 depicts schematically a plot of $\ln \sigma$ versus $1/T$ as experimentally found for alkali halides. The linear $\ln \sigma$ versus $1/T$ relationship indicates that Fig. 9.9 is an actual representation of (9.6). The slopes of the straight lines in Arrhenius plots are utilized to calculate the activation energy of the processes under consideration. We notice in Fig. 9.9 two temperature regions representing two different activation energies: at low temperatures, the activation energy is small, the thermal energy is just sufficient to allow the hopping of ions into already existing vacancy sites. This temperature range is commonly called the *extrinsic* region. On the other hand, at high temperatures, the thermal energy is large enough to create

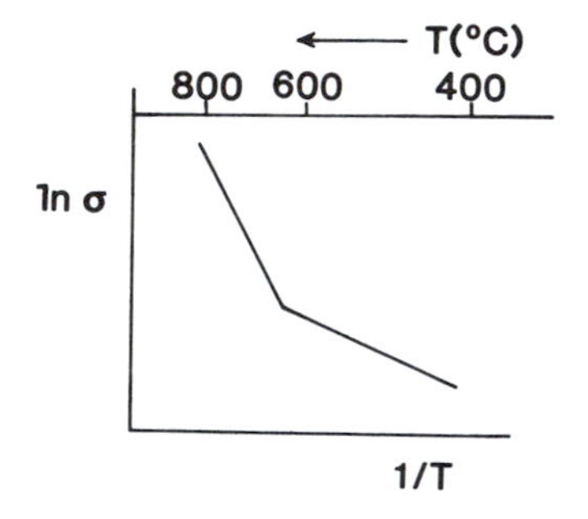

Figure 9.9. Schematic representation of $\ln \sigma$ versus $1/T$ for Na^+ ions in sodium chloride. (Arrhenius plot.)

additional vacancies. The related activation energy is thus the sum of the activation energies for vacancy creation and ion movement. This temperature range is called the *intrinsic* region.

So far, we have not been very specific in describing the circumstances of vacancy formation in an ionic crystal. Now, we have to realize that whenever vacant lattice sites are created, an overall charge neutrality needs to be maintained. The latter is the case when both a cation and an anion are removed from a lattice. Another permissible mechanism is the formation of a vacancy-interstitial pair (*Frenkel defect*). More often, however, vacancies are created as a consequence of introducing differently charged impurity atoms into an ionic lattice, i.e., by substituting, say, a monovalent metal atom with a divalent atom. In order to maintain charge neutrality in this case, a positively charged vacancy needs to be introduced. For example, if a divalent Mg^{2+}-ion substitutes for a monovalent Na^{+}-ion one extra Na^{+}-ion has to be removed to restore charge neutrality. Or, if zirconia (ZrO_2) is treated with CaO (to produce the technically important calcia-stabilized zirconia), the Ca^{2+}-ions substitute for Zr^{4+}-ions and an anion vacancy needs to be created to maintain charge neutrality. Nonstoichiometric compounds contain a high amount of vacancies even at low temperatures, whereas in stoichiometric compounds vacancies need to be formed by elevating the temperature.

In principle, both cations and anions are capable of moving simultaneously under the influence of an electric field. It turns out, however, that in most alkali halides the majority carriers are provided by the metal ions, whereas in other materials, such as the lead halides, the conduction is predominantly performed by the halide ions.

So far, it was implied that the materials under consideration are single crystals. For polycrystalline materials, however, it appears reasonable to assume that the vacant lattice sites provided by the grain boundaries would be utilized by the ions as preferred paths for migration, thus enhancing the conductivity. This has indeed been experimentally observed for alkali ions.

One piercing question remains to be answered: If ionic conduction entails the transport of ions, i.e., of matter from one electrode to the other, would this not imply some segregation of the constituents? Indeed, a pile-up of mobile ions at the electrodes has been observed for long lasting experiments with a concomitant induced electric field in the opposite direction to the externally applied field. As a consequence the conductivity decreases gradually over time. Of course, this does not happen when nonblocking electrodes are utilized which provide a source and a sink for the mobile species.

9.3. Conduction in Metal Oxides

Metal oxides do not actually represent a separate class of conducting materials on their own. Indeed, they can be insulating, such as TiO_2, have metallic conduction properties, such as TiO, or are semiconducting. For understanding the mechanisms involved in metal oxides, e.g., in the afore-mentioned titanium oxides, it is helpful to inspect the table in Appendix 3. Oxygen is seen there to have four $2p$-electrons in its outermost shell. Two more electrons will bring O^{2-} into the closed-shell configuration and four electrons are obviously needed to accomplish the same for two oxygen ions, such as in TiO_2. These four electrons are provided by the titanium from its $3d$- and $4s$-shells. Thus, in the case of TiO_2, all involved elements are in the noble gas configuration. Since ionic bonds are involved, any attempted removal of electrons would require a considerable amount of thermal energy. TiO_2 is, consequently, an insulator having a wide band gap. Not so for TiO. Since only two titanium valence electrons are needed to fill the $2p$-shell of *one* oxygen ion, two more titanium electrons are free to serve as conduction electrons. Thus, TiO has metallic properties with a σ in the $10^3\ \Omega^{-1}\ \text{cm}^{-1}$ range.

A refinement of our understanding is obtained by considering the pertinent electron bands. TiO has, according to the afore-mentioned explanations, a filled oxygen $2p$-valence band and an essentially empty titanium $4s$-conduction band. Also involved is a narrow titanium $3d$-band which is partially filled by the above-mentioned two electrons. The conduction in TiO takes place, therefore, in the titanium $3d$-band which can host, as we know, a total of 10 electrons.

We discuss zinc oxide as a next example. ZnO has two valence $4s$-electrons which transfer to the oxygen $2p$-band. ZnO, if strictly stoichiometric, has, thus, a filled valence $2p$-band and an empty zinc $4s$-band employing a gap energy of 3.3 eV. Stoichiometric ZnO is therefore an insulator or a wide band-gap semiconductor. Now, if interstitial zinc atoms (or oxygen vacancies) are introduced into the lattice (by heating ZnO in a reducing atmosphere which causes neutral oxygen to leave the crystal) then the valence electrons of these zinc interstitials are only loosely bound to their nuclei. One of these two electrons can easily be ionized (0.05 eV) and acts therefore as a donor. Nonstoichiometric ZnO is, consequently, an n-type semiconductor. The same is incidentally true for nonstoichiometric Cu_2O (see Appendix 3), an established semiconducting material from which Cu/Cu_2O Schottky-type rectifiers have been manufactured long before silicon technology was invented.

Another interesting metal oxide is SnO_2 (sometimes doped with In_2O_3) which is transparent in the visible region and which is a reasonable conductor in the $1\ \Omega^{-1}\ \text{cm}^{-1}$ range. It is used in optoelectronics to provide electrical contacts without blocking the light from reaching a device. It is known as *indium-tin-oxide* or *ITO*.

Finally, we discuss NiO. Again, a filled oxygen $2p$-band and an empty nickel $4s$-band are involved. In order to form the nickel $3d$-bands required for conduction, a substantial overlap of the $3d$-wave functions would be required by quantum mechanics. Band structure calculations show, however, that these interactions do not take place. Instead, *deep-lying localized electron states* in the forbidden band close to the upper edge of the valence band are observed. Thus, no $3d$-band conduction can take place which results in stoichiometric NiO being an insulator. Nonstoichiometry (obtained by removing some nickel atoms, thus creating cation vacancies) causes NiO to become a *p*-type semiconductor.

9.4. Amorphous Materials (Metallic Glasses)

Before we discuss electrical conduction in amorphous materials, we need to clarify what the term *amorphous* means in the present context. Strictly speaking, *amorphous* implies the random arrangement of atoms, the absence of any periodic symmetry, or the absence of any crystalline structure. One could compare the random distribution of atoms with the situation in a gas, as seen in an instantaneous picture. Now, such a completely random arrangement of atoms is seldom found even in liquids, much less in solids. In actuality, the relative positions of nearest neighbors surrounding a given atom in an amorphous solid are almost identical to the positions in crystalline solids because of the ever-present binding forces between the atoms. In short, the atomic order in amorphous materials is restricted to the nearest neighbors. Amorphous materials exhibit, therefore, only *short-range order*. In contrast to this, the exact positions of the atoms which are farther apart from a given central atom cannot be predicted. This is particularly the case when various kinds of stacking orders, i.e., if polymorphic modifications are possible. As a consequence one observes atomic disorder at long range. The term *amorphous solid* should therefore be used *cum grano salis*. We empirically define materials to be amorphous when their diffraction patterns consist of diffuse rings, rather than sharply defined Bragg rings, as are characteristic for polycrystalline solids.

So far we have discussed *positional disorder* only as it might be found in pure materials. If more than one component is present in a material, a second type of disorder is possible: The individual species might be randomly distributed over the lattice sites; i.e., the species may not be alternately positioned as is the case for, say, sodium and chlorine atoms in NaCl. This random distribution of species is called *compositional disorder*.

The best-known representative of an amorphous solid is window glass, whose major components are silicon and oxygen. Glass is usually described as a supercooled liquid.

Interestingly enough, many elements and compounds which are generally known to be crystalline under equilibrium conditions can also be obtained in the nonequilibrium amorphous state by applying rapid solidification techniques, i.e., by utilizing cooling rates of about 10^5 K/s. These cooling rates can be achieved by fast quenching, melt spinning, vapor deposition, sputtering, radiation damage, filamentary casting in continuous operation, etc. The degree of amorphousness (or, the degree of short range order) may be varied by the severity of the quench. The resulting *metallic glasses*, or *glassy metals*, have unusual electrical, mechanical, optical, magnetic, and corrosion properties and are therefore of considerable interest. Amorphous semiconductors (con-

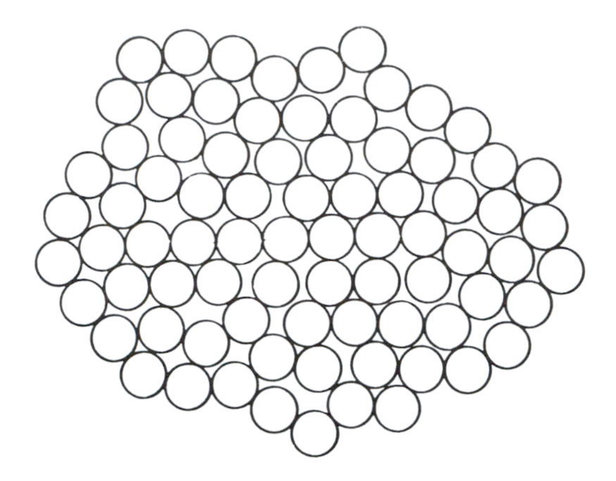

Figure 9.10. Two-dimensional schematic representation of a dense random packing of hard spheres (Bernal model).

sisting, e.g., of Ge, Si, GeTe, etc.) have also received substantial attention because they are relatively inexpensive to manufacture, have unusual switching properties, and have found applications in inexpensive photovoltaic cells.

We now turn to the atomic structure of amorphous metals and alloys. They have essentially nondirectional bonds. Thus, the short-range order does not extend beyond the nearest neighbors. The atoms must be packed together tightly, however, in order to achieve the observed density. There are only a limited number of ways of close packing. One way of arranging the atoms in amorphous metals is depicted by the *dense random packing of hard spheres* model (Fig. 9.10). This *Bernal* model is considered as the ideal amorphous state. No significant regions of crystalline order are present. In transition metal–metalloid compounds (such as Ni–P) it is thought that the small metalloid atoms occupy the holes which occur as a consequence of this packing (*Bernal–Polk model*).

The atoms in amorphous semiconductors, on the other hand, do not arrange themselves in a close-packed manner. Atoms of group IV elements are, as we know, covalently bound. They are often arranged in a *continuous random network* with correlations in ordering up to the third or fourth nearest neighbors (Fig. 9.11(b) and (c)). Amorphous pure silicon contains numerous *dangling bonds* similar to those found in crystalline silicon in the presence of vacancies (Fig. 9.11(a)).

Since amorphous solids have no long-range crystal symmetry, we can no longer apply the Bloch theorem, which led us in Section 4.4 from the distinct energy levels for isolated atoms to the broad quasi-continuous bands for crystalline solids. Thus, the calculation of electronic structures for amorphous metals and alloys has to use alternate techniques, e.g., the *cluster model* approach. This method has been utilized to calculate the electronic structure of amorphous Zr–Cu (which is a representative of a noble metal–transition

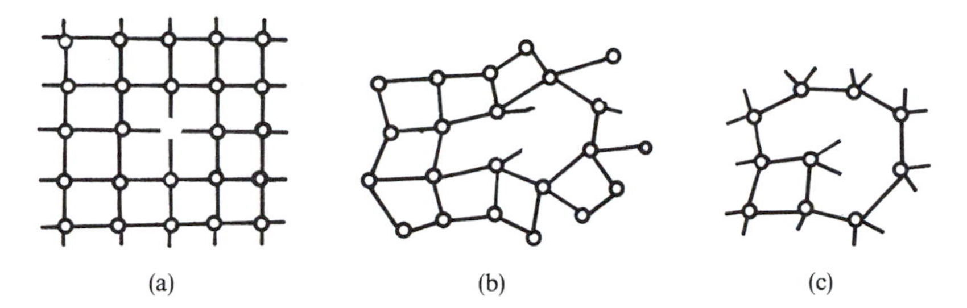

Figure 9.11. Defects in crystalline and amorphous silicon. (a) Monovacancy in a crystalline semiconductor; (b) one and (c) two dangling bonds in a continuous random network of an amorphous semiconductor. (Note the deviations in the interatomic distances and of the bond angles.)

metal metallic glass). A series of clusters were assumed which exhibit the symmetry of the close-packed lattices fcc (as for Cu) and hcp[15] (as for Zr). The energy level diagram depicted in Fig. 9.12 shows two distinct "bands" of levels. The lower band consists primarily of copper *d*-levels, while the upper band consists mainly of zirconium *d*-levels. A sort of gap separates the two bands of levels. Even though the concept of quasi-continuous energy bands is no longer meaningful for amorphous solids, the density of states concept still is, as can be seen in Fig. 9.12. We notice that the Fermi energy is located in the upper part of the zirconium levels. Further, we observe partially filled

[15] Hexagonal close-packed.

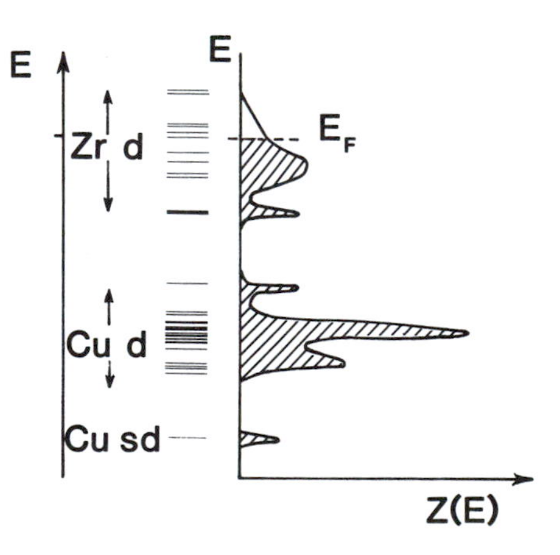

Figure 9.12. Schematic representation of the molecular orbital energy level diagram and the density of states curves for Zr–Cu clusters. The calculated density of states curves agree reasonably well with photoemission experiments.

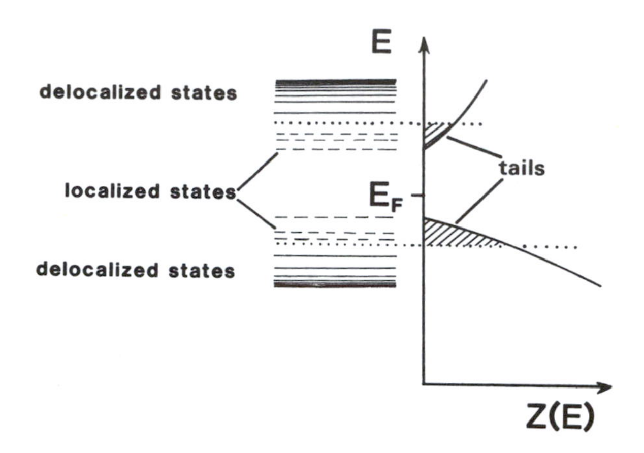

Figure 9.13. Localized and delocalized states and density of states $Z(E)$ for amorphous semiconductors. Note the band tails which are caused by the localized states.

electron states. This has two interesting consequences. First, we expect *metal-like* conduction. Second, $Z(E)$ near E_F is small which suggests relatively small values for the conductivity (see (7.26)). Indeed, σ for Cu–Zr is comparable to that of poor metallic conductors (i.e., approximately 5×10^3 $1/\Omega$ cm).

The electrical resistivity of many metallic glasses (such as $Pd_{80}Si_{20}$ or $Fe_{32}Ni_{36}Cr_{14}P_{12}B_6$[16]) stays constant over a fairly wide temperature range, up to the temperature which marks the irreversible transition from the amorphous into the crystalline state. This makes these alloys attractive as resistance standards. The mean free path for electrons in metallic glasses is estimated to be about 1 nm.

The energy level diagrams and the density of states curves for amorphous semiconductors are somewhat different from those for amorphous metals. Because of the stronger binding forces which exist between the atoms in covalently bound materials, the valence electrons are tightly bound, or *localized.* As a consequence the density of states for the localized states extend into the "band gap" (Fig. 9.13). This may be compared to the localized impurity states in doped crystalline semiconductors which are also located in the band gap. Thus, we observe density of states *tails.* These tails might extend, for some materials, so far into the gap that they partially overlap. In general, however, the density of electron and hole states for the localized levels is very small.

The electrical conductivity for amorphous semiconductors σ_A depends, as usual (8.13), on the density of carriers N_A and the mobility of these carriers, μ_A:

$$\sigma_A = N_A e \mu_A. \tag{9.7}$$

The density of carriers in amorphous semiconductors is extremely small

[16] METGLAS 2826A, trademark of Allied Chemical.

because all electrons are, as said before, strongly bound (localized) to their respective nuclei. Likewise, the mobility of the carriers is small because the absence of a periodic lattice causes substantial incoherent scattering. As a consequence, the room temperature conductivity in amorphous semiconductors is generally very low (about 10^{-7} $1/\Omega$ cm).

Some of the localized electrons might occasionally acquire sufficient thermal energy to overcome barriers which are caused by potential wells of variable depth and hop to a neighboring site. Thus, the conduction process in amorphous semiconductors involves a (temperature-dependent) activation energy, Q_A, which leads to an equation similar to (9.5) describing a so-called variable range hopping

$$\sigma_A = \sigma_0 \exp\left[-\left(\frac{Q_A(T)}{k_B T}\right)\right]. \tag{9.8}$$

Equation (9.8) states that the conductivity in amorphous semiconductors increases exponentially with increasing temperature because any increase in thermal energy provides additional free carriers.

The application of amorphous silicon for *photovoltaic devices* (see Section 8.7.6) will be discussed briefly in closing because of its commercial as well as scientific significance. If silicon is deposited out of the gas phase on relatively cold ($< 500°$ C) substrates (utilizing silane or sputtering), a structure as shown in Fig. 9.11(b) and (c) results. Doping is virtually not possible in this case since any free charge carriers recombine immediately with the dangling bonds. However, hydrogen, if added during deposition and incorporated into the solid, neutralizes the unsaturated valencies (and reduces internal strain in the lattice network). This results in *hydrogenated amorphous silicon* which is in its properties quite comparable to crystalline silicon. Doping can be accomplished right during deposition. This way, semiconducting materials can be produced which vary in their conductivity between 10^{-11} and 10^{-2} Ω^{-1} cm^{-1} depending on doping (see Fig. 7.1). Commercial flat-plate solar cells of this type have an efficiency of about 8% compared to 14% efficiency for commercial single crystalline silicon technology. The price (and the consumption of power during manufacturing) is, however, only one-half of that for crystalline silicon mainly because of the simpler way of deposition (see page 124).

The understanding of amorphous metals, alloys, and semiconductors is still in its infancy. Future developments in this field should be followed with a great deal of anticipation because of the potentially significant applications which might arise in the years to come.

9.4.1. Xerography

Xerography (from the Greek "dry writing") or *electrophotography* is an important application of amorphous semiconductors such as amorphous selenium or amorphous silicon, etc. Such a material, when deposited on a cylindri-

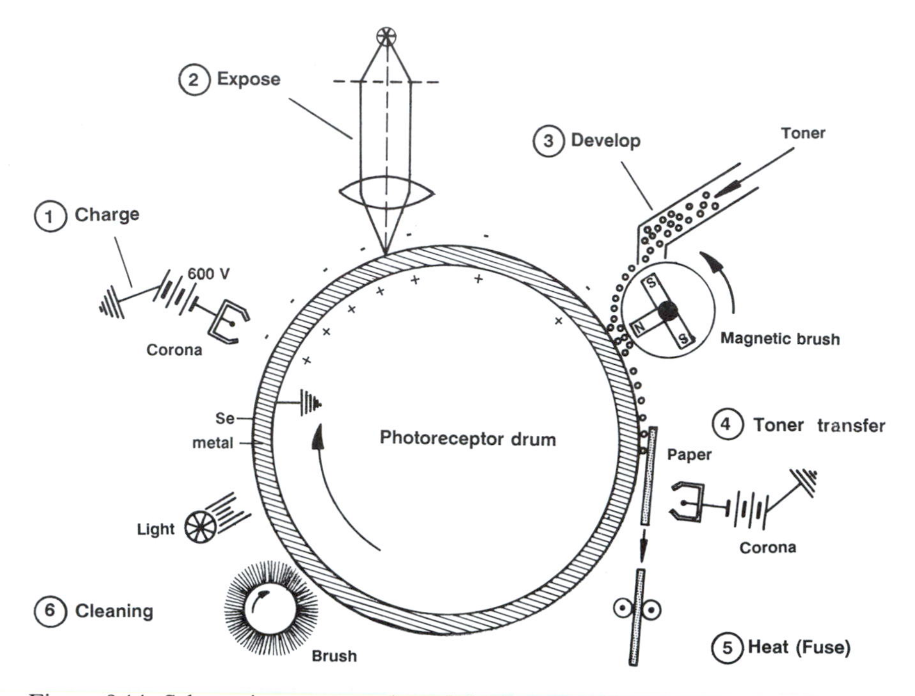

Figure 9.14. Schematic representation of the electrophotography process. The individual steps are explained in the text.

cally-shaped metallic substrate, constitutes the *photoreceptor* drum, as shown in Fig. 9.14.

Before copying, the photoreceptor is electrostatically charged by means of a corona wire to which a high voltage is applied (Step 1). Amorphous semiconductors are essentially insulators (see above) which hold this electric charge reasonably well, as long as they are kept in the dark. If, however, light which has been reflected from the document to be copied falls on the photoreceptor, electron–hole pairs are formed causing the photoreceptor to become conducting. This process discharges the affected parts on the drum, creating a latent image on the photoreceptor, i.e., a pattern consisting of charged and neutral areas. At the next step, electrostatically charged and pigmented polymer particles (called *toner*) are brought into contact with the drum. The toner clings to the charged areas only. Commonly, a two-component toner is utilized; one part consists of magnetically soft particles. They form brush-type chains under the influence of a magnetic field which is caused by permanent magnets that are rotated inside a cylinder (see Fig. 9.14, Step 3). Eventually, the toner on the photoreceptor is electrostatically transferred to a piece of paper by properly corona-charging the back of the paper. Finally, the toner is fused to the paper by heat. A cleaning and photodischarging process prepares the photoreceptor drum for the next cycle.

Laser printers use the same principle. To create the latent image, the laser light is periodically scanned across the rotating photoreceptive drum by means of a rotating multisurface mirror. The spectral sensitivity of the amorphous semiconductor has to be matched to the wavelength of the laser light. Amorphous silicon (maximal photosensitivity near 700 nm) in conjunction with a helium–neon laser (see Table 13.1) is a usable combination.

Problems

1. Calculate the mobility of the oxygen ions in UO_2 at 700 K. The diffusion coefficient of O^{2-} at this temperature is 10^{-13} cm^2/s. Compare this mobility with electron or hole mobilities in semiconductors (see Appendix 4). Discuss the difference!

2. Calculate the number of vacancy sites in an ionic conductor in which the metal ions are the predominant charge carriers. Assume the room temperature ionic conductivity to be 10^{-17} $1/\Omega$ cm and an ionic mobility of 10^{-17} m^2/V s. Does the calculated result make sense? Discuss how the vacancies might have been introduced into the crystal.

3. Calculate the activation energy for ionic conduction for a metal ion in an ionic crystal at 300 K. Take $D_0 = 10^{-3}$ m^2/s and $D = 10^{-17}$ m^2/s.

4. Calculate the ionic conductivity at 300 K for an ionic crystal. Assume 6×10^{20} Schottky defects per cubic meter, an activation energy of 0.8 eV and $D_0 = 10^{-3}$ m^2/s.

Suggestions for Further Reading (Part II)

N.W. Ashcroft and N.D. Mermin, *Solid State Physics*, Hold, Rinehart and Winston, New York (1976).

A.R. Blythe, *Electrical Properties of Polymers*, Cambridge University Press, Cambridge (1979).

M.H. Bordsky, ed., *Amorphous Semiconductors*, Springer-Verlag, Berlin, (1979).

I. Brodie and J. Muray, *The Physics of Microfabrication*, Plenum Press, New York (1982).

R.H. Bube, *Electronic Properties of Crystalline Solids*, Academic Press, New York (1974).

R.H. Bube, *Electrons in Solids*, 2nd ed., Academic Press, New York (1988).

J.R. Ferraro, and J.M. Williams, *Introduction to Synthetic Electrical Conductors*, Academic Press, Orlando (1987).

S.K. Ghandi, *The Theory and Practice of Microelectronics*, Wiley, New York (1968).

S.K. Ghandi, *VLSI Fabrication Principles*, Wiley, New York (1983).

N.J. Grant and B.C.Giessen, eds., *Rapidly Quenched Metals*, Second International Conference. The Massachusetts Institute of Technology, Boston, MA (1976).

C.R.M. Grovenor, *Materials for Semiconductor Devices*, The Institute of Metals (1987).

L.L. Hench and J.K. West, *Principles of Electronic Ceramics*, Wiley, New York (1990).

B.H. Kear, B.C. Giessen, and K.M. Cohen, eds., *Rapidly Solidified Amorphous and Crystalline Alloys*, North-Holland, Amsterdam (1982).

C. Kittel, *Introduction to Solid State Physics*, 6th ed., Wiley, New York (1986).

J. Mort and G. Pfister, eds., *Electronic Properties of Polymers*, Wiley, New York (1982).

G.W. Neudeck and R.F. Pierret, eds., *Modular Series on Solid State Devices*, Vols. I–VI, Addison-Wesley, Reading, MA (1987).

M.A. Omar, *Elementary Solid State Physics*, Addison-Wesley, Reading, MA (1978).

D.A. Seanov ed., *Electrical Properties of Polymers*, Academic Press, New York, (1982).

B.G. Streetman, *Solid State Electronic Devices*, 2nd ed., Prentice-Hall, Englewood Cliffs, NJ (1980).

S.M. Sze, *Physics of Semiconductor Devices*, 2nd ed., Wiley, New York (1981).

S.M. Sze, *Semiconductor Devices, Physics and Technology*, Wiley, New York (1985).

H.E. Talley and D.G. Daugherty, *Physical Principles of Semiconductor Devices*, Iowa University Press, Ames, IA (1976).

L.H. Van Vlack, *Physical Ceramics for Engineers*, Addison-Wesley, Reading, MA (1964).

C.A. Wert and R.M. Thompson, *Physics of Solids*, 2nd ed., McGraw-Hill, New York (1970).

PART III

OPTICAL PROPERTIES OF MATERIALS

CHAPTER 10

The Optical Constants

10.1. Introduction

The most apparent properties of metals, their luster and their color, have been known to mankind since metals were known. Because of these properties, metals were already used in ancient times for mirrors and jewelry. The color was utilized 4000 years ago by the ancient Chinese as a guide to determine the composition of the melt of copper alloys: the hue of a preliminary cast indicated whether the melt, from which bells or mirrors were to be made, already had the right tin content.

The German poet Goethe was probably the first one who explicitly spelled out 200 years ago in his *Treatise on Color* that *color* is not an absolute property of matter (such as the resistivity), but requires a living being for its perception and description. Applying Goethe's findings, it was possible to explain qualitatively the color of, say, gold in simple terms. Goethe wrote: "If the color *blue* is removed from the spectrum, then blue, violet, and green are missing and red and yellow remain." Thin gold films are bluish–green when viewed in transmission. These colors are missing in reflection. Consequently, gold appears reddish–yellow.

This chapter treats the optical properties from a completely different point of view. Measurable quantities such as the index of refraction or the reflectivity and their spectral variations are used to characterize materials. In doing so, the term "color" will almost completely disappear from our vocabulary. Instead, it will be postulated that the interactions of light with the valence electrons of a material are responsible for the optical properties. As in previous chapters, where an understanding of the electrical properties was attempted, an atomistic model and later a quantum mechanical treatment will be employed. Thus, the electron theory of metals, as introduced in the first six chapters, will serve as a foundation.

At the beginning of this century the study of the interactions of light with matter (black body radiation, etc.) laid the foundations for quantum theory. Today, optical methods are among the most important tools for elucidating the electron structure of matter. Most recently, a number of optical devices such as lasers, photodetectors, waveguides, etc., have gained considerable technological importance. They are used in communication, fiber optics, medical diagnostics, night viewing, solar applications, optical computing, or for other optoelectronic purposes. Traditional utilizations of optical materials for windows, antireflection coatings, lenses, mirrors, etc., should be likewise mentioned. All taken, it is well justified to spend a major part of this book on the optical properties of materials.

Before we start our discourse we need to define the optical constants. We make use of some elements of physics.

10.2. Index of Refraction, *n*

When light passes from an optically "thin" into an optically dense medium, one observes that in the dense medium, the angle of refraction β (i.e., the angle between the refracted light beam and a line perpendicular to the surface) is smaller than the angle of incidence, α. This well-known phenomenon is used for the definition of the refractive power of a material and is called Snell's law

$$\frac{\sin\alpha}{\sin\beta} = \frac{n_{\text{med}}}{n_{\text{vac}}} = n. \tag{10.1}$$

Commonly, the index of refraction for vacuum n_{vac} is arbitrarily set to be unity. The refraction is caused by the different velocities, c, of the light in the two media

$$\frac{\sin\alpha}{\sin\beta} = \frac{c_{\text{vac}}}{c_{\text{med}}}. \tag{10.2}$$

Thus, if light passes from vacuum into a medium, we find

$$n = \frac{c_{\text{vac}}}{c_{\text{med}}} = \frac{c}{v}. \tag{10.3}$$

The magnitude of the refractive index depends on the wavelength of the incident light. This property is called dispersion. In metals, the index of refraction varies in addition with the angle of incidence. This is particularly true when n is small.

10.3. Damping Constant, *k*

Metals damp the intensity of light in a relatively short distance. Thus, to characterize the optical properties of metals, an additional material constant is needed.

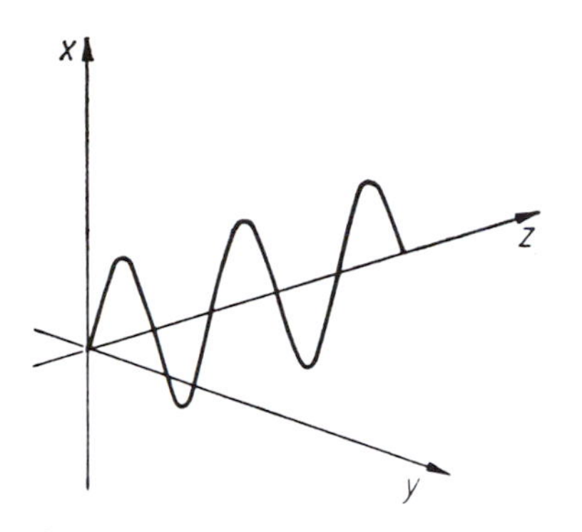

Figure 10.1. Plane polarized wave which propagates in the positive z-direction and vibrates in the x-direction.

We make use of the electromagnetic wave equation which mathematically describes the propagation of light in a medium. The derivation of this wave equation from the well-known Maxwell equations does not further our understanding of the optical properties. (The interested reader can find the derivation in specialized texts.[1])

For simplification, we consider a plane polarized wave which propagates along the positive z-axis and which vibrates in the x-direction (Fig. 10.1). We neglect possible magnetic properties. For this special case, the electromagnetic wave equation reads

$$c^2 \frac{\partial^2 \mathscr{E}_x}{\partial z^2} = \varepsilon \frac{\partial^2 \mathscr{E}_x}{\partial t^2} + 4\pi\sigma \frac{\partial \mathscr{E}_x}{\partial t}, \tag{10.4}$$

where $\mathscr{E}_x$ is the x-component of the electric field strength,[2] ε is the dielectric constant, and σ is the (a.c.) conductivity. The solution to (10.4) is commonly achieved by using the following trial solution:

$$\mathscr{E}_x = \mathscr{E}_0 \exp\left[i\omega\left(t - \frac{zn}{c}\right)\right], \tag{10.5}$$

where $\mathscr{E}_0$ is the maximal value of the electric field strength and $\omega = 2\pi\nu$ is the angular frequency. Differentiating (10.5) once with respect to time, and twice with respect to time *and* z, and inserting these values into (10.4) yields

$$\hat{n}^2 = \varepsilon - \frac{4\pi\sigma}{\omega} i = \varepsilon - \frac{2\sigma}{\nu} i. \tag{10.6}$$

Equation (10.6) leads to an important result: The index of refraction is generally a complex number as inspection of the right-hand side of (10.6) indicates. We denote for clarity the complex index of refraction by $\hat{n}$. As is true for all

[1] For example, R.E. Hummel, *Optische Eigenschaften von Metallen und Legierungen*, Springer-Verlag, Berlin (1971).

[2] We use, for the electric field strength, $\mathscr{E}$ to distinguish it from the energy.

complex quantities, the complex index of refraction consists of a real and an imaginary part

$$\hat{n} = n_1 - in_2. \tag{10.7}$$

In the literature, the imaginary part of the index of refraction n_2 is often denoted by "k" and (10.7) is then written as

$$\hat{n} = n - ik. \tag{10.8}$$

We will call n_2 or k the *damping constant.* (In some books n_2 or k is named the *absorption constant.* We will not follow this practice because the latter term is extremely misleading. Other authors call k the *attenuation index* or the *extinction coefficient* which we will not use either in this context.)

Squaring (10.8) yields, together with (10.6),

$$\hat{n}^2 = n^2 - k^2 - 2nki = \varepsilon - \frac{2\sigma}{\nu} i. \tag{10.9}$$

Equating individually the real and imaginary parts of (10.9) yields two important relations between electrical and optical constants

$$\varepsilon = n^2 - k^2, \tag{10.10}$$

$$\sigma = nk\nu. \tag{10.11}$$

Let us return to (10.9). The right-hand side is the difference between two dielectric constants (a real and an imaginary one). Thus, the left side must be a dielectric constant too and (10.9) may be rewritten as

$$\hat{n}^2 = n^2 - k^2 - 2nik \equiv \hat{\varepsilon} = \varepsilon_1 - i\varepsilon_2. \tag{10.12}$$

Equating individually the real and imaginary parts in (10.12) yields

$$\varepsilon_1 = n^2 - k^2 \tag{10.13}$$

and (with (10.11))

$$\varepsilon_2 = 2nk = \frac{2\sigma}{\nu}. \tag{10.14}$$

Similarly as above, ε_1 and ε_2 are called the real and the imaginary parts of the complex dielectric constant $\hat{\varepsilon}$, respectively. (ε_1 in (10.13) is identical to ε in (10.10).) ε_2 is often called the *absorption product* or, briefly, the *absorption.*

We consider a special case: For insulators ($\sigma \approx 0$) it follows from (10.11) that $k \approx 0$ (see also Table 10.1). Then (10.10) reduces to $\varepsilon = n^2$ (*Maxwell relation*).

From (10.10), (10.11), (10.13), and (10.14) one obtains

$$n^2 = \frac{1}{2}\left(\sqrt{\varepsilon^2 + \left(\frac{2\sigma}{\nu}\right)^2} + \varepsilon\right) = \frac{1}{2}\left(\sqrt{\varepsilon_1^2 + \varepsilon_2^2} + \varepsilon_1\right), \tag{10.15}$$

$$k^2 = \frac{1}{2}\left(\sqrt{\varepsilon^2 + \left(\frac{2\sigma}{\nu}\right)^2} - \varepsilon\right) = \frac{1}{2}\left(\sqrt{\varepsilon_1^2 + \varepsilon_2^2} - \varepsilon_1\right). \tag{10.16}$$

Table 10.1. Characteristic Penetration Depth, W, and Damping Constant, k, for Some Materials ($\lambda = 589.3$ nm).

Material	Water	Flint glass	Graphite	Gold
W(cm)	32	29	6×10^{-6}	1.5×10^{-6}
k	1.4×10^{-7}	1.5×10^{-7}	0.8	3.2

It should be emphasized that (10.10)–(10.16) are only valid if ε, σ, n, and k are measured at the same wavelength, because these "constants" are wavelength dependent. For small frequencies, however, the d.c. values for ε and σ can be used with good approximation, as will be shown later. Finally, it should be noted that the above equations are only valid for optically isotropic media; otherwise ε becomes a tensor.

We return now to (10.5) in which we replace the index of refraction by the general expression (10.8). This yields

$$\mathscr{E}_x = \mathscr{E}_0 \exp\left[i\omega\left(t - \frac{z(n - ik)}{c}\right)\right], \tag{10.17}$$

which may be rewritten to read

$$\mathscr{E}_x = \mathscr{E}_0 \underbrace{\exp\left[-\frac{\omega k}{c} z\right]}_{\text{Damped amplitude}} \cdot \underbrace{\exp\left[i\omega\left(t - \frac{zn}{c}\right)\right]}_{\text{Undamped wave}}. \tag{10.18}$$

Equation (10.18) is now the complete solution of the wave equation (10.4). It represents a damped wave and expresses that in matter the amplitude decreases exponentially with increasing z (Fig. 10.2). The constant k determines

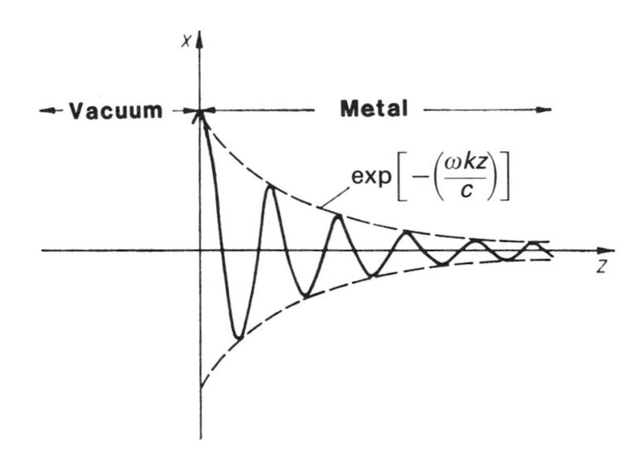

Figure 10.2. Modulated light wave. The amplitude decreases exponentially.

how much the amplitude decreases, i.e., k expresses the degree of damping of the light wave. We understand now why k is termed the *damping constant*.

The result which we just obtained is well known to electrical engineers. They observe that at high frequencies the electromagnetic waves are conducted only on the outer surface of a wire. They call this phenomenon the (*normal*) *skin effect*.

10.4. Characteristic Penetration Depth, W, and Absorbance, α

The field strength $\mathscr{E}$ is hard to measure. Thus, the intensity, I, which can be measured effortlessly with light sensitive devices (such as a photodetector, see Section 8.7.6) is commonly used. The intensity equals the square of the field strength. Thus, the damping term in (10.18) may be written as

$$I = \mathscr{E}^2 = I_0 \exp\left(-\frac{2\omega k}{c} z\right). \tag{10.19}$$

We define a characteristic penetration depth, W, as that distance at which the intensity of the light wave, which travels through a material, has decreased to $1/e$ or 37% of its original value, i.e., when

$$\frac{I}{I_0} = \frac{1}{e} = e^{-1}. \tag{10.20}$$

This definition yields in conjunction with (10.19)

$$z = W = \frac{c}{2\omega k} = \frac{c}{4\pi\nu k} = \frac{\lambda}{4\pi k}. \tag{10.21}$$

Table 10.1 presents values for k and W for some materials obtained by using sodium vapor light ($\lambda = 589.3$ nm).

The inverse of W is sometimes called the (exponential) *attenuation* or the *absorbance* which is, by making use of (10.21), (10.14), and (10.11),

$$\alpha = \frac{4\pi k}{\lambda} = \frac{2\pi\varepsilon_2}{\lambda n} = \frac{4\pi\sigma}{nc} = \frac{2\omega k}{c}. \tag{10.21a}$$

It is measured in cm^{-1}, or when multiplied by 4.3 in decibels (dB) per centimeter (1 dB $= 10 \log I/I_0$).

10.5. Reflectivity, R, and Transmissivity, T

Metals are characterized by a large reflectivity. This stems from the fact that light penetrates a metal only a short distance, as shown in Fig. 10.2 and Table 10.1. Thus, only a small part of the impinging energy is converted into heat. The major part of the energy is reflected (in some cases close to 99%, see Table

Table 10.2. Optical Constants of Some Metals ($\lambda = 600$ nm).

Metal	n	k	R [%]
Copper	0.14	3.35	95.6
Silver	0.05	4.09	98.9
Gold	0.21	3.24	92.9
Aluminum	0.97	6.0	90.3

10.2). In contrast to this, visible light penetrates into glass much farther than into metals, i.e., approximately seven orders of magnitude more, see Table 10.1. As a consequence, very little light is reflected by glass. Nevertheless, a piece of glass about one or two meters thick eventually dissipates a substantial part of the impinging light into heat. (In practical applications, one does not observe this large reduction in light intensity because windows are as a rule only a few millimeters thick.) It should be noted that typical window panes reflect the light on the front as well as on the back.

The ratio between reflected intensity I_R and incoming intensity I_0 of the light serves as a definition for the *reflectivity*

$$R = \frac{I_R}{I_0}. \tag{10.22}$$

Quite similarly, one defines the ratio between the transmitted intensity, I_T, and the impinging light intensity as the *transmissivity*

$$T = \frac{I_T}{I_0}. \tag{10.22a}$$

Experiments have shown that for insulators, R depends solely on the index of refraction. For perpendicular incidence one finds

$$R = \frac{(n-1)^2}{(n+1)^2}. \tag{10.23}$$

This equation can also be derived from the Maxwell equations.

We know already that n is generally a complex quantity. By definition, however, R has to remain real. Thus, the modulus of R becomes

$$R = \left|\frac{\hat{n}-1}{\hat{n}+1}\right|^2 \tag{10.24}$$

which yields

$$R = \frac{(n-ik-1)}{(n-ik+1)}\cdot\frac{(n+ik-1)}{(n+ik+1)} = \frac{(n-1)^2+k^2}{(n+1)^2+k^2} \tag{10.25}$$

(*Beer's equation*). The reflectivity is a unitless material constant and is often given in percent of the incoming light (see Table 10.2). R is, like the index of refraction, a function of the wavelength of the light.

The reflectivity is a function of ε_1 and ε_2. We shall derive this relationship by performing a few transformations. Equation (10.25) is rewritten as

$$R = \frac{n^2 + k^2 + 1 - 2n}{n^2 + k^2 + 1 + 2n}, \tag{10.26}$$

(1) $$n^2 + k^2 = \sqrt{(n^2 + k^2)^2} = \sqrt{n^4 + 2n^2k^2 + k^4}$$
$$= \sqrt{n^4 - 2n^2k^2 + k^4 + 4n^2k^2} = \sqrt{(n^2 - k^2)^2 + 4n^2k^2}$$
$$= \sqrt{\varepsilon_1^2 + \varepsilon_2^2}, \tag{10.27}$$

(2) $$2n = \sqrt{4n^2} = \sqrt{2(n^2 + k^2 + n^2 - k^2)} = \sqrt{2(\sqrt{\varepsilon_1^2 + \varepsilon_2^2} + \varepsilon_1)}. \tag{10.28}$$

Inserting (10.27) and (10.28) into (10.26) provides

$$R = \frac{\sqrt{\varepsilon_1^2 + \varepsilon_2^2} + 1 - \sqrt{2(\sqrt{\varepsilon_1^2 + \varepsilon_2^2} + \varepsilon_1)}}{\sqrt{\varepsilon_1^2 + \varepsilon_2^2} + 1 + \sqrt{2(\sqrt{\varepsilon_1^2 + \varepsilon_2^2} + \varepsilon_1)}}. \tag{10.29}$$

10.6. Hagen–Rubens Relation

Our next task is to find a relationship between reflectivity and conductivity. For small frequencies (i.e., $\nu < 10^{13}\ \mathrm{s}^{-1}$) the ratio σ/ν for metals is very large (d.c. conductivity[3] $\sigma_0 \approx 10^{17}\ \mathrm{s}^{-1}$, $\varepsilon \approx 10$). Thus, we obtain

$$\frac{\sigma}{\nu} \approx \frac{10^{17}}{10^{13}} \gg \varepsilon. \tag{10.30}$$

Then (10.15) and (10.16) reduce to

$$n^2 \approx \frac{\sigma}{\nu} \approx k^2. \tag{10.31}$$

The reflectivity may now be rewritten by combining the slightly modified equation (10.26) with (10.31) to read

$$R = \frac{n^2 + 2n + 1 + k^2 - 4n}{n^2 + 2n + 1 + k^2} = 1 - \frac{4n}{2n^2 + 2n + 1}. \tag{10.32}$$

If $2n + 1$ is neglected as small compared to $2n^2$ (which can be done only for small frequencies for which n is much larger than 1), then (10.32) reduces by using (10.31) to

$$R = 1 - \frac{2}{n} = 1 - 2\sqrt{\frac{\nu}{\sigma}}. \tag{10.33}$$

[3] The conductivity in cgs units is expressed in s^{-1}, see Appendix 4.

Finally, we set $\sigma = \sigma_0$ which is again only permissible for small frequencies, i.e., in the infrared region of the spectrum. This yields the Hagen–Rubens equation

$$R = 1 - 2\sqrt{\frac{\nu}{\sigma_0}} \tag{10.34}$$

which states that in the infrared (IR) region metals with large electrical conductivity are good reflectors. This equation was found empirically by Hagen and Rubens from reflectivity measurements in the IR and was derived theoretically by Drude. As stated above, the Hagen–Rubens relation is only valid at frequencies below 10^{13} s^{-1} or, equivalently, at wavelengths larger than about 30 μm.

Problems

1. Complete the intermediate steps between (10.5) and (10.6).
2. Calculate the conductivity from the index of refraction and the damping constant for copper (0.14 and 3.35, respectively; measurement at room temperature and $\lambda = 0.6$ μm). Compare your result with the conductivity of copper (see Appendix 4). You will notice a difference between these conductivities by several orders of magnitude. Why? (Compare only the same units!)
3. Express n and k in terms of ε and σ (or ε_1 and ε_2) by using $\varepsilon = n^2 - k^2$ and $\sigma = nk\nu$. (Compare with (10.15) and (10.16.)
4. The intensity of Na light passing through a gold film was measured to be about 15% of the incoming light. What is the thickness of the gold film? ($\lambda = 589$ nm; $k = 3.2$. *Note*: $I = \mathscr{E}^2$.)
5. Calculate the reflectivity of silver and compare it with the reflectivity of flint glass ($n = 1.59$). Use $\lambda = 0.6$ μm.
6. Calculate the characteristic penetration depth in aluminum for Na light ($\lambda =$ 589 nm; $k = 6$).
7. Derive the Hagen–Rubens relation from (10.29). (*Hint*: In the IR region $\varepsilon_2^2 \gg \varepsilon_1^2$ can be used. Justify this approximation.)
8. The transmissivity of a piece of glass of thickness $d = 1$ cm was measured at $\lambda =$ 589 nm to be 89%. What would the transmissivity of this glass be if the thickness were reduced to 0.5 cm?

CHAPTER 11

Atomistic Theory of the Optical Properties

11.1. Survey

In the preceding chapter, the optical constants and their relationship to electrical constants were introduced by employing the "continuum theory." The continuum theory considers only macroscopic quantities and interrelates experimental data. No assumptions are made about the structure of matter when formulating equations. Thus, the conclusions which have been drawn from the empirical laws in Chapter 10 should have general validity as long as nothing is neglected in a given calculation. The derivation of the Hagen–Rubens equation has served as an illustrative example for this.

The validity of equations derived from the continuum theory is, however, often limited to frequencies for which the atomistic structure of solids does not play a major role. Experience shows that the atomistic structure does not need to be considered in the far infrared (IR) region. Thus, the Hagen–Rubens equation reproduces the experimental results of metals in the far IR quite well. It has been found, however, that proceeding to higher frequencies (i.e., in the near IR and visible spectrum), the experimentally observed reflectivity of metals decreases faster than predicted by the Hagen–Rubens equation (Fig. 11.1). For the visible and near IR region an atomistic model needs to be considered to explain the optical behavior of metals. Drude did this important step at the turn of last century. He postulated that some electrons in a metal can be considered to be free, i.e., they can be separated from their respective nuclei. He further assumed that the free electrons can be accelerated by an external electric field. This preliminary Drude model was refined by considering that the moving electrons collide with certain metal atoms in a nonideal lattice.

The free electrons are thought to perform periodic motions in the alternating electric field of the light. These vibrations are restrained by the above-

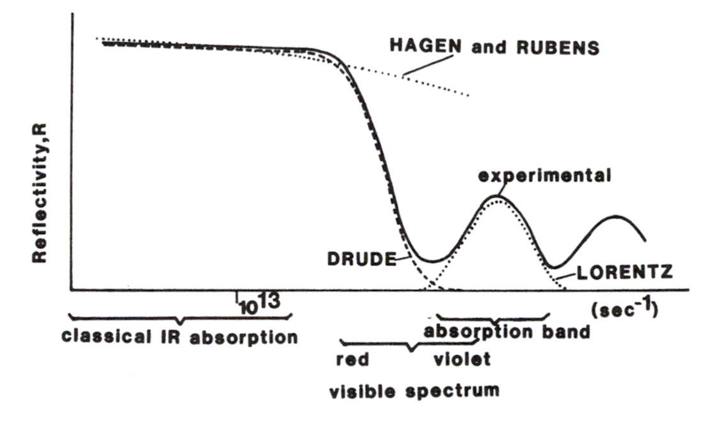

Figure 11.1. Schematic frequency dependence of the reflectivity of metals, experimentally (solid line) and according to three models.

mentioned interactions of the electrons with the atoms of a nonideal lattice. Thus, a *friction force* is introduced which takes this interaction into consideration. The calculation of the frequency dependence of the optical constants is accomplished by using the well-known equations for vibrations, whereby the interactions of electrons with atoms are taken into account by a damping term which is assumed to be proportional to the velocity of the electrons. The free electron theory describes, to a certain degree, the dispersion of the optical constants quite well. This is schematically shown in Fig. 11.1, in which the spectral dependence of the reflectivity is plotted for a specific case. The Hagen–Rubens relation reproduces the experimental findings only up to 10^{13} s^{-1}. In contrast to this, the Drude theory correctly reproduces the spectral dependence of R even in the visible spectrum. Proceeding to yet higher frequencies, however, the experimentally found reflectivity eventually rises and then decreases again. Such an *absorption band* cannot be explained by the Drude theory. For its interpretation, a new concept needs to be applied.

Lorentz postulated that the electrons should be considered to be bound to their nuclei and that an external electric field displaces the positive charge of an atomic nucleus against the negative charge of its electron cloud. In other words, he represented each atom as an electric dipole. Retracting forces were thought to occur which try to eliminate the displacement of charges. Lorentz postulated further that the centers of gravity of the electric charges are identical if no external forces are present. However, if one shines light onto a solid, i.e., if one applies an alternating electric field to the atoms, then the dipoles are thought to perform forced vibrations. Thus, a dipole is considered to behave similarly as a mass which is suspended on a spring, i.e., the equations for a harmonic oscillator may be applied. An oscillator is known to absorb a maximal amount of energy when excited near its resonance frequency (Fig. 11.2). The absorbed energy is thought to be dissipated mainly by diffuse radiation. Figure 11.2 resembles an absorption band as shown in Fig. 11.1.

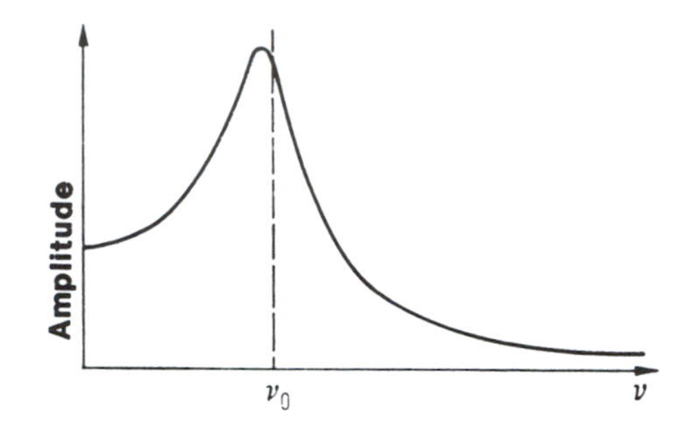

Figure 11.2. Frequency dependence of the amplitude of a harmonic oscillator which was excited to perform forced vibrations, assuming weak damping. ν_0 is the resonance frequency.

Thirty or forty years ago, many scientists considered the electrons in metals to behave at low frequencies as if they were free and at higher frequencies as if they were bound. In other words, electrons in a metal under the influence of light were described to behave as a series of classical free electrons and a series of classical harmonic oscillators. Insulators, on the other hand, were described by harmonic oscillators only.

We shall now treat the optical constants of materials by applying the above-mentioned theories.

11.2. Free Electrons Without Damping

We consider the simplest case at first and assume that the free electrons are excited to perform forced but undamped vibrations under the influence of an external alternating field, i.e., under the influence of light. As explained in Section 11.1, the damping of the electrons is thought to be caused by collisions between electrons and atoms of a nonideal lattice. Thus, we neglect in this section the influence of lattice defects. For simplicity, we treat the one-dimensional case because the result obtained this way does not differ from the general case. Thus, we consider the interaction of plane polarized light with the electrons. The momentary value of the field strength of a plane polarized light wave is given by

$$\mathscr{E} = \mathscr{E}_0 \exp(i\omega t), \tag{11.1}$$

where $\omega = 2\pi\nu$ is the angular frequency, t is the time, and $\mathscr{E}_0$ is the maximal value of the field strength. The equation describing the motion of an electron which is excited to perform forced, harmonic vibrations under the influence of light is (see Appendix 1 and (7.6))

$$m\frac{d^2x}{dt^2} = e\mathscr{E} = e\mathscr{E}_0 \exp(i\omega t), \tag{11.2}$$

where e is the electron charge, m is the electron mass, and $e \cdot \mathscr{E}$ is the modulus of the excitation force. The stationary solution of this vibrational equation is obtained by forming the second derivative of the trial solution $x = x_0 \exp(i\omega t)$ and inserting it into (11.2). This yields

$$x = -\frac{e\mathscr{E}}{m4\pi^2\nu^2}. \tag{11.3}$$

The vibrating electrons carry an electric dipole moment which is the product of the electron charge e and displacement x. The polarization P is defined to be the sum of the dipole moments of all N_f free electrons per cubic centimeter

$$P = exN_f. \tag{11.4}$$

It is known from electrodynamics that the dielectric constant may be calculated from polarization and electric field strength as follows:[4]

$$\varepsilon = 1 + 4\pi\frac{P}{\mathscr{E}}. \tag{11.5}$$

Inserting (11.3) and (11.4) into (11.5) yields

$$\hat{\varepsilon} = 1 - \frac{e^2 N_f}{\pi m \nu^2}. \tag{11.6}$$

(It is appropriate to use in the present case the *complex* dielectric constant, see below.) The dielectric constant equals the square of the index of refraction n (see (10.12)). Equation (11.6) thus becomes

$$\hat{n}^2 = 1 - \frac{e^2 N_f}{\pi m \nu^2}. \tag{11.7}$$

We consider two special cases:

(a) For small frequencies, the term $e^2 N_f/\pi m \nu^2$ is larger than one. Then $\hat{n}^2$ is negative and $\hat{n}$ imaginary. An imaginary $\hat{n}$ means that the real part of $\hat{n}$ disappears. Equation (10.25) becomes, for $n = 0$,

$$R = \frac{(n-1)^2 + k^2}{(n+1)^2 + k^2} = \frac{1 + k^2}{1 + k^2} = 1,$$

i.e., the reflectivity is 100% (see Fig. 11.3).

(b) For large frequencies (UV light), the term $e^2 N_f/\pi m \nu^2$ becomes smaller than one. Thus, $\hat{n}^2$ is positive and $\hat{n} \equiv n$ real (but smaller than one). The reflectivity for real values of $\hat{n}$, i.e., for $k = 0$, becomes

$$R = \frac{(n-1)^2}{(n+1)^2}.$$

[4] Equation (11.5) reads in the SI system: $\varepsilon = 1 + P/\varepsilon_0\mathscr{E}$ (see also Appendix 4).

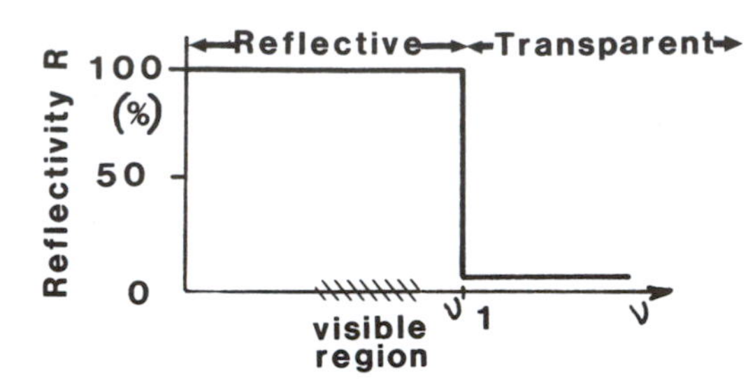

Figure 11.3. Schematic frequency dependence of an alkali metal according to the free electron theory without damping. ν_1 is the plasma frequency.

i.e., the material is essentially transparent for these wavelengths (and perpendicular incidence) and therefore behaves optically like an insulator, see Fig. 11.3.

We define a characteristic frequency ν_1, often called the *plasma frequency*, which separates the reflective from the transparent region (Fig. 11.3). The plasma frequency is defined by setting $e^2 N_f / \pi m \nu_1^2$ equal to unity which yields

$$\nu_1^2 = \frac{e^2 N_f}{\pi m}. \tag{11.8}$$

We conclude from (11.6) that the dielectric constant becomes zero at the plasma frequency because of $e^2 N_f / \pi m \nu_1^2 = 1$. $\hat{\varepsilon} = 0$ is the condition for a *plasma oscillation*, i.e., a fluidlike oscillation of the entire electron gas. We will discuss this phenomenon in detail in Section 13.2.2.

The alkali metals behave essentially as shown in Fig. 11.3. They are transparent in the near UV and reflect the light in the visible region. This result indicates that the *s*-electrons[5] of the outer shell of the alkali metals can be considered to be free.

Table 11.1 contains some measured, as well as some calculated, plasma frequencies. For the calculations, applying (11.8), *one* free electron per atom was assumed. This means that N_f was set equal to the number of atoms per volume, N_a. (The latter quantity is obtained by using

$$N_a = \frac{N_0 \cdot \delta}{M}, \tag{11.9}$$

where N_0 is the Avogadro constant, δ = density, and M = atomic mass.)

We note in Table 11.1 that the calculated and the observed values for ν_1 are only identical for sodium. This may be interpreted to mean that only in sodium exactly *one* free electron per atom is contributing to the electron gas. For other metals an "effective number of free electrons" is commonly introduced which is defined to be the ratio between the observed and calculated ν_1^2

[5] See Appendix 3.

Table 11.1. Plasma Frequencies and Effective Numbers of Free Electrons for Some Alkali Metals.

Metal	Li	Na	K	Rb	Cs
ν_1 (10^{14} s^{-1}), observed	14.6	14.3	9.52	8.33	6.81
ν_1 (10^{14} s^{-1}), calculated	19.4	14.3	10.34	9.37	8.33
λ_1 nm ($=c/\nu_1$), observed	150	210	290	320	360
N_{eff}[free electrons/atom]	0.57	1.0	0.8	0.79	0.67

values

$$\frac{\nu_1^2\,(\text{observed})}{\nu_1^2\,(\text{calculated})} = N_{\text{eff}}. \tag{11.10}$$

The effective number of free electrons is a parameter which is of great interest because it is contained in a number of nonoptical equations (such as the Hall constant, electrotransport, superconductivity, etc.). Since for most metals the plasma frequency, ν_1, cannot be measured as readily as for the alkalis, another avenue for determining N_{eff} has to be found. For reasons which will become clear later, N_{eff} can be obtained by measuring n and k in the red or IR spectrum (i.e., in a frequency range without absorption bands, Fig. 11.1) and by applying

$$N_{\text{eff}} = \frac{(1 - n^2 + k^2)\nu^2 \pi m}{e^2}. \tag{11.10a}$$

Equation (11.10a) follows by combining (11.6) with (10.10) and replacing N_f by N_{eff}.

11.3. Free Electrons With Damping (Classical Free Electron Theory of Metals)

The simple reflectivity spectrum as depicted in Fig. 11.3 is seldom found for metals. We need to refine our model. We postulate that the motion of electrons in metals is damped. More specifically, we postulate that the velocity is reduced by collisions of the electrons with atoms of a nonideal lattice. Lattice defects may be introduced into a solid by interstitial atoms, vacancies, impurity atoms, dislocations, grain boundaries, or thermal motion of the atoms.

To take account of the damping, we add to the vibration equation (11.2) a damping term $\gamma(dx/dt)$ which is proportional to the velocity (See Appendix 1 and (7.7))

$$m\frac{d^2x}{dt^2} + \gamma\frac{dx}{dt} = e\mathscr{E} = e\mathscr{E}_0 \exp(i\omega t). \tag{11.11}$$

We determine first the damping factor γ. For this we write a particular solution of (11.11) which is obtained by assuming that the electrons drift under the influence of a steady or slowly varying electric field (see Section 7.3) with a velocity $v' = \text{const.}$ through the crystal. (The drift velocity of the electrons, which is caused by an external field, is superimposed on the random motion of the electrons.) The damping is depicted to be a friction force which counteracts the electron motion. $v' = \text{const.}$ yields

$$\frac{d^2x}{dt^2} = 0. \tag{11.12}$$

By using (11.12), equation (11.11) becomes

$$\frac{e\mathscr{E}}{\gamma} = \frac{dx}{dt} = v'. \tag{11.13}$$

The drift velocity is

$$v' = \frac{j}{eN_f} \tag{11.14}$$

(see (7.4)), where j is the current density (i.e., that current which passes in one second through an area of one square centimeter). N_f is the number of free electrons per cubic centimeter. The current density is connected with the d.c. conductivity σ_0 and the field strength $\mathscr{E}$ by Ohm's law (7.2)

$$j = \sigma_0 \mathscr{E}. \tag{11.15}$$

Inserting (11.14) and (11.15) into (11.13) yields

$$\gamma = \frac{N_f e^2}{\sigma_0}. \tag{11.16}$$

Thus, (11.11) becomes

$$m\frac{d^2x}{dt^2} + \frac{N_f e^2}{\sigma_0}\frac{dx}{dt} = e\mathscr{E} = e\mathscr{E}_0 \exp(i\omega t). \tag{11.17}$$

We note that the damping term in (11.17) is inversely proportional to the conductivity, i.e., proportional to the resistivity. This result makes sense.

The stationary solution of (11.17) is obtained, similarly as in Section 11.2, by differentiating the trial solution $x = x_0 \exp(i\omega t)$ by the time, and inserting first and second derivatives into (11.17) which yields

$$-m\omega^2 x + \frac{N_f e^2}{\sigma_0} x\omega i = \mathscr{E} e. \tag{11.18}$$

Rearranging (11.18) provides

$$x = \frac{\mathscr{E}}{\frac{N_f e\omega}{\sigma_0} i - \frac{m\omega^2}{e}}. \tag{11.19}$$

Inserting (11.19) into (11.4) yields the polarization

$$P = \frac{eN_f \mathscr{E}}{\frac{N_f e\omega}{\sigma_0} i - \frac{m\omega^2}{e}}. \tag{11.20}$$

With (11.20) and (11.5) the complex dielectric constant becomes

$$\hat{\varepsilon} = 1 + 4\pi \frac{P}{\mathscr{E}} = 1 + \frac{1}{\frac{\nu}{2\sigma_0} i - \frac{m\pi}{N_f e^2} \nu^2}. \tag{11.21}$$

The term $N_f e^2/m\pi$ is set, as in (11.8), equal to ν_1^2, which reduces (11.21) to

$$\hat{\varepsilon} = 1 + \frac{1}{\frac{\nu}{2\sigma_0} i - \frac{\nu^2}{\nu_1^2}} = 1 + \frac{\nu_1^2}{i\nu \frac{\nu_1^2}{2\sigma_0} - \nu^2}. \tag{11.22}$$

The term $\nu_1^2/2\sigma_0$ in (11.22) has the unit of a frequency. Thus, for abbreviation, we define a *damping frequency*

$$\nu_2 = \frac{\nu_1^2}{2\sigma_0}. \tag{11.23}$$

(Table 11.2 lists values for ν_2 which were calculated using experimental σ_0 and ν_1 values.) Now (11.22) becomes

$$\hat{\varepsilon} = 1 + \frac{\nu_1^2}{i\nu\nu_2 - \nu^2}, \tag{11.24}$$

where $\hat{\varepsilon}$ is, as usual, identical to $\hat{n}^2$

$$(\hat{n})^2 = n^2 - 2nki - k^2 = 1 - \frac{\nu_1^2}{\nu^2 - \nu\nu_2 i}. \tag{11.25}$$

Multiplying the numerator and denominator of the fraction in (11.25) by the complex conjugate of the denominator $(\nu^2 + \nu\nu_2 i)$ allows us to equate individually real and imaginary parts. This provides the Drude equations for the optical constants.

$$\boxed{n^2 - k^2 = \varepsilon_1 = 1 - \frac{\nu_1^2}{\nu^2 + \nu_2^2}} \tag{11.26}$$

and

$$\boxed{2nk = \varepsilon_2 = \frac{2\sigma}{\nu} = \frac{\nu_2}{\nu} \frac{\nu_1^2}{\nu^2 + \nu_2^2},} \tag{11.27}$$

Table 11.2. Resistivities, Conductivities, and Damping Frequencies for Some Metals.

Metal	Li	Na	K	Rb	Cs	Cu	Ag	Au
ρ_0 ($\mu\Omega$ cm)[a]	8.55	4.2	6.15	12.5	20	1.67	1.59	2.35
σ_0 (10^{17} s^{-1})	1.05	2.14	1.46	0.72	0.45	5.37	5.66	3.83
ν_2 (10^{12} s^{-1})	10.1	4.8	3.1	4.82	5.15	4.7	4.35	5.9

[a] *Handbook of Chemistry and Physics*, 1977; room temperature values.

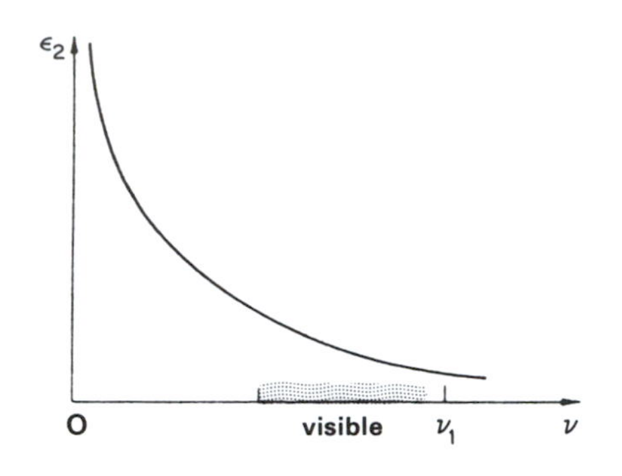

Figure 11.4. The absorption $\varepsilon_2 = 2nk$ versus frequency ν according to the free electron theory (schematic).

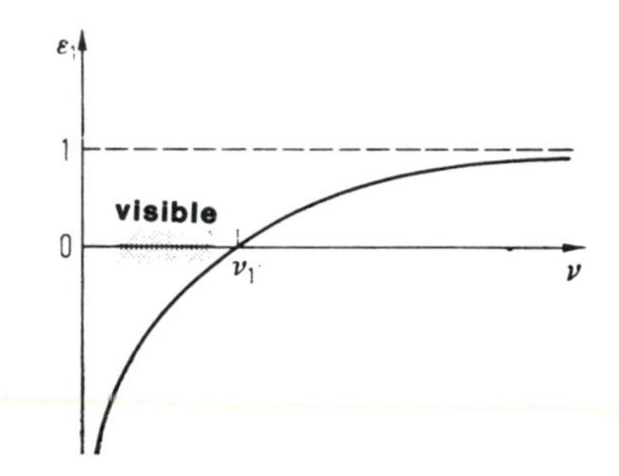

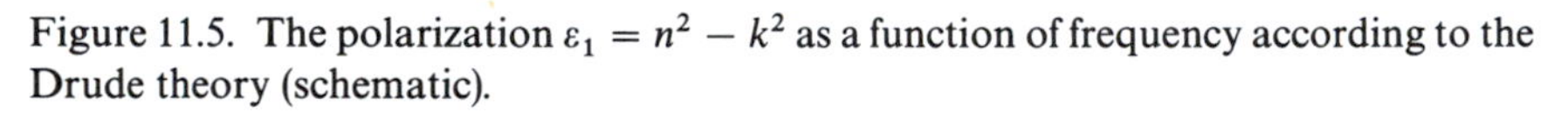

Figure 11.5. The polarization $\varepsilon_1 = n^2 - k^2$ as a function of frequency according to the Drude theory (schematic).

with the characteristic frequencies

$$\nu_1 = \sqrt{\frac{e^2 N_f}{\pi m}} \tag{11.8}$$

and

$$\nu_2 = \frac{\nu_1^2}{2\sigma_0}. \tag{11.23}$$

The absorption ε_2 and the polarization ε_1 are plotted in Figs. 11.4 and 11.5 as a function of frequency, making use of (11.27) and (11.26).

11.4. Special Cases

For the *UV*, *visible*, and *near IR* regions, the frequency varies between 10^{14} and 10^{15} s^{-1}. The average damping frequency ν_2 is 5×10^{12} s^{-1} (Table 11.2). Thus, $\nu^2 \gg \nu_2^2$. Equation (11.27) then reduces to

$$\varepsilon_2 = \frac{\nu_2}{\nu}\frac{\nu_1^2}{\nu^2}. \tag{11.28}$$

With $\nu \approx \nu_1$ (Table 11.1) we obtain

$$\varepsilon_2 = \frac{\nu_2}{\nu}. \tag{11.29}$$

Equation (11.29) confirms that ε_2 plotted versus the frequency yields a hyperbola with ν_2 as parameter (Fig. 11.4). For *very small frequencies* ($\nu^2 \ll \nu_2^2$), we may neglect ν^2 in the denominator of (11.27). This yields with (11.23)

$$nk\nu = \sigma = \frac{1}{2}\frac{\nu_1^2}{\nu_2} = \sigma_0. \tag{11.30}$$

Thus, in the far IR the a.c. conductivity σ and the d.c. conductivity σ_0 may be considered to be identical. We have already made use of this condition in Section 10.6. In general, however, σ is *not* identical to the d.c. conductivity σ_0. (The same is true for the dielectric constant ε.)

11.5. Reflectivity

The reflectivity of metals is calculated using (10.29) in conjunction with (11.26) and (11.27), see Fig. 11.6. We notice that the experimental behavior for not too high frequencies (Fig. 11.1) is essentially reproduced. See also in this

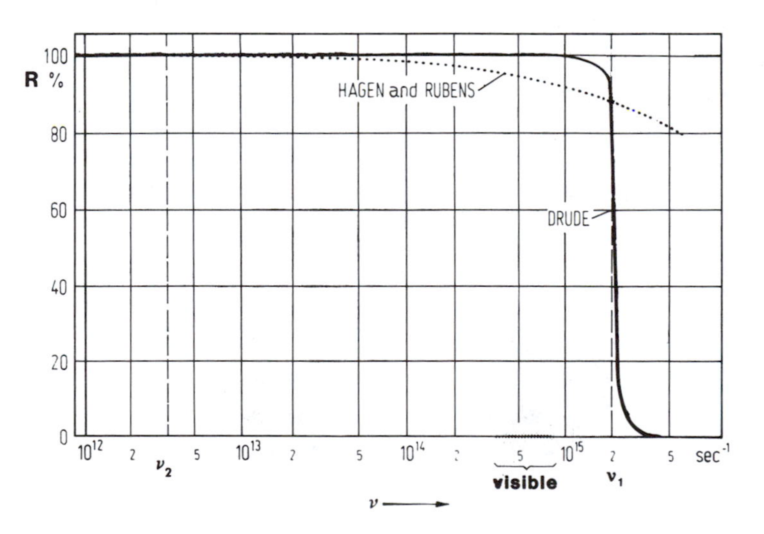

Figure 11.6. Calculated spectral reflectivity for a metal using the exact Drude equation (solid line), and the Hagen–Rubens equation (10.34) using $\nu_1 = 2 \times 10^{15}$ s^{-1} and $\nu_2 = 3.5 \times 10^{12}$ s^{-1}.

context the experimentally obtained reflectivities in Figs. 13.7, 13.10, and 13.12. For higher frequencies, however, we need to resort to a different model than the one discussed so far. This will be done in the next chapter.

11.6. Bound Electrons (Classical Electron Theory of Dielectric Materials)

The preceding sections have shown that the optical properties of metals can be described and calculated quite well in the low-frequency range by applying the free electron theory. We mentioned already that this theory has its limits at higher frequencies at which we observe that light is absorbed by metals as well as by nonmetals in a narrow frequency band. To interpret these absorption bands, Lorentz postulated that the electrons are bound to their respective nuclei. He assumed that under the influence of an external electric field, the positively charged nucleus and the negatively charged electron cloud are displaced with respect to each other (Fig. 11.7). An electrostatic force tries to counteract this displacement. For simplicity, we describe the negative charge of the electrons to be united in one point. Thus, we describe the atom in an electric field as consisting of a positively charged core which is bound quasi-elastically to *one* electron (electric dipole, Fig. 11.8). A bound electron, thus, may be compared to a mass which is suspended from a spring. Under the influence of an alternating electric field (i.e., by light), the electron is thought to perform forced vibrations. For the description of these vibrations, the well-known equations of mechanics dealing with a harmonic oscillator may be applied. This will be done now.

We first consider an isolated atom, i.e., we neglect the influence of the surrounding atoms upon the electron. An external electric field with force

$$e\mathscr{E} = e\mathscr{E}_0 \exp(i\omega t) \tag{11.31}$$

periodically displaces an electron from its rest position by a distance x. This

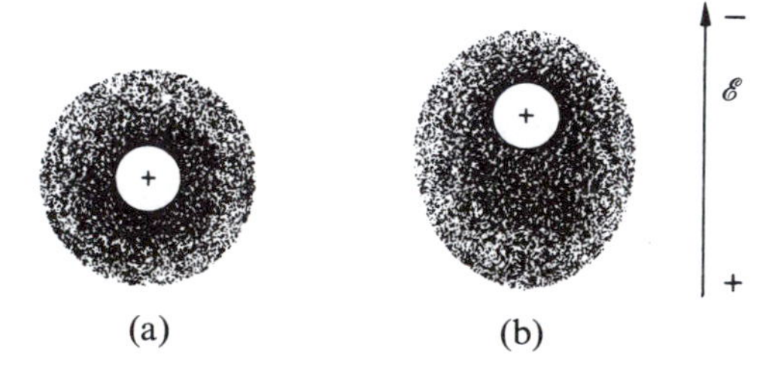

Figure 11.7. An atom is represented by a positively charged core and a surrounding, negatively charged electron cloud (a) in equilibrium and (b) in an external electric field.

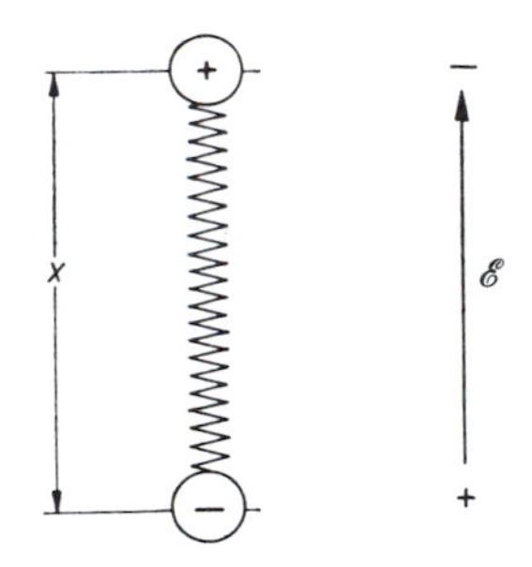

Figure 11.8. Quasi-elastic bound electron in an external electric field (harmonic oscillator).

displacement is counteracted by a restoring force $\kappa \cdot x$ which is proportional to the displacement x. Then, the vibration equation becomes (see Appendix 1)

$$m\frac{d^2x}{dt^2} + \gamma'\frac{dx}{dt} + \kappa x = e\mathscr{E}_0 \exp(i\omega t). \tag{11.32}$$

The factor κ is the *spring constant* which determines the binding strength between the atom and electron. Each vibrating dipole (e.g., an antenna) loses energy by radiation. Thus, $\gamma'(dx/dt)$ represents the damping of the oscillator by radiation (γ' = damping parameter). The stationary solution of (11.32) for weak damping is (see Appendix 1)

$$x = \frac{e\mathscr{E}_0}{\sqrt{m^2(\omega_0^2 - \omega^2)^2 + \gamma'^2\omega^2}} \exp[i(\omega t - \phi)], \tag{11.33}$$

where

$$\omega_0 = 2\pi\nu_0 = \sqrt{\frac{\kappa}{m}} \tag{11.34}$$

is called the resonance frequency of the oscillator, i.e., that frequency at which the electron vibrates freely without an external force. ϕ is the phase difference between forced vibration and the excitation force of the light wave. It is defined to be (see Appendix 1)

$$\tan\phi = \frac{\gamma'\omega}{m(\omega_0^2 - \omega^2)} = \frac{\gamma'\nu}{2\pi m(\nu_0^2 - \nu^2)}. \tag{11.35}$$

As in the previous sections, we calculate the optical constants starting with the polarization P which is the product of the dipole moment $e \cdot x$ of *one* dipole times the number of all dipoles (oscillators) N_a. As before, we assumed *one* oscillator per atom. Thus, N_a is identical to the number of atoms per unit volume. We obtain

$$P = exN_a. \tag{11.36}$$

Inserting (11.33) yields

$$P = \frac{e^2 N_a \mathscr{E}_0 \exp[i(\omega t - \phi)]}{\sqrt{m^2(\omega_0^2 - \omega^2)^2 + \gamma'^2\omega^2}}. \tag{11.37}$$

With

$$\exp[i(\omega t - \phi)] = \exp(i\omega t)\cdot\exp(-i\phi) \tag{11.38}$$

we obtain

$$P = \frac{e^2 N_a \mathscr{E}}{\sqrt{m^2(\omega_0^2 - \omega^2)^2 + \gamma'^2\omega^2}} \exp(-i\phi), \tag{11.39}$$

which yields with (11.5) and (10.12)

$$\hat{\varepsilon} = n^2 - k^2 - 2nki = 1 + \frac{4\pi e^2 N_a}{\sqrt{m^2(\omega_0^2 - \omega^2)^2 + \gamma'^2\omega^2}} \exp(-i\phi). \tag{11.40}$$

Equation (11.40) becomes with[6]

$$\exp(-i\phi) = \cos\phi - i\sin\phi, \tag{11.41}$$

$$n^2 - k^2 - 2nki = 1 + \frac{4\pi e^2 N_a}{\sqrt{m^2(\omega_0^2 - \omega^2)^2 + \gamma'^2\omega^2}} \cos\phi - i\frac{4\pi e^2 N_a}{\sqrt{m^2(\omega_0^2 - \omega^2)^2 + \gamma'^2\omega^2}} \sin\phi. \tag{11.42}$$

The trigonometric terms in (11.42) are replaced, using (11.35), as follows:

$$\cos\phi = \frac{1}{\sqrt{1 + \tan^2\phi}} = \frac{m(\omega_0^2 - \omega^2)}{\sqrt{m^2(\omega_0^2 - \omega^2)^2 + \gamma'^2\omega^2}}, \tag{11.43}$$

$$\sin\phi = \frac{\tan\phi}{\sqrt{1 + \tan^2\phi}} = \frac{\gamma'\omega}{\sqrt{m^2(\omega_0^2 - \omega^2)^2 + \gamma'^2\omega^2}}. \tag{11.44}$$

Separating the real and imaginary parts in (11.42) finally provides the optical constants

$$\varepsilon_1 = n^2 - k^2 = 1 + \frac{4\pi e^2 m N_a(\omega_0^2 - \omega^2)}{m^2(\omega_0^2 - \omega^2)^2 + \gamma'^2\omega^2},$$

that is,

$$\boxed{\varepsilon_1 = 1 + \frac{4\pi e^2 m N_a(\nu_0^2 - \nu^2)}{4\pi^2 m^2(\nu_0^2 - \nu^2)^2 + \gamma'^2\nu^2},} \tag{11.45}$$

and

$$\varepsilon_2 = 2nk = \frac{4\pi e^2 N_a \gamma'\omega}{m^2(\omega_0^2 - \omega^2)^2 + \gamma'^2\omega^2}$$

---

[6] See Appendix 2.

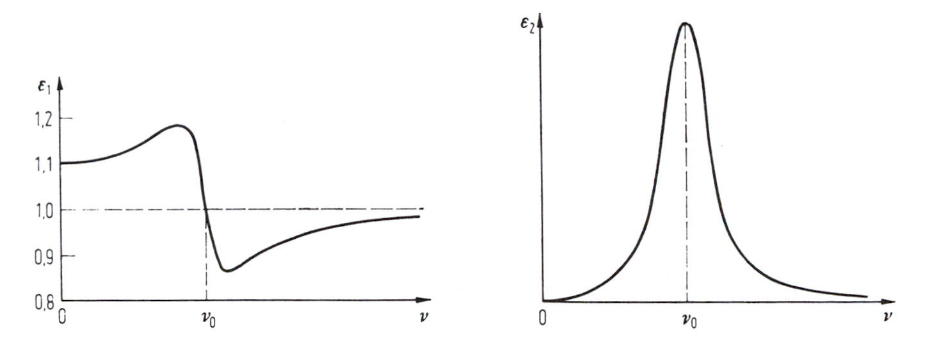

Figures 11.9 and 11.10. Frequency dependence of polarization $\varepsilon_1 = n^2 - k^2$ and absorption $\varepsilon_2 = 2nk$, as calculated with (11.45) and (11.46), respectively, using characteristic values for N_a and γ'.

or

$$\varepsilon_2 = \frac{2e^2 N_a \gamma' \nu}{4\pi^2 m^2 (\nu_0^2 - \nu^2)^2 + \gamma'^2 \nu^2}. \tag{11.46}$$

The frequency dependencies of ε_1 and ε_2 are plotted in Figs. 11.9 and 11.10. Figure 11.9 resembles the dispersion curve for the index of refraction as it is experimentally obtained for dielectrics. Figure 11.10 depicts the absorption product ε_2 in the vicinity of the resonance frequency ν_0 (absorption band) as experimentally observed for dielectrics. Equations (11.45) and (11.46) reduce to the Drude equations for $\nu_0 \to 0$ (no oscillators).

*11.7. Discussion of the Lorentz Equations for Special Cases

11.7.1. High Frequencies

We observe in Fig. 11.10 that ε_2 approaches zero at high frequencies and far away from any resonances (absorption bands). In the same frequency region, $\varepsilon_1 = n^2 - k^2$ and thus essentially n assumes the constant value 1 (Fig. 11.9). This is consistent with experimental observations that X-rays are not refracted and are not absorbed by many materials. (Note, however, that high energetic X-rays interact with the *inner* electrons, i.e., they may be absorbed by the K, L, ..., etc. electrons. Metals are, therefore, opaque for high energetic X-rays).

11.7.2. Small Damping

We consider the case for which the radiation-induced energy loss of the oscillator is very small. Then, γ' is small. With $\gamma'^2\nu^2 \ll 4\pi^2 m^2 (\nu_0^2 - \nu^2)^2$ (which is only valid for $\nu \neq \nu_0$), equation (11.45) reduces to

$$\varepsilon_1 = n^2 - k^2 = 1 + \frac{e^2 N_a}{\pi m(\nu_0^2 - \nu^2)}. \tag{11.47}$$

Figure 11.11 depicts a sketch of (11.47). We observe that for small damping, ε_1 (and thus essentially n^2) approaches infinity near the resonance frequency. A dispersion curve, such as Fig. 11.11, is indeed observed for many dielectrics (glass, etc.).

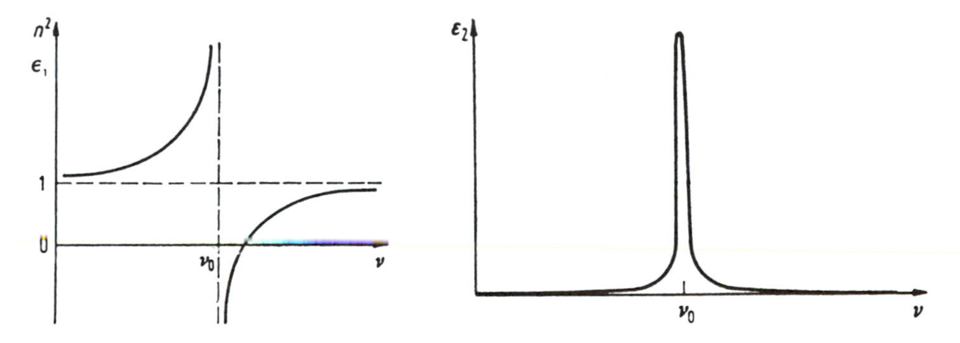

Figures 11.11 and 11.12. The functions $\varepsilon_1(n^2)$ and ε_2, respectively, versus frequency according to the bound electron theory for the special case of small damping.

11.7.3. Absorption Near ν_0

Electrons absorb most energy from light at the resonance frequency, i.e., ε_2 has a maximum near ν_0. For small damping, the absorption band becomes an absorption line (see Fig. 11.12). Inserting $\nu = \nu_0$ into (11.46) yields

$$\varepsilon_2 = \frac{2e^2 N_a}{\gamma' \nu_0}, \tag{11.48}$$

which shows that the absorption becomes large for small damping (γ').

11.7.4. More Than One Oscillator

At the beginning of Section 11.6 we assumed that *one* electron is quasi-elastically bound to a given nucleus; in other words, we assumed *one* oscillator per atom. This assumption is certainly a gross simplification, as one can deduce from the occurrence of multiple absorption bands in experimental optical spectra. Thus, each atom has to be associated with a number of i oscillators each having an oscillator strength, f_i. The ith oscillator vibrates with its resonance frequency ν_{0i}. The related damping constant is γ_i'. (This description has its equivalent in the mechanics of a system of mass points having one basic frequency and higher harmonics.) If all oscillators are taken into account, (11.45) and (11.46) become

$$\varepsilon_1 = n^2 - k^2 = 1 + 4\pi e^2 m N_a \sum_i \frac{f_i(\nu_{0i}^2 - \nu^2)}{4\pi^2 m^2(\nu_{0i}^2 - \nu^2)^2 + \gamma_i'^2 \nu^2}, \tag{11.49}$$

$$\varepsilon_2 = 2nk = 2e^2 N_a \sum_i \frac{f_i \nu \gamma_i'}{4\pi^2 m^2(\nu_{0i}^2 - \nu^2)^2 + \gamma_i'^2 \nu^2}. \tag{11.50}$$

Equations (11.49) and (11.50) reduce for weak damping (see above) to

$$\varepsilon_1 = n^2 - k^2 \approx n^2 = 1 + \frac{e^2 N_a}{\pi m} \sum_i \frac{f_i}{\nu_{0i}^2 - \nu^2}, \tag{11.51}$$

$$\varepsilon_2 = 2nk = \frac{e^2 N_a}{2\pi^2 m^2} \sum_i \frac{f_i \nu \gamma_i'}{(\nu_{0i}^2 - \nu^2)^2}. \tag{11.52}$$

11.8. Contributions of Free Electrons and Harmonic Oscillators to the Optical Constants

In the previous section, we ascribed two different properties to the electrons of a solid. In Section 11.4 we postulated that N_f electrons move freely in metals under the influence of an electric field and that this motion is damped by collisions of the electrons with vibrating lattice atoms and lattice defects. In Section 11.6 we postulated that a certain number of electrons are quasi-elastically bound to N_a atoms which are excited by light to perform forced vibrations. The energy loss was thought to be by radiation.

The optical properties of metals may be described by postulating a certain number of free electrons and a certain number of harmonic oscillators. Both the free electrons and the oscillators contribute to the polarization. Thus, the equations for the optical constants may be rewritten, by combining (11.26), (11.27), (11.49), and (11.50),

$$\varepsilon_1 = 1 - \frac{\nu_1^2}{\nu^2 + \nu_2^2} + 4\pi e^2 m N_a \sum_i \frac{f_i(\nu_{0i}^2 - \nu^2)}{4\pi^2 m^2(\nu_{0i}^2 - \nu^2)^2 + \gamma_i'^2 \nu^2}, \tag{11.53}$$

$$\varepsilon_2 = 2nk = \frac{\nu_2}{\nu} \frac{\nu_1^2}{\nu^2 + \nu_2^2} + 2e^2 N_a \sum_i \frac{f_i \nu \gamma_i'}{4\pi^2 m^2(\nu_{0i}^2 - \nu^2)^2 + \gamma_i'^2 \nu^2}. \tag{11.54}$$

Figures 11.13 and 11.14 depict schematically the frequency dependence of

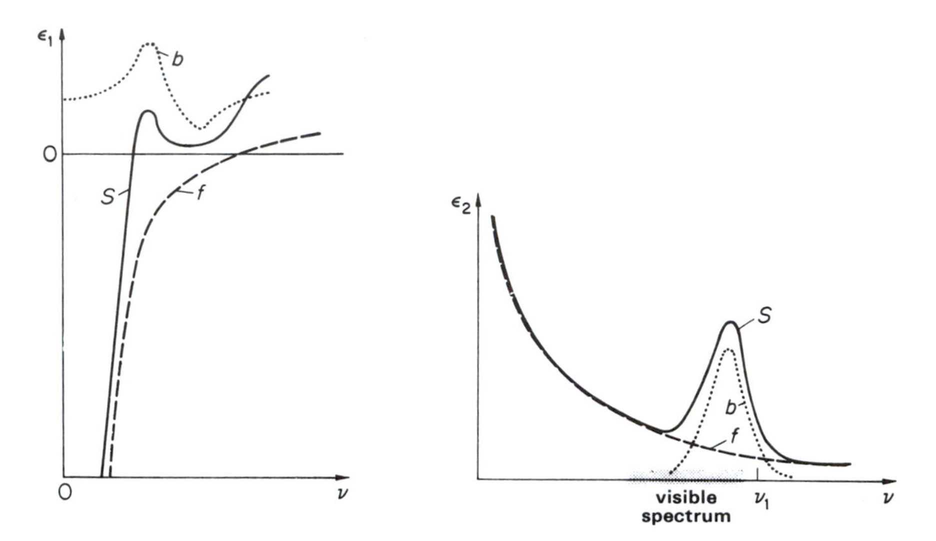

Figures 11.13 and 11.14. Frequency dependence of ε_1 and ε_2 according to (11.53) and (11.54). $(i = 1) \cdot f$ = free electron theory; b = bound electron theory; S = summary curve (schematic).

ε_1 and ε_2 as obtained by using (11.53) and (11.54). These figures also show the contributions of free and bound electrons on the optical constants. The experimentally found frequency dependence of ε_1 and ε_2 resembles these calculated spectra quite well. We will elaborate on this in Chapter 13 in which experimental results are presented.

Problems

1. Calculate the reflectivity of sodium in the frequency ranges $\nu > \nu_1$ and $\nu < \nu_1$ using the theory for free electrons without damping. Sketch R versus frequency.
2. The plasma frequency, ν_1, can be calculated for the alkali metals by assuming *one* free electron per atom, i.e., by substituting N_f by the number of atoms per unit volume (atomic density N_a). Calculate ν_1 for potassium and lithium.
3. Calculate N_{eff} for sodium and potassium. For which of these two metals is the assumption of *one* free electron per atom justified?
4. What is the meaning of the frequencies ν_1 and ν_2? In which frequency ranges are they situated compared to visible light?
5. Calculate the reflectivity of gold at $\nu = 9 \times 10^{12}\ \text{s}^{-1}$ from its conductivity. Is the reflectivity increasing or decreasing at this frequency when the temperature is increased? Explain.
6. Calculate ν_1 and ν_2 for silver (0.5×10^{23} free electrons per cubic centimeter).

7. The experimentally found dispersion of NaCl is as follows:

λ [μm]	0.3	0.4	0.5	0.7	1	2	5
n	1.607	1.568	1.552	1.539	1.532	1.527	1.519

   Plot these results along with calculated values obtained by using the equations of the "bound electron theory" assuming small damping. Let

$$\frac{e^2 N_a}{\pi m} = 1.81 \times 10^{30}\ \mathrm{s}^{-2} \qquad \text{and} \qquad \nu_0 = 1.47 \times 10^{15}\ \mathrm{s}^{-1}.$$

8. The optical properties of an absorbing medium can be characterized by various sets of parameters. One such set is the index of refraction and the damping constant. Explain the physical significance of those parameters, and indicate how they are related to the complex dielectric constant of the medium. What other set of parameters are commonly used to characterize the optical properties? Why are there always "sets" of parameters?

9. Describe the damping mechanisms for free electrons and bound electrons, respectively.

10. Why does it make sense that we assume *one* free electron per atom for the alkali metals?

11. Derive the Drude equations from (11.45) and (11.46) by setting $\nu_0 \to 0$.

12. Calculate the effective number of free electrons per cubic centimeter and per atom for silver from its optical constants ($n = 0.05$ and $k = 4.09$ at 600 nm). (*Hint*: Use the free electron mass.) How many free electrons per atom would you expect? Does the result make sense? Why may we use the free electron theory for this wavelength?

13. *Computer problem.* Plot (11.26), (11.27), and (10.29) for various values of ν_1 and ν_2. Start with $\nu_1 = 2 \times 10^{15}\ \mathrm{s}^{-1}$ and $\nu_2 = 3.5 \times 10^{12}\ \mathrm{s}^{-1}$.

14. *Computer problem.* Plot (11.45), (11.46), and (10.29) for various values of N_a, γ', and ν_0. Start with $\nu_0 = 1.5 \times 10^{15}\ \mathrm{s}^{-1}$ and $N_a = 2.2 \times 10^{22}\ \mathrm{cm}^{-3}$ and vary γ' between 100 and 0.1.

15. *Computer problem.* Plot (11.51), (11.52), and (10.29) by varying the parameters as in the previous problems. Use one, two or three oscillators. Try to "fit" an experimental curve such as the ones in Figs. 13.10 or 13.11.

CHAPTER 12

Quantum Mechanical Treatment of the Optical Properties

12.1. Introduction

We assumed in the preceding chapter that the electrons behave like particles. This working hypothesis provided us (at least for small frequencies) with equations which reproduce the optical spectra of solids reasonably well. Unfortunately, the treatment had one flaw: For calculation and interpretation of the infrared (IR) absorption we used the concept that electrons in metals are free; whereas the absorption bands in the visible and ultraviolet (UV) spectrum could only be explained by postulating harmonic oscillators. From the classical point of view, however, it is not immediately evident why the electrons should behave freely at low frequencies and respond as if they would be bound at higher frequencies. An unconstrained interpretation for this is only possible by applying wave mechanics. This will be done in the present chapter. We make use of the material presented in Chapters 5 and 6.

12.2. Absorption of Light by Interband and Intraband Transitions

When light (photons) having sufficiently large energy impinges on a solid, the electrons in this crystal are thought to be excited into a higher energy level, provided that unoccupied higher energy levels are available. For these transitions the total momentum of electrons and photons must remain constant (conservation of momentum). For optical frequencies, the momentum of a photon, and thus its wave vector $\mathbf{k}_{phot} = p/h$ (see (4.8)) is much smaller than that of an electron. Thus, $\mathbf{k}_{phot}$ is much smaller than the diameter of the Brillouin zone (Fig. 12.1). Electron transitions at which $\mathbf{k}$ remains constant

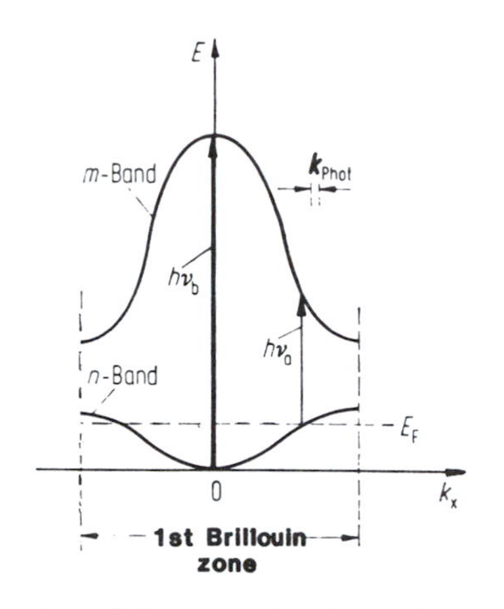

Figure 12.1. Electron bands and direct interband transitions in a reduced zone. (Compare with Fig. 5.4.)

(vertical transitions) are called "*direct interband transitions.*" Optical spectra for metals are dominated by direct interband transitions.

Another type of interband transition is possible however. It involves the absorption of a light quantum under participation of a *phonon* (lattice vibration quantum, see Chapter 20). To better understand these "*indirect interband transitions*" (Fig. 12.2) we have to know that a phonon can only absorb very small energies, but is able to absorb a large momentum which is comparable to that of an electron. During an indirect interband transition, the excess momentum (i.e., the wave number vector) is transferred to the lattice (or is absorbed from the lattice). In other words, a phonon is exchanged with the solid. Indirect interband transitions may be disregarded for the interpretation of metal spectra because they are generally weaker than direct transitions

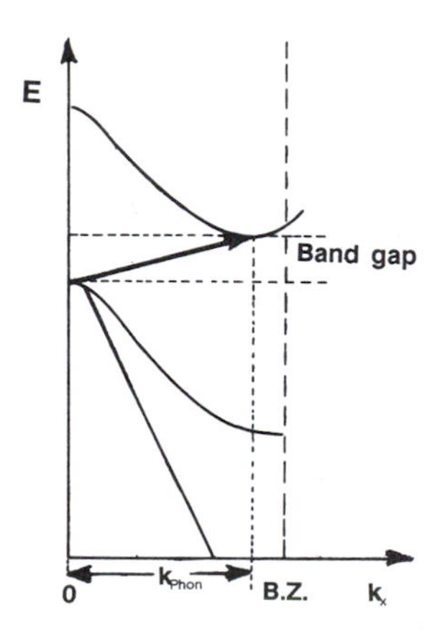

Figure 12.2. Indirect interband transition. (The properties of phonons are explained in Chapter 20.)

by two or three orders of magnitude. They are only observed in the absence of direct transitions. In the case of semiconductors, however, and for the interpretation of photoemission, indirect interband transitions play an important role.

We now make use of the simplified model depicted in Fig. 12.1 and consider direct interband transitions from the n to the m band. The smallest photon energy in this model is absorbed by those electrons whose energy equals the Fermi energy E_F, i.e., by electrons which already possess the highest possible energy at $T = 0$ K. This energy is marked in Fig. 12.1 by $h\nu_a$. Similarly, $h\nu_b$ is the *largest* energy, which leads to an interband transition from the n to the m band. In the present case, a variety of interband transitions may take place between the energy interval $h\nu_a$ and $h\nu_b$.

Interband transitions are also possible by skipping one or more bands, which occur by involving photons with even larger energies. Thus, a multitude of absorption bands are possible. These bands may partially overlap.

As an example for interband transitions in an actual case, we consider the band diagram for copper. In Fig. 12.3, a portion of Fig. 5.22 is shown, i.e., the pertinent bands around the L-symmetry point are depicted. The interband transition having the smallest possible energy difference is shown to occur between the upper d-band and the Fermi energy. This smallest energy is called the "*threshold energy for interband transitions*" (or the "*fundamental edge*") and is marked in Fig. 12.3 by a solid arrow. We mention in passing that this transition, which can be stimulated by a photon energy of 2.2 eV, is responsible for the red color of copper. At slightly higher photon energies, a second transition takes place, which originates *from* the Fermi energy. It is marked in Fig. 12.3 by a dashed arrow. Needless to say, many more transitions are

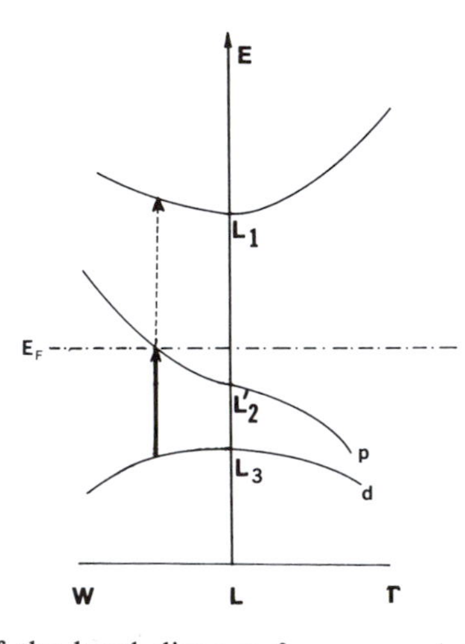

Figure 12.3. Section of the band diagram for copper (schematic). Two pertinent interband transitions are shown with arrows. The smallest possible interband transition occurs from a filled d-state to an unfilled state just above the Fermi energy.

possible. They can take place over a wide range in the Brillouin zone. This will become clearer in Chapter 13 when we return to the optical spectra of materials and their interpretation.

We now turn to another photon-induced absorption mechanism. Under certain conditions photons may excite electrons into a higher energy level *within the same band*. This occurs with participation of a phonon. We call such a transition appropriately an *intra*band transition (Fig. 12.4). It should be kept in mind, however, that because of the Pauli principle, electrons can only be excited into *empty* states. Thus, intraband transitions can only be observed in metals because only metals have unfilled electron bands.

*Intra*band transitions are equivalent to the behavior of *free* electrons in classical physics, i.e., to the "classical infrared absorption." Insulators and semiconductors have no classical infrared absorption. This explains why some insulators (such as glass) are transparent in the visible spectrum or, more precisely, why they are transparent in the absence of interband transitions. The largest photon energy, E_{max}, which can be absorbed by means of an intraband transition, corresponds to an excitation from the lower to the upper band edge, see Fig. 12.4. All energies which are smaller than E_{max} are absorbed continuously.

In summary, at low photon energies, *intra*band transitions (if possible) are the prevailing absorption mechanism. Intraband transitions are not quantized and occur in metals only. Above a critical light energy *inter*band transitions set in. Only certain energies or energy intervals are absorbed in this case. The onset of this absorption mechanism depends on the energy difference between the bands in question. *Interband* transitions occur in metals

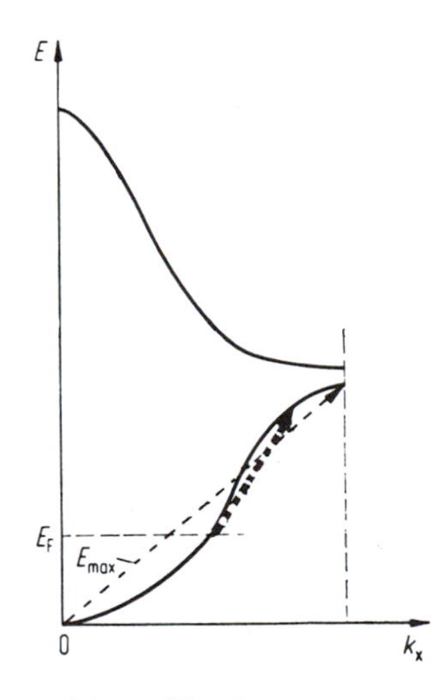

Figure 12.4. Intraband transitions. The largest energy which can be absorbed by intraband transitions is obtained by projecting the arrow marked "E_{max}" onto the energy axis.

as well as in insulators or semiconductors. They are analogous to optical excitations in solids with *bound* electrons. In an intermediate frequency range interband as well as intraband transitions may take place.

12.3. Optical Spectra of Materials

Optical spectra are the principal means to obtain experimentally the band gaps and energies for interband transitions. For isolated atoms and ions the absorption and emission spectra are known to be extremely sharp. Thus, absorption and emission energies for atoms can be determined with great accuracy. The same is basically true for molecular spectra. In contrast to this, the optical spectra of solids are rather broad. This stems from the high particle density in solids and from the interatomic interactions which split the atomic levels into quasi-continuous bands. The latter extend through the three-dimensional momentum space of a Brillouin zone.

A further factor has to be considered too. Plain *reflection* spectra of solids are, in general, not too useful for the deduction of transition energies, mainly because R is a rather involved function of ε_1 and ε_2 (see (10.29)). Thus, ε_2 (i.e., absorption) spectra are often utilized instead. The characteristic features in the ε_2-spectra of solids stem from discontinuities in the energy profile of the density of states. However, relatively sharp features in ε_2-spectra are superimposed on noncharacteristic transitions from other parts of the Brillouin zone. In other words, the ε_2-spectra derive their shape from a summation over extended, rather than localized, regions in the Brillouin zone. Modulated optical spectra (see Section 13.1.3) separate the small contributions stemming from points of high symmetry (such as the centers and edges of a Brillouin zone) from the general, much larger background. This will become clearer in the next chapter.

*12.4. Dispersion

To calculate the behavior of electrons in a periodic lattice we used, in Section 4.4, the periodic potential shown in Fig. 4.9. We implied at that time that the potential does not vary with time. This proposition needs to be dropped when the interaction of light with a solid is considered. The alternating electric field of the light which impinges on the solid perturbs the potential field of the lattice periodically. Thus, we need to add to the potential energy a correction term, the so-called perturbation potential V',

$$V = V_0 + V' \tag{12.1}$$

(V_0 = unperturbed potential energy). It goes without saying that this perturbational potential oscillates with the frequency ν of the light.

We consider, as always, plane polarized light. The momentary value of the field strength $\mathscr{E}$ is

$$\mathscr{E} = A \cos \omega t, \tag{12.2}$$

where A is the maximal value of the field strength. Then the perturbation potential (potential energy of the perturbation, or force times displacement x) is

$$V' = e\mathscr{E}x = eA \cos(\omega t) \cdot x. \tag{12.3}$$

Since the potential now varies with time, we need to make use of the time-dependent Schrödinger equation (3.8)

$$\nabla^2 \Psi - \frac{2m}{\hbar^2} V\Psi - \frac{2im}{\hbar} \frac{\partial \Psi}{\partial t} = 0, \tag{12.4}$$

which reads, with (12.1) and (12.3),

$$\nabla^2 \Psi - \frac{2m}{\hbar^2}(V_0 + eAx \cos \omega t)\Psi - \frac{2im}{\hbar} \frac{\partial \Psi}{\partial t} = 0. \tag{12.5}$$

Our goal is to calculate the optical constants from the polarization, in a similar way as it was done in Sections 11.2, 11.3, and 11.6. We have to note, however, the following: In wave mechanics, the electron is not considered to be a point, but instead is thought to be "smeared" about the space $d\tau$. The locus of the electron in classical mechanics is thus replaced by the probability $\Psi\Psi^*$ of finding an electron in space (see (2.12)). The classical polarization

$$P = Nex$$

(11.4) is replaced in wave mechanics by

$$P = Ne \int x\Psi\Psi^* \, d\tau. \tag{12.6}$$

We seek to find a solution Ψ of the *perturbed Schrödinger equation* (12.5) and calculate from that the *norm* $\Psi\Psi^*$; then by using (12.6) we can calculate the polarization P. The equation for the optical constants thus obtained is given in (12.31).

The detailed calculation of this approach will be given below. The first step is to transform the space- *and* time-dependent Schrödinger equation into a Schrödinger equation which is only space-dependent. The perturbed Schrödinger equation (12.5) is rewritten, using the Euler equation[7] $\cos \rho = \frac{1}{2}(e^{i\rho} + e^{-i\rho})$, as

$$\nabla^2 \Psi - \frac{2m}{\hbar^2} V_0 \Psi - \frac{2im}{\hbar} \frac{\partial \Psi}{\partial t} = \frac{2m}{\hbar^2} eAx\tfrac{1}{2}[e^{i\omega t} + e^{-i\omega t}]\Psi. \tag{12.7}$$

Now, the left side of (12.7) has the form of the unperturbed Schrödinger equation (12.4). We assume that the perturbation is very small. Then we can insert in the perturbation term (right side of (12.7)) the expression (3.4), and

[7] See Appendix 2.

get

$$\Psi_i^0(x, y, z, t) = \psi_i^0(x, y, z)e^{i\omega_i t} \tag{12.8}$$

for the unperturbed ith eigenfunction. This yields

$$\nabla^2\Psi - \frac{2m}{\hbar^2}V_0\Psi - \frac{2im}{\hbar}\frac{\partial\Psi}{\partial t} = \frac{m}{\hbar^2}eAx\psi_i^0[e^{i(\omega_i+\omega)t} + e^{i(\omega_i-\omega)t}]. \tag{12.9}$$

The right-hand side will be contracted to simplify the calculation

$$\nabla^2\Psi - \frac{2m}{\hbar^2}V_0\Psi - \frac{2im}{\hbar}\frac{\partial\Psi}{\partial t} = \frac{m}{\hbar^2}eAx\psi_i^0 e^{i(\omega_i\pm\omega)t}. \tag{12.10}$$

To solve (12.10), we seek a trial solution which consists of an unperturbed solution and two terms with the angular frequencies $(\omega_i + \omega)$ and $(\omega_i - \omega)$

$$\Psi = \Psi_i^0 + \psi_+ e^{i(\omega_i+\omega)t} + \psi_- e^{i(\omega_i-\omega)t}. \tag{12.11}$$

This trial solution is condensed as before

$$\Psi = \Psi_i^0 + \psi_\pm e^{i(\omega_i\pm\omega)t}. \tag{12.12}$$

Equation (12.12) is differentiated twice with respect to space and once with respect to time, and the results are inserted into (12.10). This yields

$$\underline{\nabla^2\Psi_i^0} + \nabla^2\psi_\pm e^{i(\omega_i\pm\omega)t} - \underline{\frac{2m}{\hbar^2}V_0\Psi_i^0} - \frac{2m}{\hbar^2}V_0\psi_\pm e^{i(\omega_i\pm\omega)t}$$

$$\underline{-\frac{2im}{\hbar}\frac{\partial\Psi_i^0}{\partial t}} + \frac{2m}{\hbar}(\omega_i \pm \omega)\psi_\pm e^{i(\omega_i\pm\omega)t} = \frac{m}{\hbar^2}eAx\psi_i^0 e^{i(\omega_i\pm\omega)t}. \tag{12.13}$$

The underlined terms in (12.13) vanish according to (12.4) if Ψ_i^0 is the solution to the unperturbed Schrödinger equation. In the remaining terms, the exponential factors can be cancelled which yields, with $\hbar\omega = h\nu = E$,

$$\nabla^2\psi_\pm + \frac{2m}{\hbar^2}\psi_\pm(E_i \pm h\nu - V_0) = \frac{m}{\hbar^2}eAx\psi_i^0. \tag{12.14}$$

In writing (12.14) we have reached our first goal, i.e., to obtain a time-independent, perturbed Schrödinger equation. We solve this equation with a procedure which is common in perturbation theory. We develop the function $x\psi_i^0$ (the right side of (12.14)) in a series of eigenfunctions

$$x\psi_i^0 = a_{1i}\psi_1^0 + a_{2i}\psi_2^0 + \cdots + a_{ni}\psi_n^0 + \cdots = \sum a_{ni}\psi_n^0, \tag{12.15}$$

multiply (12.15) by ψ_n^{0*}, and integrate over the entire space $d\tau$. Then, due to $\int\psi\psi^*\,d\tau = 1$ (3.15) and $\int\psi_m\psi_n^*\,d\tau = 0$ (for $m \neq n$), we obtain

$$\int x\psi_i^0\psi_n^{0*}\,d\tau = a_{1i}\underbrace{\int\psi_1^0\psi_n^{0*}\,d\tau}_{0} + \underbrace{\cdots}_{0} + a_{ni}\underbrace{\int\psi_n^0\psi_n^{0*}\,d\tau}_{1} + \cdots = a_{ni}. \tag{12.16}$$

Similarly, we develop the function $\psi_\pm$ in a series of eigenfunctions

$$\psi_\pm = \sum b_{\pm n}\psi_n^0. \tag{12.17}$$

Inserting (12.15) and (12.17) into (12.14) yields

$$\sum b_{\pm n}\left(\underline{\nabla^2\psi_n^0} + \frac{2m}{\hbar^2}E_i\psi_n^0 \pm \frac{2m}{\hbar^2}h\nu\psi_n^0 - \underline{\frac{2m}{\hbar^2}V_0\psi_n^0}\right) = \frac{m}{\hbar^2}eA\sum a_{ni}\psi_n^0. \quad (12.18)$$

Rewriting the unperturbed time-independent Schrödinger equation (3.1) yields

$$\nabla^2\psi_n^0 - \frac{2m}{\hbar^2}V_0\psi_n^0 = -\frac{2m}{\hbar^2}E_n\psi_n^0. \quad (12.19)$$

Equation (12.19) shows that the underlined terms in (12.18) may be equated to the right side of (12.19). Thus, (12.18) may be rewritten as

$$\frac{2m}{\hbar^2}\sum \psi_n^0 b_{\pm n}(E_i - E_n \pm h\nu) = \frac{2m}{\hbar^2}\frac{eA}{2}\sum \psi_n^0 a_{ni}. \quad (12.20)$$

Comparing the coefficients in (12.20) yields, with

$$E_i - E_n = E_{ni} = h\nu_{ni}, \quad (12.21)$$

the following expression:

$$b_{\pm n} = \frac{eAa_{ni}}{2(E_i - E_n \pm h\nu)} = \frac{eAa_{ni}}{2h(\nu_{ni} \pm \nu)}. \quad (12.22)$$

Now we are able to determine the functions ψ_+ and ψ_- by using (12.22) and (12.17). We insert these functions together with (3.4) into the trial solution (12.11) and obtain a solution for the time-dependent, perturbed Schrödinger equation (12.5)

$$\Psi = \psi_i^0 e^{i\omega_i t} + \frac{1}{2h}\sum eAa_{ni}\psi_n^0\left[\frac{e^{i(\omega_i+\omega)t}}{\nu_{ni}+\nu} + \frac{e^{i(\omega_i-\omega)t}}{\nu_{ni}-\nu}\right], \quad (12.23)$$

and thus

$$\Psi^* = \psi_i^{0*} e^{-i\omega_i t} + \frac{1}{2h}\sum eAa_{ni}^*\psi_n^{0*}\left[\frac{e^{-i(\omega_i+\omega)t}}{\nu_{ni}+\nu} + \frac{e^{-i(\omega_i-\omega)t}}{\nu_{ni}-\nu}\right]. \quad (12.24)$$

In order to write the polarization (12.6) we have to form the product $\Psi\Psi^*$. As can be seen from (12.23) and (12.24), this calculation yields time-dependent as well as time-independent terms. The latter ones need not be considered here since they provided only an additive constant to the polarization (light scattering). The time-dependent part of the norm $\Psi\Psi^*$ is

$$\Psi\Psi^* = \frac{eA}{2h}\Bigg[\sum a_{ni}^*\psi_n^{0*}\psi_i^0\underbrace{\left(\frac{e^{-i\omega t}}{\nu_{ni}+\nu} + \frac{e^{i\omega t}}{\nu_{ni}-\nu}\right)}_{Q}$$

$$+ \sum a_{ni}\psi_n^0\psi_i^{0*}\underbrace{\left(\frac{e^{i\omega t}}{\nu_{ni}+\nu} + \frac{e^{-i\omega t}}{\nu_{ni}-\nu}\right)}_{R}\Bigg]. \quad (12.25)$$

To simplify, we abbreviate the terms in parentheses by Q and R, respectively. The polarization (12.6) is then

$$P = \frac{Ne^2 A}{2h}\left[\sum a_{ni}^* Q \underbrace{\int x\psi_n^{0*}\psi_i^0 \, d\tau}_{a_{ni}} + \sum a_{ni} R \underbrace{\int x\psi_n^0\psi_i^{0*} \, d\tau}_{a_{ni}^*}\right], \tag{12.26}$$

which reduces, with (12.16),

$$\int x\psi_i^0\psi_n^{0*} \, d\tau = a_{ni}$$

and

$$a_{ni} \cdot a_{ni}^* = |a_{ni}|^2 \equiv a_{ni}^2 \tag{12.27}$$

to

$$P = \frac{Ne^2 A}{2h}\sum a_{ni}^2 (Q + R). \tag{12.28}$$

A numerical calculation applying the above-quoted Euler equation yields

$$Q + R = \frac{2\nu_{ni}e^{-i\omega t}}{\nu_{ni}^2 - \nu^2} + \frac{2\nu_{ni}e^{i\omega t}}{\nu_{ni}^2 - \nu^2} = \frac{4\nu_{ni}\cos \omega t}{\nu_{ni}^2 - \nu^2}, \tag{12.29}$$

which gives with (12.2)

$$P = \frac{Ne^2 \mathscr{E}}{\pi\hbar}\sum a_{ni}^2 \frac{\nu_{ni}}{\nu_{ni}^2 - \nu^2}. \tag{12.30}$$

Finally, we make use of (10.13) and (11.5) and obtain with (12.30)

$$\boxed{\varepsilon_1 = n^2 - k^2 = 1 + \frac{4Ne^2}{\hbar}\sum a_{ni}^2 \frac{\nu_{ni}}{\nu_{ni}^2 - \nu^2}.} \tag{12.31}$$

Equation (12.31) is the sought-after relation for the optical properties of solids, obtained by wave mechanics. It is similar in form to the classical dispersion equation (11.51). A comparison of classical and quantum mechanical results might be helpful to better understand the meaning of the empirically introduced oscillator strength f_i. We obtain

$$\boxed{f_i = \frac{4\pi m}{\hbar} a_{ni}^2 \nu_{ni}.} \tag{12.32}$$

We know that $h\nu_{ni}$ is that energy which an electron absorbs when it is excited from the n-band into the i-band (e.g., the m-band). Thus, the resonance frequency ν_{0i} of the ith oscillator introduced in Section 11.6 is replaced in wave mechanics by a frequency ν_{ni} which corresponds to an allowed electron transition from the nth into the ith band. Furthermore, we see from (12.16) that a_{ni} is proportional to the probability of an electron transition from the nth into the ith band. The oscillator strength f_i is therefore, essentially, the probability for a certain interband transition.

Problems

1. What information can be gained from the quantum mechanical treatment of the optical properties of metals which cannot be obtained by the classical treatment?
2. What can we conclude from the fact that the spectral reflectivity of a metal (e.g., copper) has "structure"?
3. Below the reflection spectra for two materials A and B are given.
   (a) What type of material belongs to reflection spectrum A, what type to B? (Justify.) Note the scale difference!
   (b) For which colors are these (bulk) materials transparent?
   (c) What is the approximate threshold energy for interband transitions for these materials?
   (d) For which of the materials would you expect intraband transitions in the infrared region? (Justify.)
   (e) Why do these intraband transitions occur in this region?

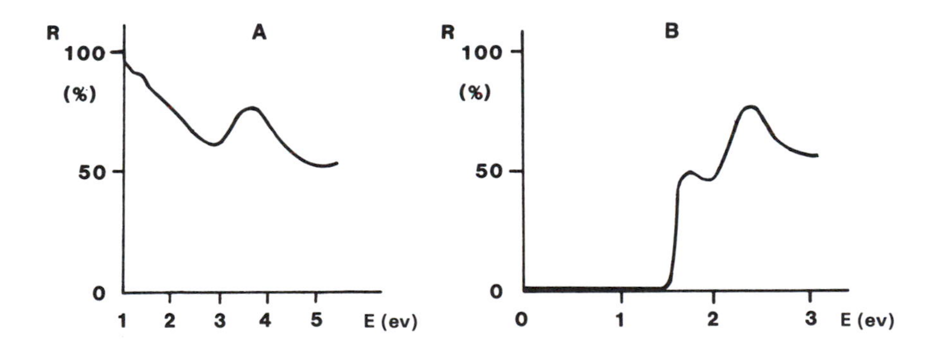

4. What is the smallest possible energy for interband transitions for aluminum? (*Hint*: Consult the band diagram in Fig. 5.21.)
5. Are intraband transitions possible in semiconductors at high temperatures?

CHAPTER 13

Applications

13.1. Measurement of the Optical Properties

The measurement of the optical properties of solids is simple in principle, but can be involved in practice. This is so because many bulk solids (particularly metals) are opaque so that the measurements have to be taken in reflection. Light penetrates about 10 nm into a metal (see Table 10.1). As a consequence, the optical properties are basically measured near the surface which is susceptible to oxidation, deformation (polishing), or contamination by adsorbed layers. One tries to alleviate the associated problems by utilizing ultrahigh vacuum, vapor deposition, sputtering, etc. Needless to say, the method by which a given sample was prepared may have an effect on the numerical value of its optical properties.

Let us assume that the surface problems have been resolved. Then, still another problem remains. The most relevant optical properties, namely, n, k, ε_1, ε_2, and the energies for interband transitions cannot be easily deduced by simply measuring the reflectivity, i.e., the ratio between reflected and incident intensity. Thus, a wide range of techniques have been developed in the past century to obtain the above-mentioned parameters. Only three methods will be briefly discussed here. It should be mentioned, however, that thirty or forty other techniques could be easily presented. They all have certain advantages for some specific applications and disadvantages for others. The reader who is not interested in the measurement of optical properties may skip the next three sections for the time being and return to them at a later time.

*13.1.1. Kramers–Kronig Analysis (Dispersion Relations)

This method was very popular in the 1960s and involves the measurement of the reflectivity over a wide spectral range. A relationship exists between real and imaginary terms of any complex function, which enables one to calculate *one* component of a complex quantity if the other one is known. In the present case, one calculates the phase jump δ' (between the reflected and incident ray) from the reflectivity which was measured at a given frequency ν. This is accomplished by the Kramers–Kronig relation

$$\delta'(\nu_x) = \frac{1}{\pi}\int_0^\infty \frac{d\ln\rho}{d\nu}\ln\left|\frac{\nu+\nu_x}{\nu-\nu_x}\right| d\nu, \tag{13.1}$$

where

$$\rho = \sqrt{R} = \sqrt{\frac{I_R}{I_0}} \tag{13.2}$$

is obtained from the reflected intensity I_R and the incident intensity I_0 of the light. The optical constants are calculated by applying

$$n = \frac{1-\rho^2}{1+\rho^2+2\rho\cos\delta'} \tag{13.3}$$

and

$$k = \frac{2\rho\sin\delta'}{1+\rho^2+2\rho\cos\delta'}. \tag{13.4}$$

Equation (13.1) shows that the reflectivity should be known in the entire frequency range (i.e., between $\nu = 0$ and $\nu = \infty$). Since measured values can hardly be obtained for such a large frequency range, one usually extrapolates the reflectivity beyond the experimental region using theoretical or phenomenological considerations. Such an extrapolation would not cause a substantial error if one could assume that no interband transitions exist beyond the measured spectral range. This assumption is probably valid only on rare occasions. (For details, see specialized books listed at the end of Part III.)

*13.1.2. Spectroscopic Ellipsometry

This technique was developed in its original form at the turn of the century. The underlying idea is as follows: If plane polarized light impinges under an angle α on a metal, the reflected light is generally elliptically polarized. The analysis of this elliptically polarized light yields two parameters, the *azimuth* and the *phase difference* from which the optical properties are calculated.

We consider plane polarized light whose vibrational plane is inclined by 45° towards the plane of incidence (Fig. 13.1). This angle is called azimuth ψ_e

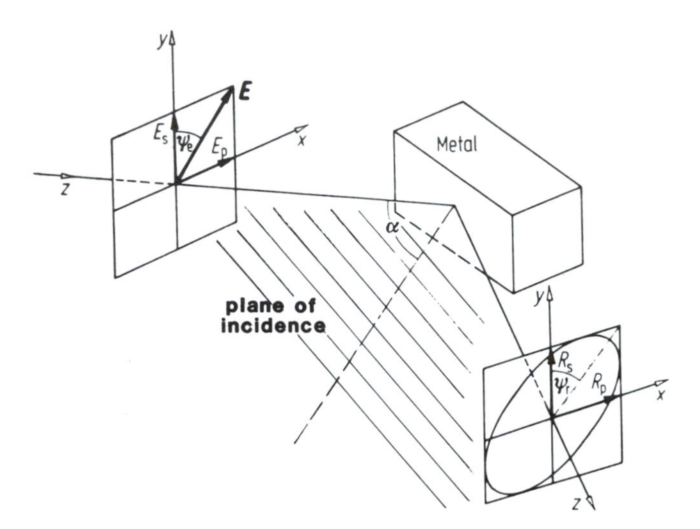

Figure 13.1. Reflection of plane polarized light on a metal surface. (*Note*: In the figure $\mathscr{E}_{Rp} \equiv R_p$ and $\mathscr{E}_{Rs} = R_s$.)

in contrast to the azimuth of the reflected light ψ_r which is defined as

$$\tan \psi_r = \frac{\mathscr{E}_{Rp}}{\mathscr{E}_{Rs}}, \tag{13.5}$$

see Fig. 13.1, where $\mathscr{E}_{Rp}$ and $\mathscr{E}_{Rs}$ are parallel and perpendicular components of the reflected electric field strength $|\mathscr{E}|$, i.e., the amplitudes of the reflected light wave.

In elliptically polarized light the length and direction of the light vector is altered periodically. The tip of the light vector moves along a continuous screw, having the direction of propagation as an axis (Fig. 13.2(a)). The projection of this screw onto the x–y plane is an ellipse (Fig. 13.1). Elliptically polarized light can be thought of as composed of two mutually perpendicular, plane polarized waves, having a phase difference δ between them (expressed in fractions of 2π) (see Fig. 13.2(b)).

For the actual measurement of ψ_r and δ, one needs two polarizers (consisting of a birefringent material which allows only plane polarized light to pass), and a compensator (also consisting of birefringent material which allows one to measure the phase difference δ; see Fig. 13.3). In Fig. 13.4, the light reflected from a metal is represented by two light vectors pointing in the x- and the y-directions, respectively. They have a phase difference δ between them. By varying the thickness of the birefringent materials in the compensator, one eventually accomplishes that the light which leaves the compensator is plane polarized (i.e., $\delta = 0°$). The resulting vector R_{res} is then tilted by an angle ψ_r against the normal to the plane of incidence. One determines ψ_r by turning the analyzer to a position at which its axis is perpendicular to R_{res}. In short, δ and ψ_r are measured by simultaneously altering the thickness of the

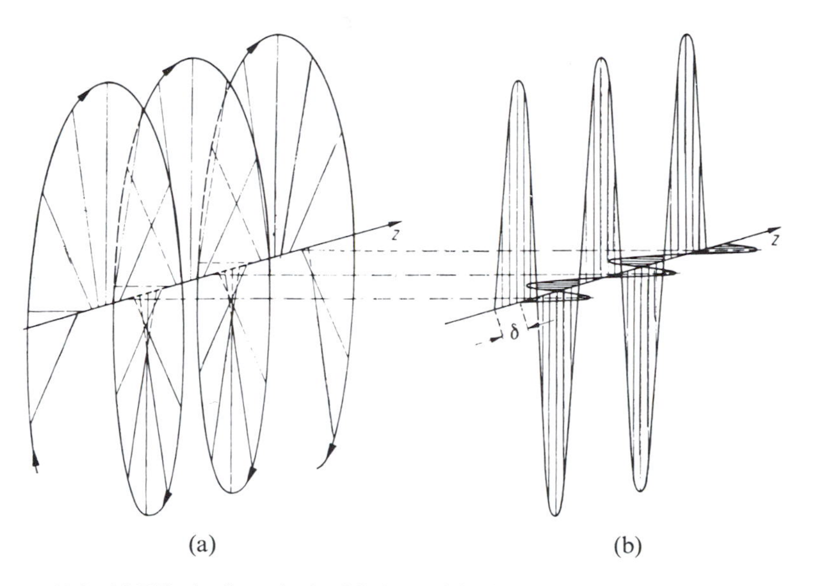

Figure 13.2. (a) Elliptically polarized light and (b) decomposition of elliptically polarized light into two mutually perpendicular plane polarized waves with phase difference δ. Adapted from R.W. Pohl, *Optik und Atomphysik*. Springer-Verlag, Berlin (1958).

compensator and turning the analyzer until no light leaves the analyzer. It is evident that this method is cumbersome and time-consuming, particularly in cases in which an entire spectrum needs to be measured point by point. Thus, in recent years automated and computerized ellipsometers have been developed.

The optical constants are calculated using

$$n^2 = \tfrac{1}{2}[\sqrt{(a^2 - b^2 + \sin^2 \alpha)^2 + 4a^2b^2} + a^2 - b^2 + \sin^2 \alpha], \quad (13.6)$$

$$k^2 = \tfrac{1}{2}[\sqrt{(a^2 - b^2 + \sin^2 \alpha)^2 + 4a^2b^2} - (a^2 - b^2 + \sin^2 \alpha)], \quad (13.7)$$

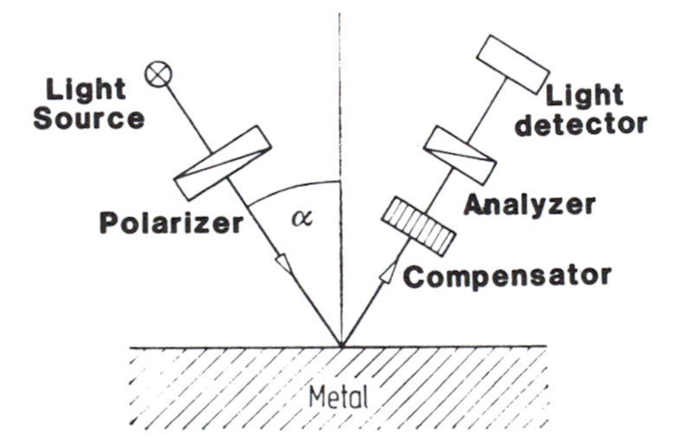

Figure 13.3. Schematic of an ellipsometer (polarizer and analyzer are identical devices).

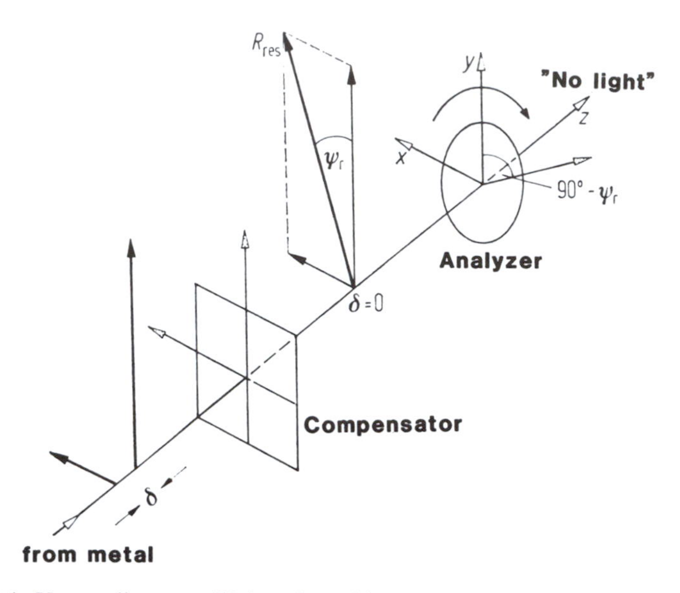

Figure 13.4. Vector diagram of light reflected from a metal surface. The vectors having solid arrowheads give the vibrational direction and magnitude of the light.

with

$$a = \frac{\sin \alpha \tan \alpha \cos 2\psi_r}{1 - \cos \delta \sin 2\psi_r} \tag{13.8}$$

and

$$b = -a \sin \delta \tan 2\psi_r. \tag{13.9}$$

Alternatively, one obtains, for the polarization ε_1 and absorption ε_2,

$$\varepsilon_1 = n^2 - k^2 = \sin^2 \alpha \left[1 + \frac{\tan^2 \alpha(\cos^2 2\psi_r - \sin^2 2\psi_r \sin^2 \delta)}{(1 - \sin 2\psi_r \cos \delta)^2}\right], \tag{13.10}$$

$$\varepsilon_2 = 2nk = -\frac{\sin 4\psi_r \sin \delta \tan^2 \alpha \sin^2 \alpha}{(1 - \sin 2\psi_r \cos \delta)^2}. \tag{13.11}$$

*13.1.3. Differential Reflectometry

The information gained by differential reflectometry is somewhat different from that obtained by the afore-mentioned techniques. A "differential reflectogram" allows the direct measurement of the energies which electrons absorb from photons as they are raised into higher allowed energy states. The differential reflectometer measures the normalized difference between the reflectivities of two similar specimens which are mounted side by side (Fig. 13.5). For example, one specimen might be pure copper and the other copper with, say, 1% zinc. Unpolarized light coming from a monochromator is alternately

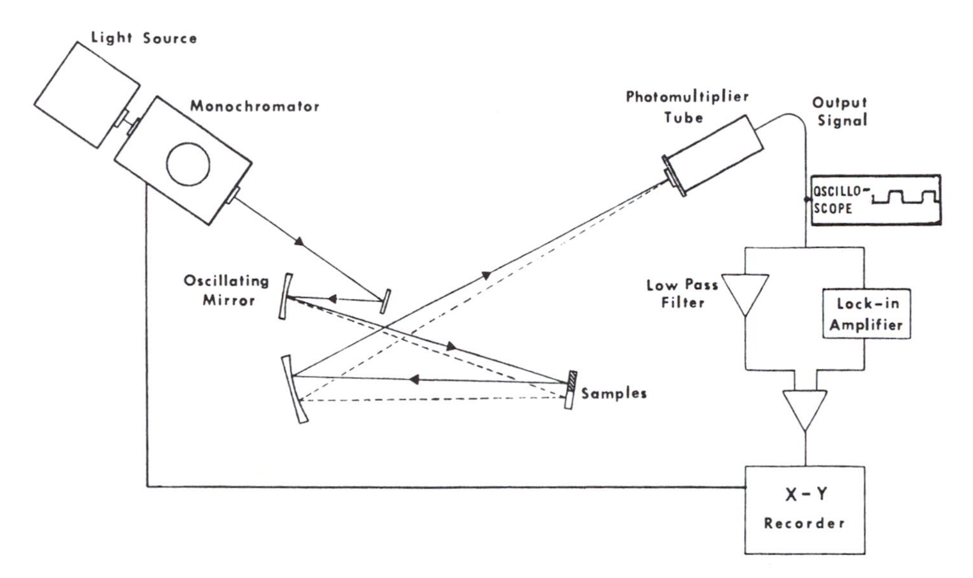

Figure 13.5. Schematic diagram of the differential reflectometer. (For clarity, the angle of incidence of the light beam impinging on the samples is drawn larger than it is in reality.) From R.E. Hummel, *Phys. Stat. Sol.* (a) **76**, 11 (1983).

deflected under near normal incidence to one or the other sample by means of a vibrating mirror. The reflected light is electronically processed to yield $\Delta R/\bar{R} = 2(R_1 - R_2)/(R_1 + R_2)$. A complete differential reflectogram, i.e., a scan from the near IR through the visible into the near UV, is generated automatically and takes about two minutes. The main advantage of differential reflectometry over conventional optical techniques lies in its ability to eliminate any undesirable influences of oxides, deformations, windows, electrolytes (for corrosion studies), or instrumental parameters upon a differential reflectogram, owing to the differential nature of the technique. No vacuum is needed. Thus, the formation of a surface layer due to environmental interactions can be studied *in situ*. Finally, the data can be taken under near normal incidence.

Differential reflectometry belongs to a family of techniques, called *modulation spectroscopy*, in which the derivative of the unperturbed reflectivity (or ε_2) with respect to an external parameter is measured. Modulation techniques restrict the action to so-called *critical points* in the band structure, i.e., they emphasize special electron transitions from an essentially featureless background. This background is caused by the allowed transitions at practically all points in the Brillouin zone. Most modulation techniques, such as differential reflectometry, wavelength modulation, thermoreflectance, or piezoreflectance, are *first-derivative* techniques (Fig. 13.6(a)). In semiconductor research (Section 13.6) another modulation technique, called electroreflectance, is often used, which provides the third derivative of R or ε_2. (It utilizes an

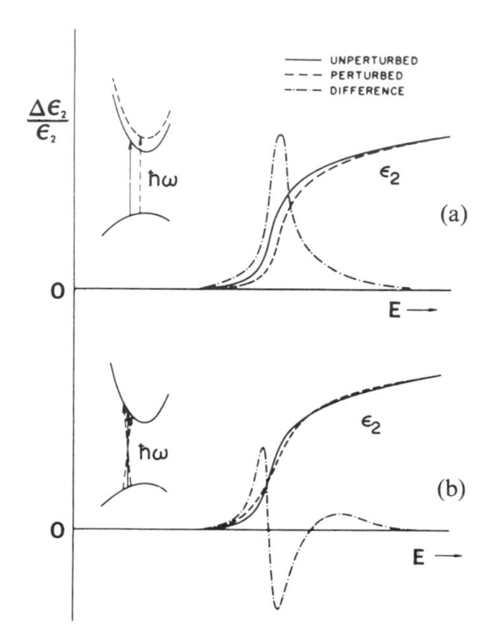

Figure 13.6. Schematic representation of (a) the first derivative and (b) the third derivative of an ε_2-spectrum. The equivalent interband transitions at a so-called M_0 symmetry point are shown in the inserts. Adapted from D.E. Aspnes, *Surface Science* **37**, 418 (1973).

alternating electric field which is applied to the semiconducting material during the reflection measurement.) The third derivative provides sharper and more richly structured spectra than the first derivative techniques (Fig. 13.6(b)). In a first derivative modulation spectrum, the lattice periodicity is retained, the optical transitions remain vertical and the interband transition energy changes with the perturbation (see insert of Fig. 13.6(a)). In electromodulation, the formerly sharp vertical transitions are spread over a finite range of initial and final momenta (see insert of Fig. 13.6(b)). A relatively involved line-shape analysis of electroreflectance spectra eventually yields the interband transition energies.

We shall make use of reflection, absorption, first and third derivative spectra in the sections to come.

13.2. Optical Spectra of Pure Metals

13.2.1. Reflection Spectra

The spectral dependence of the optical properties of metals was described and calculated in Chapter 11 by postulating that light interacts with a certain

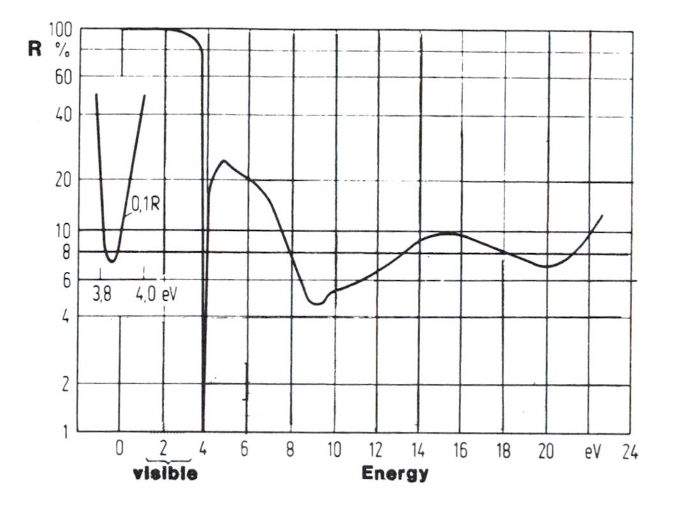

Figure 13.7. Reflectivity spectrum for silver. Adapted from H. Ehrenreich et al., *IEEE Spectrum* **2**, 162 (1965). © 1965 IEEE.

number of *free electrons* and a certain number of *classical harmonic oscillators*, or equivalently, by *intraband* and *interband* transitions. In the present section we shall inspect experimental reflection data and see what conclusions can be drawn from these results with respect to the electron band structure.

Figure 13.7 depicts the spectral reflectivity for silver. From this diagram, the optical constants (i.e., the real and imaginary parts of the complex dielectric constant, $\varepsilon_1 = n^2 - k^2$ and $\varepsilon_2 = 2nk$) have been calculated by means of a Kramers–Kronig analysis (Section 13.1.1). Comparing Fig. 13.8 with Fig.

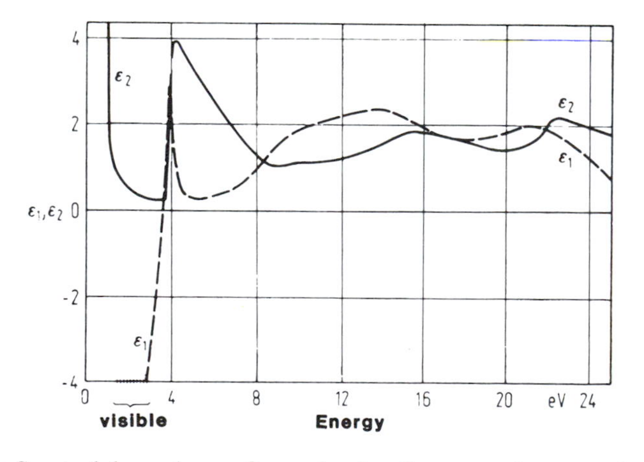

Figure 13.8. Spectral dependence of ε_1 and ε_2 for silver. ε_1 and ε_2 were obtained from Fig. 13.7 by a Kramers–Kronig analysis. Adapted from H. Ehrenreich et al., *IEEE Spectrum* **2**, 162 (1965). © 1965 IEEE.

11.5 shows that for small photon energies, i.e., for $E < 3.8$ eV, the spectral dependences of ε_1 and ε_2 have the characteristic curve shapes for free electrons. In other words, the optical properties of silver can be described in this region by the concept of free electrons. Beyond 3.8 eV, however, the spectral dependence of ε_1 and ε_2 deviates considerably from the free electron behavior. In this range, classical oscillators, or equivalently, interband transitions need to be considered.

Now it is possible to separate the contributions of free and bound electrons in ε_1- and ε_2-spectra. For this, one *fits* the theoretical ε_2 to the experimental ε_2 curves in the low-energy region. The theoretical spectral dependence of ε_2 is obtained by the Drude equation (11.27). An "effective mass" and the damping frequency ν_2 are used as adjustable parameters. With these parameters, the free electron part of ε_1 (denoted by ε_1^f) is calculated in the *entire* spectral range by using (11.26). Next, ε_1^f is subtracted from the experimental ε_1, which yields the bound electron contribution ε_1^b. Figure 13.9 depicts an absorption band thus obtained which resembles a calculated absorption band quite well (Fig. 11.9).

We now turn to the optical spectra for copper (Figs. 13.10 and 13.11). We notice immediately one important feature: Copper possesses an absorption

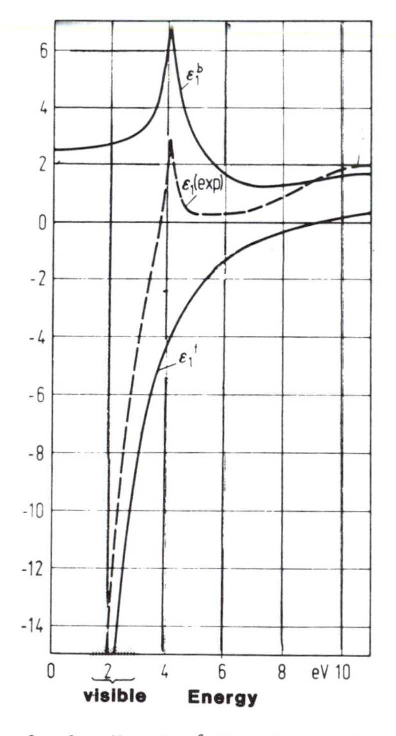

Figure 13.9. Separation of ε_1 for silver in ε_1^f (free electrons) and ε_1^b (bound electrons). Adapted from H. Ehrenreich et al., *IEEE Spectrum* **2**, 162 (1965). © 1965 IEEE.

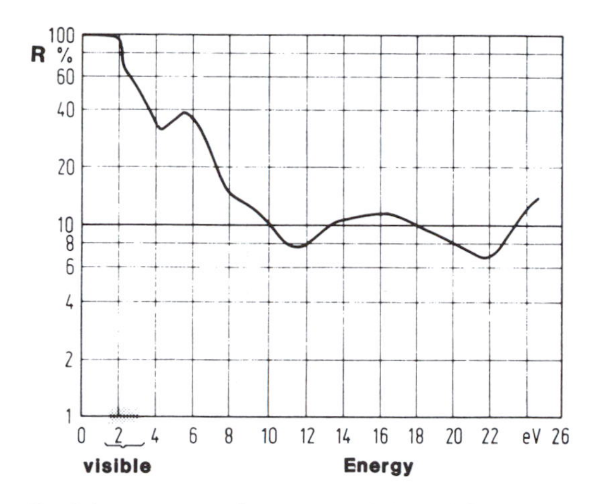

Figure 13.10. Reflectivity spectrum for copper. Adapted from H. Ehrenreich et al., *IEEE Spectrum* **2**, 162 (1965). © 1965 IEEE.

band in the *visible* spectrum which is, as already mentioned, responsible for the characteristic color of copper. We defined above a threshold energy at which interband transitions set in. In copper the threshold energy is about 2.2 eV (Fig. 13.11) which is assigned to the d-band $\rightarrow E_F$ transition near the L-symmetry point. (This is marked by an arrow in Fig. 5.22.) Another peak is observed at slightly above 4 eV which is ascribed to interband transitions *from* the Fermi energy near the L-symmetry point, as depicted in Figs. 12.3 and 5.22.

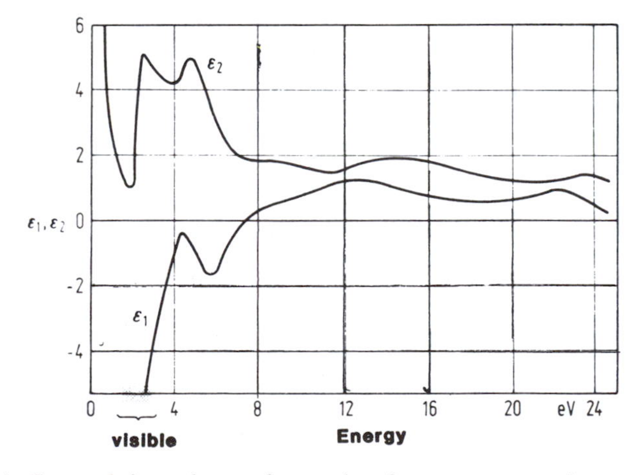

Figure 13.11. Spectral dependence of ε_1 and ε_2 for copper. ε_1 and ε_2 were obtained from Fig. 13.10 by a Kramers–Kronig analysis. Adapted from H. Ehrenreich et al., *IEEE Spectrum* **2**, 162 (1965). © 1965 IEEE.

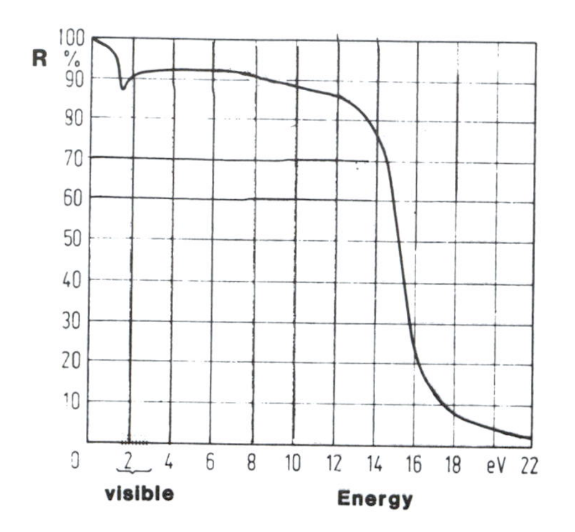

Figure 13.12. Reflection spectrum for aluminum. Adapted from H. Ehrenreich et al., *IEEE Spectrum* **2**, 162 (1965). © 1965 IEEE.

As a final example, we inspect the reflection spectrum of aluminum. Figures 13.12 and 13.13 show that the spectral dependences of ε_1 and ε_2 resemble those shown in Fig. 11.5, except in the small energy region around 1.5 eV. Thus, the behavior of aluminum may be described essentially by the free electron theory. This free electronlike behavior of aluminum can also be deduced from its band structure (Fig. 5.21) which has essential characteristics of free electron bands for fcc metals (Fig. 5.20). Interband transitions which contribute to the ε_2-peak near 1.5 eV occur between the W_2' and W_1 symmetry points and the closely spaced and almost parallel Σ_3 and Σ_1 bands. A small contribution stems from the $W_3 \rightarrow W_1$ transition near 2 eV.

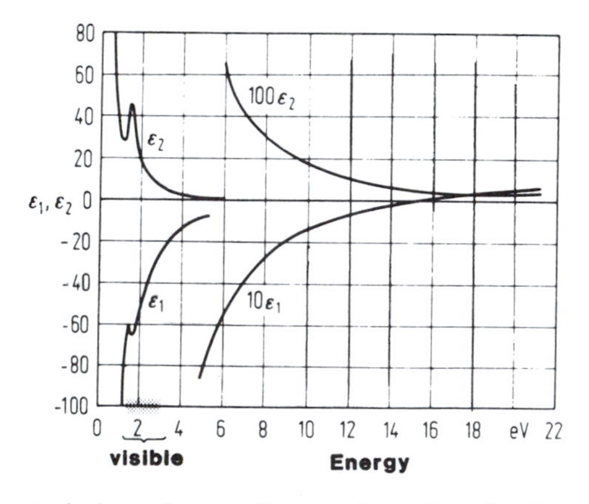

Figure 13.13. Spectral dependence of ε_1 and ε_2 for aluminum. Adapted from H. Ehrenreich et al., *IEEE Spectrum* **2**, 162 (1965). © 1965 IEEE.

*13.2.2. Plasma Oscillations

We postulate now that the free electrons of a metal interact electrostatically, thus forming an electron "plasma" which can be excited by light of proper photon energy to collectively perform fluidlike oscillations. This plasma possesses, just as an oscillator, a resonance frequency, often called the plasma frequency. We already introduced in Section 11.2 the plasma frequency ν_1 and noted that the dielectric constant $\hat{\varepsilon}$ becomes zero at ν_1. Thus, (10.12) reduces to

$$\hat{\varepsilon} = \varepsilon_1 - i\varepsilon_2 = 0, \tag{13.12}$$

from which we conclude that at the plasma frequency ε_1 as well as ε_2 must be zero. Experience shows that oscillations of the electron plasma already occur when ε_1 and ε_2 are close to zero.

The frequency dependence of the imaginary part of the reciprocal dielectric constant peaks at the plasma frequency, as we will see shortly. We write

$$\frac{1}{\hat{\varepsilon}} = \frac{1}{\varepsilon_1 - i\varepsilon_2} = \frac{\varepsilon_1 + i\varepsilon_2}{\varepsilon_1^2 + \varepsilon_2^2} = \frac{\varepsilon_1}{\varepsilon_1^2 + \varepsilon_2^2} + i\frac{\varepsilon_2}{\varepsilon_1^2 + \varepsilon_2^2}. \tag{13.13}$$

The imaginary part of the reciprocal dielectric constant, i.e.,

$$\mathrm{Im}\frac{1}{\hat{\varepsilon}} = \frac{\varepsilon_2}{\varepsilon_1^2 + \varepsilon_2^2}, \tag{13.14}$$

is called the "energy loss function" which is large for $\varepsilon_1 \to 0$ and $\varepsilon_2 < 1$, i.e., at the plasma frequency. We will now inspect the energy loss functions for some metals. We begin with aluminum because its behavior may well be interpreted by the free electron theory. We observe in Fig. 13.14 a pronounced maximum of $\mathrm{Im}(1/\hat{\varepsilon})$ near 15.2 eV. The real part of the dielectric constant (ε_1)

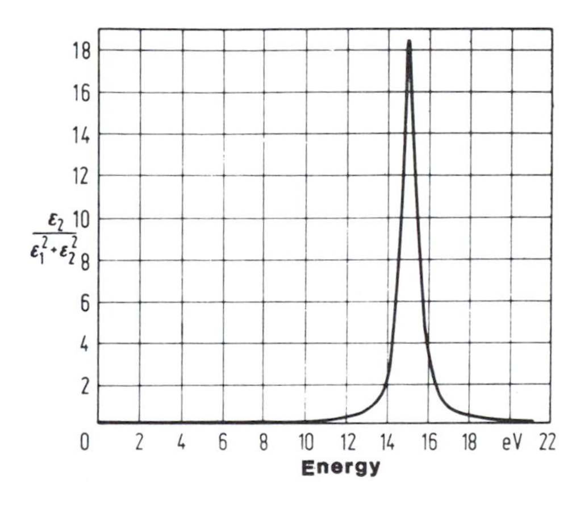

Figure 13.14. Energy loss function for aluminum. Adapted from H. Ehrenreich et al., *IEEE Spectrum* **2**, 162 (1965). © 1965 IEEE.

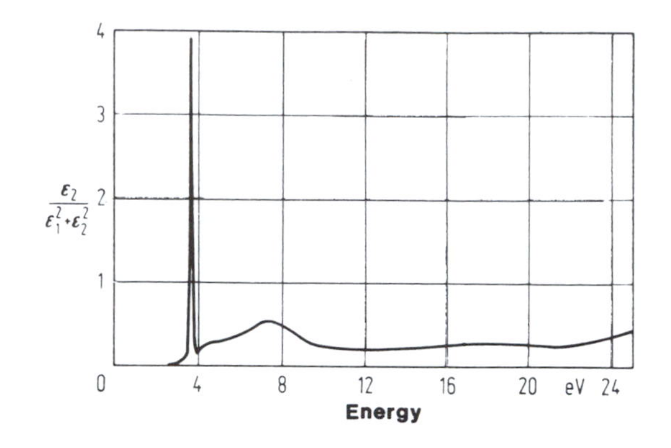

Figure 13.15. Energy loss function for silver. Adapted from H. Ehrenreich et al., *IEEE Spectrum* **2**, 162 (1965). © 1965 IEEE.

is zero at this frequency and ε_2 is small (see Fig. 13.13). Thus, we conclude that aluminum has a plasma resonance at 15.2 eV.

Things are slightly more complicated for silver. Here, the energy loss function has a steep maximum near 4 eV (Fig. 13.15) which cannot be solely attributed to free electrons, since ε_1^f is only zero at 9.2 eV (see Fig. 13.9). The plasma resonance near 4 eV originates by cooperation of the *d*- as well as the conduction electrons. The loss function for silver has another, but much weaker, resonance near 7.5 eV. This maximum is essentially caused by the conduction electrons, but is perturbed by interband transitions which occur at higher energies.

The reflection spectrum for silver (Fig. 13.7) can now be completely interpreted. The sharp decrease in R near 4 eV by almost 99% within a fraction of an electron volt is caused by a weakly damped plasma resonance. The sudden increase, only 0.1 eV above the plasma resonance takes place because of interband transitions which commence at this energy. Such a dramatic change in optical constants is unparalleled.

13.3. Optical Spectra of Alloys

It was demonstrated in the previous sections that knowledge of the spectral dependence of the optical properties contributes to the understanding of the electronic structure of metals. We will now extend our discussion to alloys. Several decades ago, N.F. Mott suggested that when a small amount of metal A is added to a metal B, the Fermi energy would simply assume an average value, while leaving the electron bands of the solvent intact. It was eventually recognized, however, that this "rigid band model" needed some modification and that the electron bands are somewhat changed for an alloy. We use copper–zinc as an example. Figure 13.16 shows a series of differential reflecto-

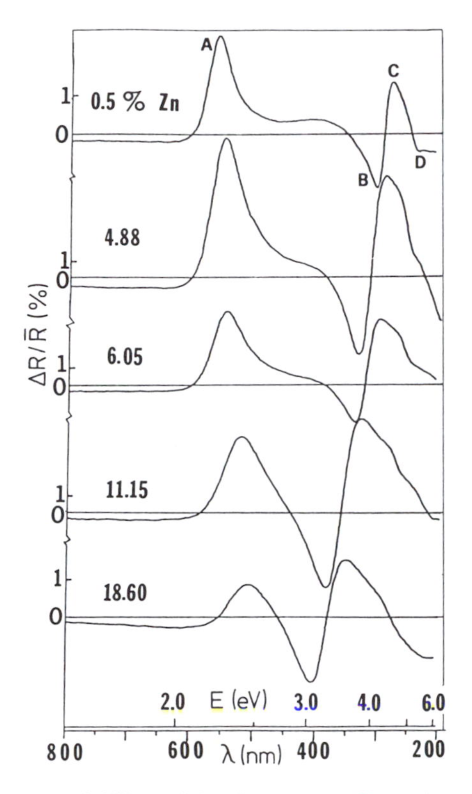

Figure 13.16. Experimental differential reflectograms for various copper–zinc alloys. The parameter on the curves is the average zinc concentration of the two alloys in at.%. The curve marked 0.5%, e.g., resulted by scanning the light beam between pure copper and a Cu–1% Zn alloy. From R.J. Nastasi-Andrews and R.E. Hummel, *Phys. Rev. B* **16**, 4314 (1977).

grams (see Section 13.1.3) from which the energies for interband transitions, E_T, can be taken. Peak A represents the threshold energy for interband transitions which can be seen to shift to higher energies with increasing zinc content. E_T is plotted in Fig. 13.17 as a function of solute (X). Essentially, a linear increase in E_T with increasing X is observed. The threshold energy for copper has been identified in Section 12.2 to be associated with electron transitions from the upper d-band to the conduction band, just above the Fermi surface (see Fig. 12.3). The rise in energy difference between the upper d-band and Fermi level, caused by solute additions, can be explained in a first approximation by suggesting a rise in the Fermi energy which results when extra electrons are introduced into the copper matrix from the higher valent solute. Gallium, which has three valence electrons would thus raise the Fermi energy more than zinc, which is indeed observed in Fig. 13.17. The slope of the $E_T = f(X)$ curve in Fig. 13.17 for zinc (as well as for other solutes) is considerably smaller than that predicted by the rigid band model. This suggests that

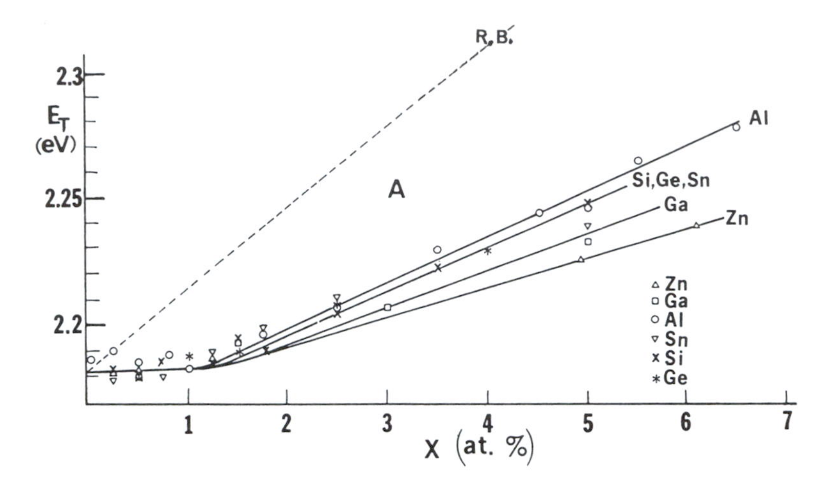

Figure 13.17. Threshold energies E_T for interband transitions for various copper-based alloys as a function of solute content. The E_T values are taken from differential reflectograms similar to those shown in Fig. 13.16. The rigid band line (R.B.) for Cu–Zn is added for comparison. From R.J. Nastasi-Andrews and R.E. Hummel, *Phys. Rev. B* **16**, 4314 (1977).

the *d*-bands are likewise raised with increasing solute content and/or that the Fermi level is shifted up much less than anticipated. Band calculations substantiate this suggestion. They reveal that upon solute additions to copper, the *d*-bands become narrower (which results from a reduction in Cu–Cu interactions) and that the *d*-bands are lifted up as a whole. Furthermore, the calculations show that solute additions to copper cause a rise in E_F and a downward shift of the bottom of the *s*-band. Figure 13.18 reflects these results. Because of the lowering of the bottom of the *s*-band (Γ_1 in Fig. 5.22), the Fermi energy rises much less than predicted had $E_{\Gamma 1}$ remained constant.

An unexpected characteristic of all $E_T = f(X)$ curves is that the threshold energy for interband transitions, E_T, does not vary appreciably for solute concentrations up to slightly above 1 at.% (Fig. 13.17). Friedel predicted just this type of behavior and related it to "screening" effects. He argued that for the first few atomic percent solute additions to copper, the additional charge from the higher valent solute is effectively screened and the copper matrix behaves as if the impurities were not present. The matrix remains essentially unperturbed as long as the impurities do not mutually interact.

The differential reflectograms shown in Fig. 13.16 suggest two additional interband transitions, one of which corresponds to feature 'D' near 5 eV and is assigned to electron transitions from the *lower d*-bands to the Fermi surface. An E_T versus X plot for peak 'D' resembles Fig. 13.17.

The third transition in the chosen energy region occurs at about 4 eV and involves the structural features 'B' and 'C'. The associated transition energy is seen to decrease with increasing solute content (Fig. 13.19). Features 'B' and 'C' are ascribed to transitions near the *L*-symmetry point, originating near the

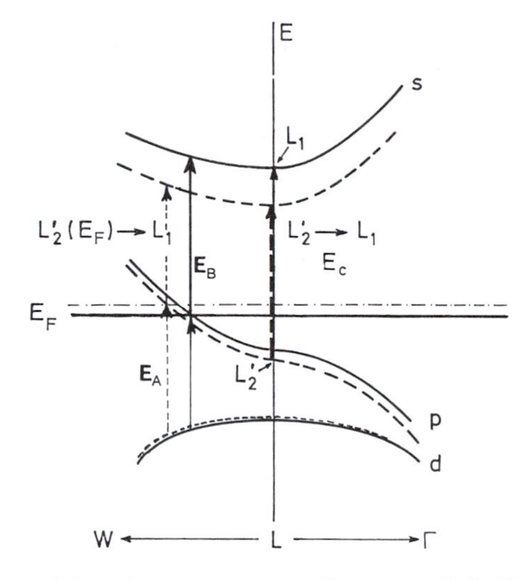

Figure 13.18. Schematic band structure near L for copper (solid lines) and an assumed dilute copper-based alloy (dashed lines). Compare with Figs. 12.3 and 5.22.

Fermi energy and terminating at the conduction band. It can be seen in Fig. 13.18 that the transition energy just mentioned is smaller for copper-based alloys than for pure copper, quite in agreement with the experimental findings. The reader is asked at this point to compare Figs. 13.10 and 13.11 with Fig. 13.16 and convince himself how different optical techniques complement each other in revealing the electronic structure of solids.

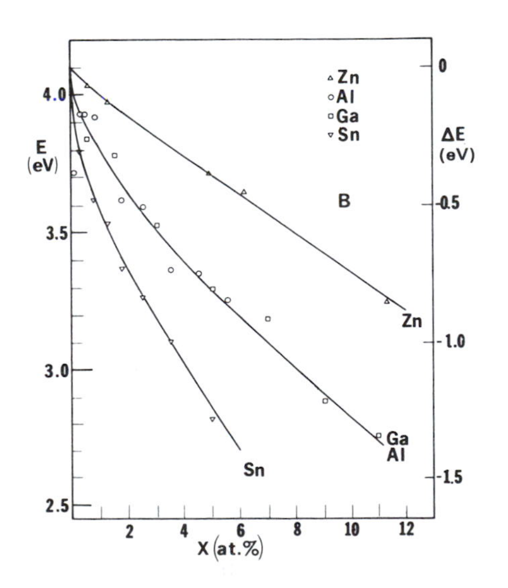

Figure 13.19. Energy of peak B for various dilute copper-based alloys. From R.J. Nastasi-Andrews, and R.E. Hummel, *Phys. Rev. B* **16**, 4314 (1977).

*13.4. Ordering

It was shown in Section 7.5.3 that the resistivity decreases when solute atoms of an alloy are periodically arranged on the regular lattice sites. Thus, we conclude that ordering has an effect on the electronic structure and hence on the optical properties of alloys. The best way to study ordering is to compare two specimens of the same alloy when one of them is ordered and the other is in the disordered state. This way, peaks occur in a differential optical spectrum whenever the ordered state causes *extra* interband transitions comparable to superlattice lines in X-ray spectroscopy. As an example, Fig. 13.20 depicts an optical spectrum for the intermetallic phase Cu_3Au. We note several transitions, among them an ε_2-type structure with a peak energy at 2.17 eV and an ε_1-type structure with a transition energy around 3.6 eV (median between 3.29 eV and 3.85 eV, see Fig. 11.9). We shall explain them by referring to Fig. 13.21 which depicts the first Brillouin zone of the disordered fcc lattice in which a simple cubic Brillouin zone, representing the superlattice, is inscribed. The $\Gamma - X$ direction of the fcc Brillouin zone is bisected by the face of the cubic Brillouin zone at the point $\bar{X}$. The point X is then thought to be folded back to the point Γ. A new transition from the d-bands (e.g., at Γ_{12}) to the point X'_4 (unfolded) can now take place (see Fig. 5.22). Folding along Γ–M–K, and possibly along other directions, explains the other transitions.

Short-range ordering shows comparatively smaller effects than long-range ordering (Fig. 13.22). The reflectivity difference between ordered and disordered alloys is about 3% for long-range ordering compared to 0.5% in the case of short-range ordering. Still, even in the latter case, a superlattice transition is observed, which is attributed to the periodic arrangement of solute atoms in small domains (about 1–2 nm in diameter).

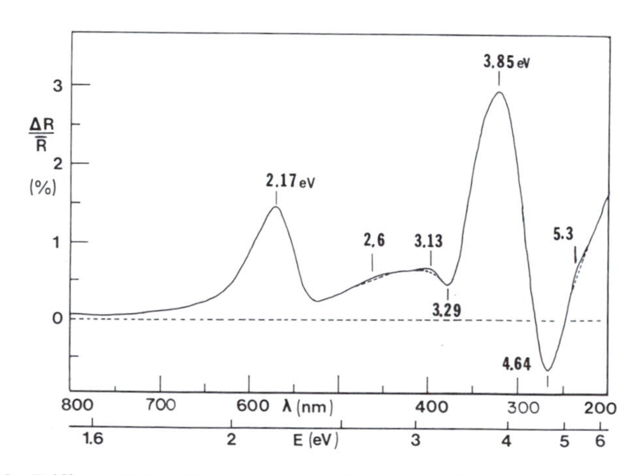

Figure 13.20. Differential reflectogram of (long-range) ordered versus disordered Cu_3Au. From R.E. Hummel, *Phys. Stat. Sol.* (*a*) **76**, 11 (1983).

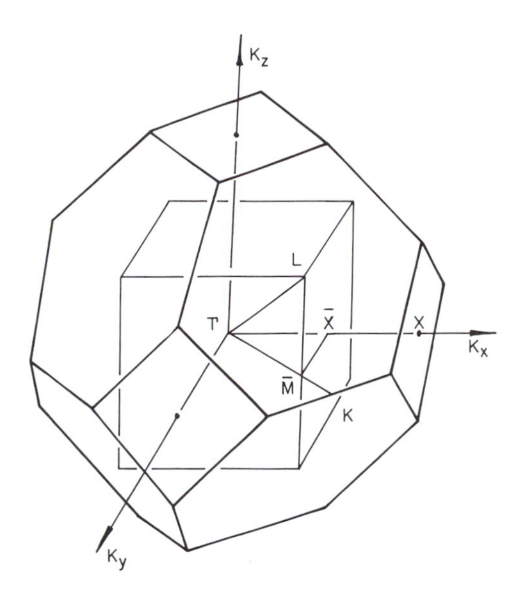

Figure 13.21. First Brillouin zone of an fcc lattice with inscribed Brillouin zone representing a cubic primitive superlattice.

Interestingly enough, optical investigations provide a further piece of information which enables us to look upon the short-range ordered state from a different perspective. It has been observed that certain peaks in a differential reflectogram shift due to ordering, exactly as they would do when a solute is added to a solvent (see Section 13.3). From this we conclude that in the short-range ordered state, the interaction between dissimilar atoms is slightly larger than that for similar atoms.

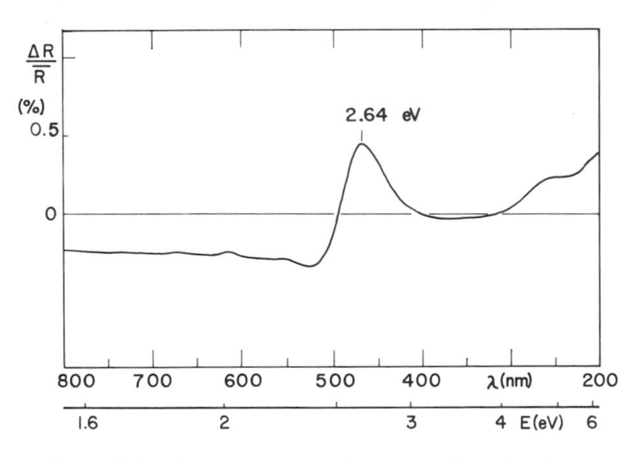

Figure 13.22. Differential reflectogram of (short-range) ordered versus disordered Cu–17 at.% Al. From J.B. Andrews, R.J. Andrews, and R.E. Hummel, *Phys. Rev. B* **22**, 1837 (1980).

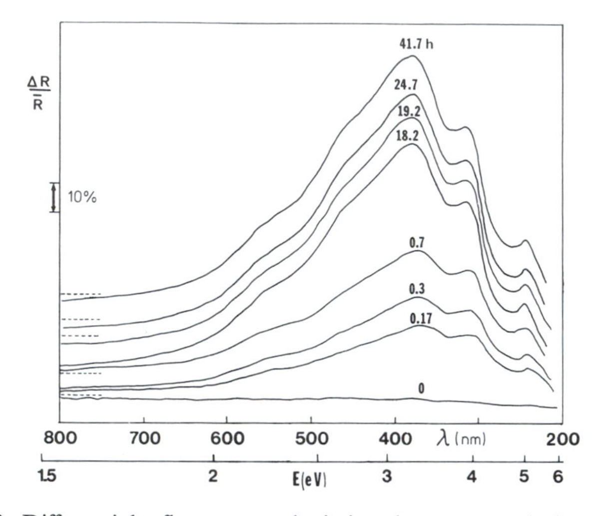

Figure 13.23. Differential reflectograms depicting the *in situ* evolution of Cu_2O on a copper substrate in a buffered electrolyte of pH 9. One sample half was held potentiostatically at −200 mV (SCE) for various times, the other at the protective potential (−500 mV (SCE)). From R.E. Hummel, *Phys. Stat. Sol.* (*a*) **76**, 11 (1983).

*13.5. Corrosion

Studies of the optical properties have been used for many decades for the investigation of environmentally induced changes of surfaces. Optical studies are *nondestructive*, simple, and allow the investigation of oxides during their formation. No vacuum is required in contrast to many other surface techniques. We use as an example the electrochemical corrosion of copper in an aqueous solution. A copper disc is divided into two parts which are electrically insulated from each other by a thin polymer film. One half is held electrically at the protective potential (as reference) and the other at the corrosion potential. No artifacts from the electrolyte, the corrosion cell window, or the metal substrate are experienced since the only difference in the light path of a differential reflectometer is the corrosion film itself. Figure 13.23 depicts a series of differential reflectograms demonstrating the evolution of Cu_2O on a copper substrate. We observe that the peak height near 3.25 eV and thus the corrosion film thickness initially grows rapidly. The growth rate slows down as the film becomes thicker. The growth kinetics has been observed to obey a logarithmic relationship.

13.6. Semiconductors

Intrinsic semiconductors have, at low temperatures, a completely filled valence band and an empty conduction band (see Chapter 8). Consequently, no *intra*band transition, or classical infrared (IR) absorption, is possible at low temperatures. Thus, the optical behavior of an intrinsic semiconductor is similar to that of an insulator, i.e., it is transparent in the low energy (far IR) region. Once the energy of the photons is increased and eventually reaches the gap energy, then the electrons are excited from the top of the valence band to the bottom of the conduction band. The semiconductor becomes opaque like a metal. The onset for *inter*band transitions is thus determined by the gap energy, which characteristically has values between 0.5 eV and 3 eV (see Table 8.1). The corresponding wavelength lies in the near IR or visible region.

As an example, the ε_2-spectrum for germanium is shown along with its band diagram in Figs. 13.24 and 13.25. The threshold energy for electron excitations across the gap from $\Gamma_{25'}$ to $\Gamma_{2'}$ can be clearly identified in both figures. Of particular interest are the nearly parallel bands Σ_4 and Σ_1 including the symmetry points X_4 and X_1 which cause a strong peak in ε_2 around 4.5 eV. (This transition is said to occur at an "M_1 critical point.") The designations of some further peaks are marked in Fig. 13.24 and should be compared to the band diagram.

We already mentioned in Section 13.1.3 that electroreflectance spectroscopy can enhance a relatively featureless part of an ε_2-spectrum and gives, in addition, some information about critical point transitions. Figure 13.26 depicts the electroreflectance spectrum for germanium between 2.8 eV and 3.6 eV. The ε_2-spectrum (Fig. 13.24) shows only a shoulder in this region, i.e., for the $\Gamma_{25'}$–Γ_{15} transition. It should be kept in mind, however, that

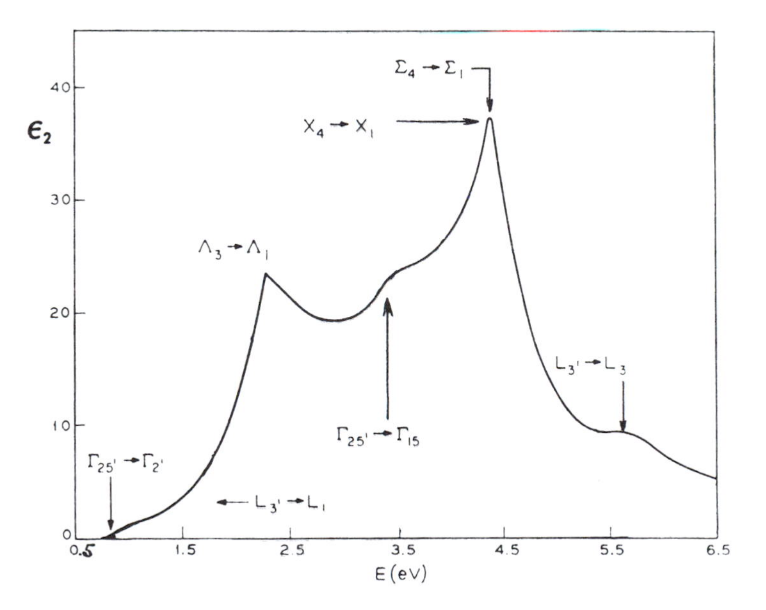

Figure 13.24. Experimental ε_2-spectrum for germanium. Adapted from J.C. Phillips, *Optical Properties and Electronic Structure of Metals and Alloys* (F. Abelès, ed.) North-Holland, Amsterdam (1966).

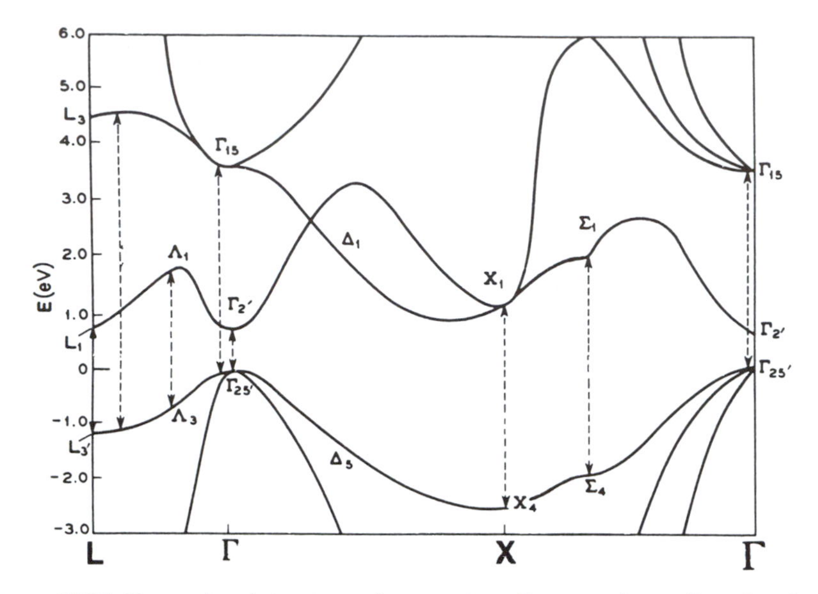

Figure 13.25. Energy band structure of germanium. Some pertinent direct interband transitions are marked. Adapted from D. Brust et al., *Phys. Rev A* **134**, 1337 (1964).

electroreflectance is a third derivative technique (Fig. 13.6(b)) which means that a sequence of peaks represents only *one* interband transition.

As a second example we consider silicon. By inspecting its band diagram (see Fig. 5.23 or Fig. 8.2) we notice that the maximum of the valence band and the minimum of the conduction band are not at the same point in **k**-space. Vertical transitions are thus not permissible at energies below about 3.4 eV

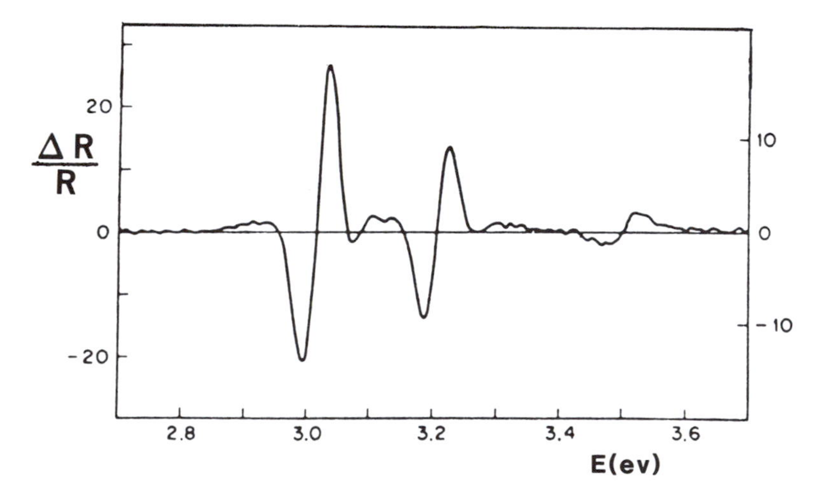

Figure 13.26. Electroreflectance spectrum of germanium at 78 K. Compare with Fig. 13.6(b). Adapted from D.E. Aspnes, *Phys. Rev. Lett.* **28**, 913 (1972).

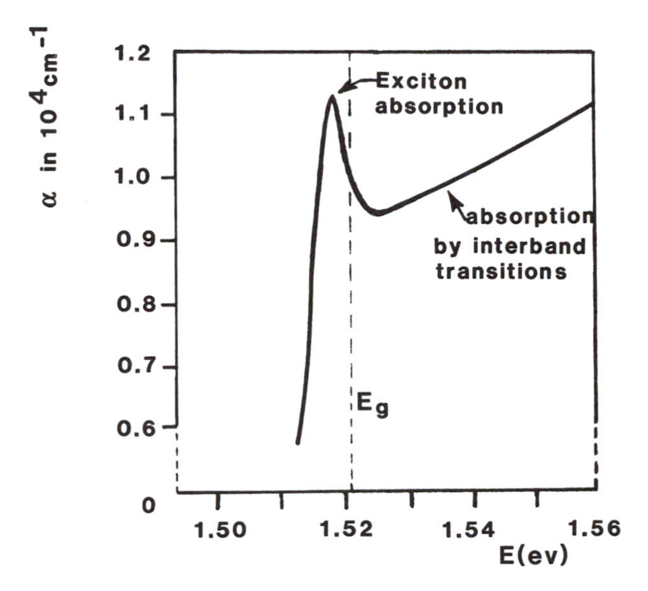

Figure 13.27. Spectral dependence of the absorbance α (10.21a) for gallium arsenide at 21 K. Adapted from M.D. Sturge, *Phys. Rev.* **127**, 768 (1962).

($\Gamma_{25'} \rightarrow \Gamma_{15}$). However, indirect transitions between the top of the valence band and the bottom of the conduction band may be possible provided the change in the wave vector **k** (or the momentum) is furnished by a phonon. We have already discussed phonon-assisted transitions in Section 12.2 and explained there that indirect interband transitions are particularly observed in the absence of direct transitions.

Our discussion of the optical spectra of semiconductors is not complete by considering only direct or indirect interband transitions. Several other absorption mechanisms may occur. It has been observed, for example, that the absorption spectra for semiconductors show a structure for photon energies slightly *below* the gap energy (Fig. 13.27). Frenkel explained this behavior by postulating that a photon may excite an electron so that it remains in the vicinity of its nucleus, thus forming an electron–hole pair, called an *exciton*. Electrons and holes are thought to be bound together by electrostatic forces and revolve around their mutual center of mass. The electrons may hop through the crystal and change their respective partner. This motion can also be described as an exciton wave. One depicts the excitons by introducing "exciton levels" into the forbidden band (Fig. 13.28). They are separated from the conduction band by the "binding energy" E_x whose position can be calculated by an equation similar to (4.18a) (see also Problem 8/10):

$$E_x = -\frac{m^* e^4}{2n^2\hbar^2\varepsilon^2}, \tag{13.15}$$

where n is an integer, m^* is the effective mass of the exciton (which is the

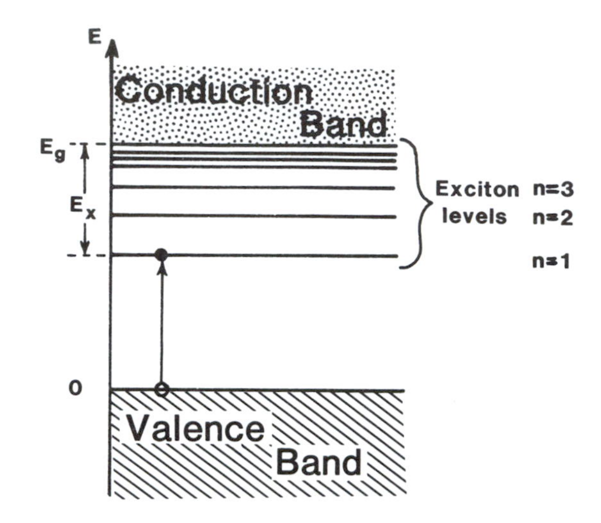

Figure 13.28. Schematic representation of exciton energy levels and an exciton in a semiconductor (or insulator).

average of m_e and m_h), and ε is the a.c. dielectric constant. E_x is characteristically about 0.01 eV. The exciton levels are broadened by interactions with impurities or phonons.

Finally, *extrinsic* semiconductors have, as we know, donor or acceptor states near the conduction or the valence band, respectively (Section 8.3). At sufficiently high temperatures, optical transitions from and to these states can take place which also cause weak absorption peaks below the gap energy.

It should be noted that the temperature slightly influences the absorption characteristics of a semiconductor. The change in gap energy is about -2×10^{-4} eV/K (see Appendix, page 375) which stems from an apparent broadening of valence and conduction levels with increasing temperature due to transitions with simultaneous emission and absorption of photons. Another temperature-enhanced effect should be considered too. Once electrons have been excited from the valence into the conduction band (either by photons or thermal excitation), holes are present in the upper part of the valence bands. Then, photons having energies well below E_g can be absorbed by *intra*band transitions. These transitions are, however, relatively weak.

Optical absorption measurements are widely used in semiconductor research since they provide the most accurate way to determine the gap energies and the energies of the localized states. Measurements are normally performed at low temperatures so that the thermal excitations of the electrons do not mask the transitions to be studied. Optical measurements are capable of discriminating between direct and indirect transitions, based on the magnitude of the absorption peaks.

13.7. Insulators (Dielectric Materials and Glass Fibers)

As we know, insulators are characterized by completely filled valence bands and empty "conduction" bands. Thus, no *intra*band transitions, i.e., no classical IR absorption takes place. Furthermore, the gap energy for insulators is fairly large (typically 5 eV or larger) so that *inter*band transitions do not occur in the IR and visible spectrum either. They take place, however, in the ultraviolet (UV) region. Third, *excitons* may be created which cause absorption peaks somewhat below the gap energy. For example, the lowest energy for an exciton level (and thus for the first exciton absorption peak) for NaCl has been found to be at about 7 eV, i.e., in the vacuum UV region. (Other alkali halides have very similar exciton energies.) We suspect, therefore, that insulators are transparent from the far IR throughout the visible up to the UV region. This is indeed essentially observed. However, in the IR region a new absorption mechanism may take place which we have not yet discussed. It is caused by the light-induced vibrations of the lattice atoms, i.e., by the excitation of *phonons* by photons. We need to explain this in some more detail.

Let us first consider a monatomic crystal (one kind of atom). The individual atoms are thought to be excited by light of appropriate frequency to perform oscillations about their points of rest. Now, the individual atoms are surely not vibrating independently. They interact with their neighbors which causes them to move simultaneously. For simplicity, we model the atoms to be interconnected by elastic springs. Thus, the interaction of light with the lattice can be mathematically represented in quite a similar manner to the one used when we discussed and calculated the classical *electron* theory of dielectric materials. (In Section 11.6 we represented *one* atom in an electric field as consisting of a positively charged core which is bound quasi-elastically to an electron.) A differential equation similar to (11.32) may be written for the present case

$$m\frac{d^2x}{dt^2} + \gamma'\frac{dx}{dt} + \kappa x = e\mathscr{E}_0 \exp(i\omega t), \tag{13.16}$$

which represents the oscillations of atoms under the influence of light whose excitation force is $e\mathscr{E}_0 \exp(i\omega t)$. As before, the factor $\kappa \cdot x$ is the restoring force which contains the displacement x and an interatomic force constant κ (i.e., a "spring constant," or a "binding strength" between the atoms). The damping of the oscillations is represented by the second term in (13.16). Damping is thought to be caused by interactions of the phonons with lattice imperfections, or with external surfaces of the crystal, or with other phonons. The oscillators possess a resonance frequency, ω_0, which depends on the mass of the atoms and on the restoring force (see (11.34)). The solution of the differential equation (13.16) yields a spectral dependence of ε_1 and ε_2 which is very similar to that shown in Figs. 11.9–11.12.

The situation becomes slightly more complicated when diatomic solids, such as ionic crystals, are considered. In this case, two differential equations of the type of (13.16) need to be written. They have to be solved simultaneously. Actually, one needs to solve $2N$ coupled differential equations, where N is the number of unit cells in the lattice. The result is, however, qualitatively still the same. The resonance frequency for diatomic crystals is

$$\omega_0 = 2\kappa\left(\frac{1}{m_1} + \frac{1}{m_2}\right), \quad (13.17)$$

where m_1 and m_2 are the masses of the two ion species. Figure 13.29 depicts the spectral reflectivity of NaCl in the IR. Sodium chloride is transmissive between 0.04 eV and 7 eV. At the upper boundary energy, exciton absorption sets in.

Fused quartz (depending on the method of manufacturing) is essentially transparent between 0.29 eV and 6.9 eV (4.28 μm and 0.18 μm) having, however, two pronounced absorption peaks near 1.38 μm and 2.8 μm. Window glass has a similar transmission spectrum as fused quartz, with the exception that its UV cut-off wavelength is already near 0.38 μm (3.3 eV). In recently developed sol–gel silica "glasses" the absorption peaks near 1.38 μm and 2.8 μm are virtually suppressed which causes this material to be transparent from 0.16 μm to 4 μm. The absorption spectrum for the commercially important borosilicate/phosphosilicate glass, used for optical fibers, is shown in Fig. 13.30. We notice the afore-mentioned peak near 1.38 μm which is caused by oscillations of OH^- ions. We shall refer to this spectrum in a later section.

A word should be added about the opacity of some dielectric materials, such as enamels, opal glasses, glazes, or porcelains, which should be transparent in the visible region according to our discussion above. This opacity is caused by the scattering of light on small particles which are contained in the matrix. Part of the light is diffusely transmitted and part of it is diffusely reflected. The larger the specular part of the reflected light, the higher the

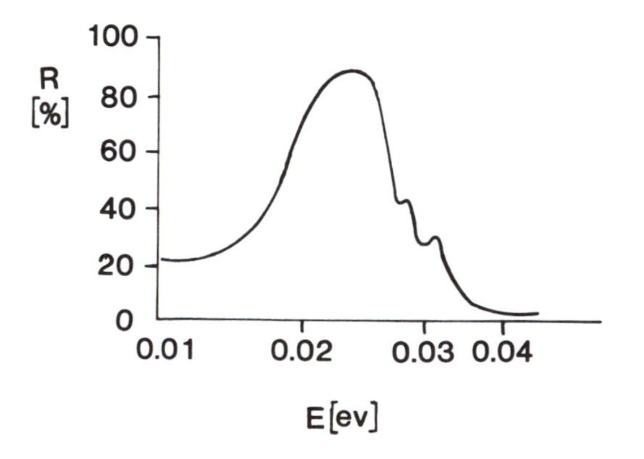

Figure 13.29. Spectral reflectivity of NaCl at room temperature in the far IR region.

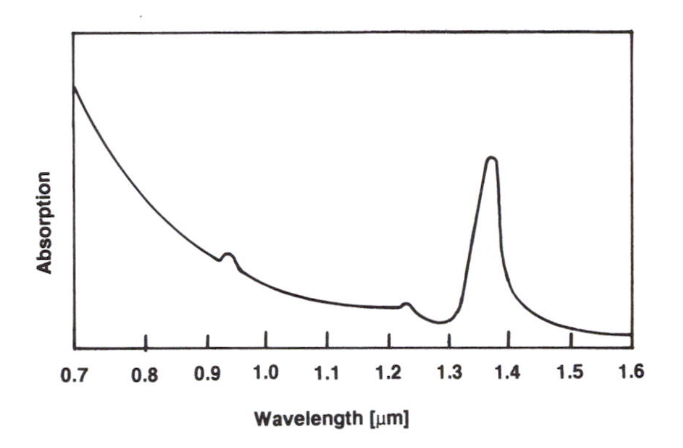

Figure 13.30. Absorption spectrum of highly purified glass for fiber-optic applications which features a phosphosilicate core surrounded by a borosilicate cladding.

gloss. Very often, *opacifiers* are purposely added to a dielectric material to cause wanted effects. The particle size should be nearly the same as the wavelength of the light, and the index of refraction should be largely different from that of the material, to obtain maximal scattering.

13.8. Lasers

13.8.1. Principles

So far we have discussed only the *absorption* of light by matter. We learned that due to the interaction of photons with electrons, the electrons are excited into higher energy states. The present section deals with the *emission* of photons, particularly with stimulated emission. A brief description of the light-emitting diode (LED) will also be given at the end of this section.

An electron, once excited, must eventually revert back into a lower, empty energy state. This occurs, as a rule, spontaneously within a fraction of a second, e.g. nanoseconds, and is accompanied by the emission of a photon and/or the dissipation of heat. This fast luminescence process is called fluorescence. (In some materials, the emission takes place after microseconds or milliseconds. This slower process is referred to as phosphorescence.) Spontaneous emission possesses none of the characteristic properties of laser light: the radiation is emitted through a wide angular region in space, the light is phase *in*coherent and is often polychromatic.

The situation changes considerably, however, when stimulated emission is induced. Let us consider two energy levels E_1 and E_2, and let us assume for a moment that the higher energy level E_2 contains more electrons than the lower level E_1, i.e., let us assume a *population inversion* of electrons (Fig. 13.31(a)). We further assume that by some means (which we shall discuss in

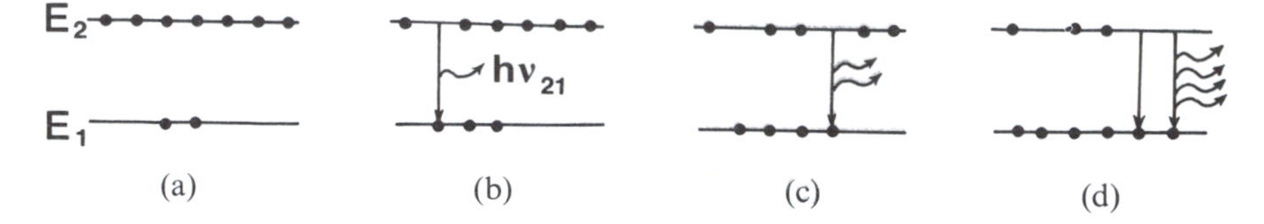

Figure 13.31. Schematic representation of stimulated emission between two energy levels E_2 and E_1. The dots symbolize electrons.

a moment) the electrons in E_2 are made to stay there for an appreciable amount of time. Nevertheless, one electron will eventually revert to the lower state. As a consequence a photon with energy $E_{21} = h\nu_{21}$ is emitted (Fig. 13.31(b)). This photon might stimulate a second electron to *descend in step* to E_1, thus causing the emission of another photon which vibrates in phase with the first one. The two photons are consequently *phase coherent* (Fig. 13.31(c)). They might stimulate two more electrons to descend in step (Fig. 13.31(d)) and so on until an avalanche of photons is created. In short, stimulated emission of light occurs when electrons are forced by incident radiation to add more photons to an incident beam. The acronym LASER can now be understood, it stands for *light amplification by stimulated emission of radiation.*

Laser light is highly monochromatic because it is generated by electron transitions between two narrow energy levels. (As a consequence, laser light can be focused to a spot less than 1 μm in diameter.) Another outstanding feature of laser light is its strong collimation, i.e., the parallel emergence of light from a laser window. (The cross section of a laser beam transmitted to the moon is only 3 km in diameter!) We understand the reason for the collimation best by knowing the physical setup of a laser.

The lasing material is embodied in a long narrow container called the cavity; the two faces at opposite ends of this cavity must be absolutely parallel to each other. One of the faces is silvered and acts as a perfect mirror, whereas the other face is partially silvered and thus transmits some of the light (Fig.

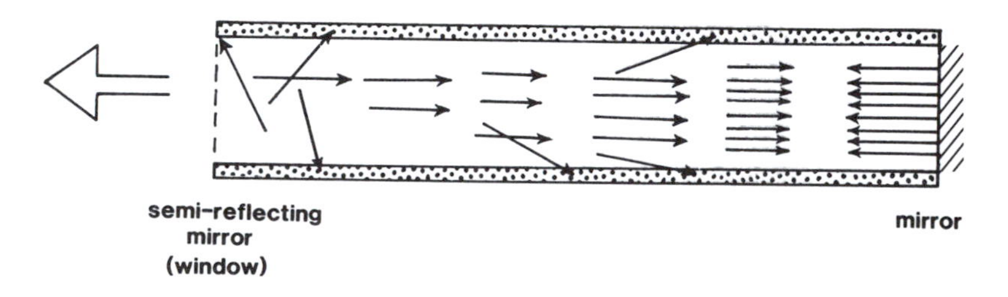

Figure 13.32. Schematic representation of a laser cavity and the buildup of laser oscillations. The stimulated emission eventually dominates over the spontaneous emission. The light leaves the cavity at the left side.

13.32). The laser light is reflected back and forth by these mirrors, thus increasing the number of photons during each pass. After the laser has been started, the light is initially emitted in all possible directions (left part of Fig. 13.32). However, only photons which travel strictly parallel to the cavity axis will remain in action whereas the photons traveling at an angle will eventually be absorbed by the cavity walls (center part in Fig. 13.32). A fraction of the photons escape through the partially transparent mirror. They constitute the emitted beam.

We now need to explain how the electrons arrive at the higher energy level, i.e., we need to discuss how they are *pumped* from E_1 into E_2. One of the methods is, of course, *optical pumping*, i.e., the absorption of light stemming from a polychromatic light source. (Xenon flashlamps for pulsed lasers, or tungsten–iodine lamps for continuously operating lasers, are often used for pumping. The lamp is either wrapped in helical form around the cavity, or the lamp is placed in one of the focal axes of a specularly reflecting elliptical cylinder, whereas the laser rod is placed along the second focal axis.) Other pumping methods involve collisions in an electric discharge, chemical reactions, nuclear reactions, or external electron beam injection.

The *pumping efficiency* is large if the bandwidth δE of the upper electron state is broad. This way, an entire frequency range (rather than a single wavelength) leads to excited electrons (Fig. 13.33(a)).

Next we discuss how population inversion can be achieved. For this we need to quote Heisenberg's uncertainty principle

$$\delta E \cdot \delta t \propto h, \tag{13.18}$$

which states that the time span δt, for which an electron remains at the higher energy level E_2, is large when the bandwidth δE of E_2 is narrow. In other words, a sharp energy level (δE small, δt large) supports the population inversion, Fig. 13.33(b)). On the other hand, a large pumping efficiency requires a large δE (Fig. 13.33(a)) which results in a small δt and a small

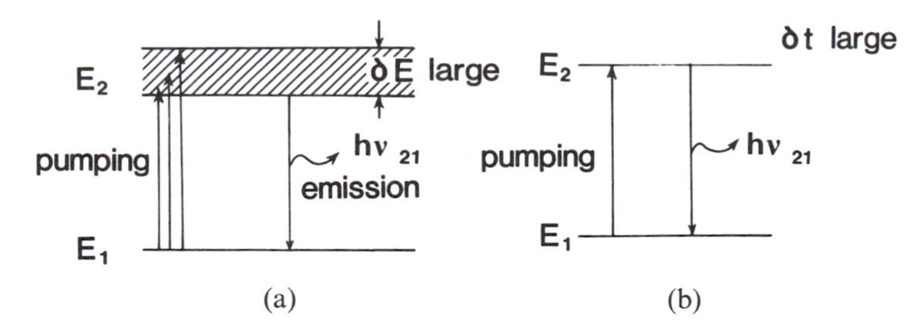

Figure 13.33. Examples of possible energy states in a two level configuration. (a) δE large, i.e., large pumping efficiency but little or no population inversion. (b) Potentially large population inversion (δt large) but small pumping efficiency. (*Note:* Two-level lasers do not produce a population inversion because absorption and emission compensate each other.)

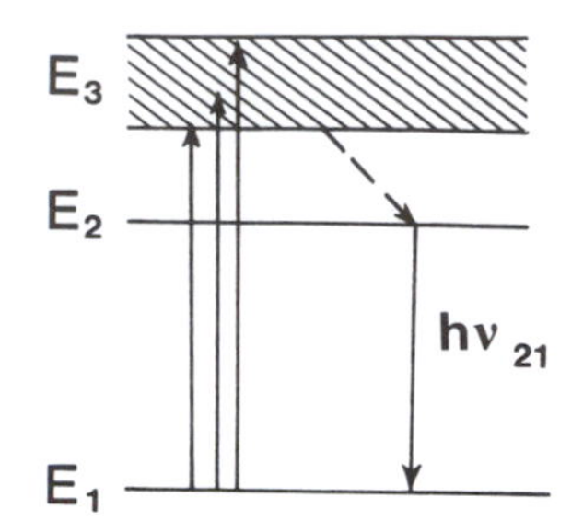

Figure 13.34. Three-level laser. The nonradiative, phonon-assisted decay is marked by a dashed line. Lasing occurs between levels E_2 and E_1. High pumping efficiency to E_3. High population inversion at E_2.

population inversion. Thus, high pumping efficiency and large population inversion mutually exclude each other in a two-level configuration. In essence, a two-level configuration as depicted in Fig. 13.33 does not yield laser action.

The three-level laser (Fig. 13.34) provides improvement. There, the "pump band" E_3 is broad, which enables a good pumping efficiency. The electrons revert after 10^{-14} s into an intermediate level E_2 via a nonradiative, phonon-assisted process. Since E_2 is sharp, the electrons remain much longer, i.e., for some milliseconds on this level. This provides the required population inversion.

An even larger population inversion is obtained using a four-level laser. In this configuration the energy level E_2 is emptied rapidly by electron transitions into a lower level E_1 (Fig. 13.35). It should be added that some three- and four-level lasers have several closely spaced pumping bands which, of course, increases the pumping efficiency.

The highest population inversion is achieved by adding *Q-switching*. For this method, the mirror in Fig. 13.32 is turned sideways during pumping to reduce stimulated emission, i.e., to build up a substantial population inversion. After some time the mirror is turned back into its original vertical position which results in a burst of light lasting 10–20 ns.

Laser materials cannot be created at will in, say, three- or four-level con-

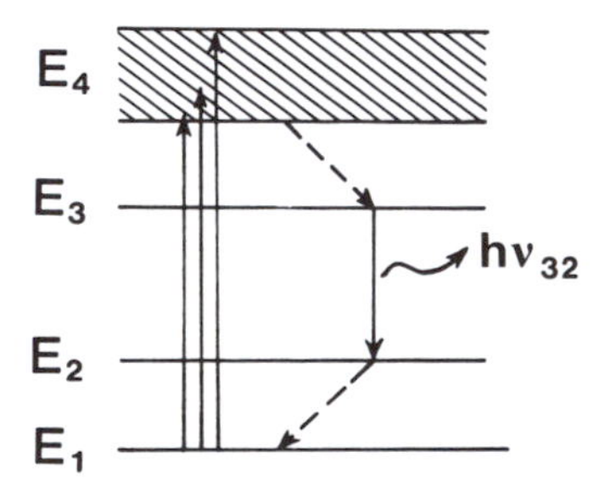

Figure 13.35. Four-level laser.

figurations. They can, however, be selected from hundreds of substances to suit a specific purpose. Laser materials include *crystals* (such as ruby), *glasses* (such as neodymium-doped glass), *gases* (such as helium, argon, xenon), *metal vapors* (such as cadmium, zinc, or mercury), *molecules* (such as carbon dioxide), or *liquids* (solvents which contain organic dye molecules). Table 13.1 lists the properties of some widely used lasers. We observe that most lasers (except the tunable dye lasers) emit their light in the red or IR spectrum. Lasers can be operated in a continuous mode (CW), or with a higher power output, in the pulsed mode. The power output varies over many orders of magnitude (and it can even be increased if Q-switching is applied). A few important laser types need special mention.

13.8.2. Helium–Neon Laser

A cavity about 2 mm in diameter is filled with 0.1 Torr Ne and 1 Torr He (Fig. 13.36(a)). A current which passes through the gas produces free electrons (and ions). The electrons are accelerated by the electric field and excite the helium gas by electron–atom collisions. Some of the helium levels are resonant with neon levels so that the neon gas also becomes excited by resonant energy transfer (Fig. 13.36(b)). This constitutes a very efficient pumping into

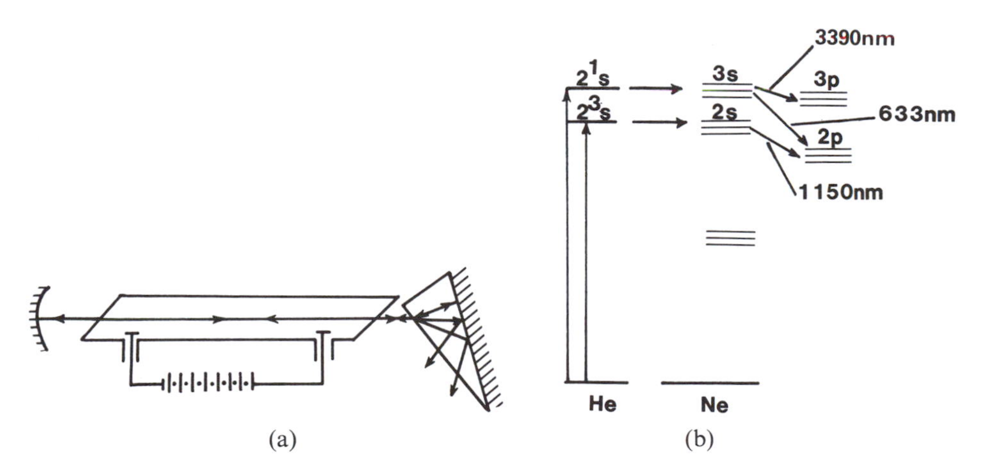

Figure 13.36. Helium–neon laser. (a) Schematic diagram of the laser cavity with Littrow prism to obtain preferred oscillation at one wavelength. (The end windows are inclined at the Brewster angle for which plane polarized light suffers no reflection losses.) (b) Energy level diagram for helium and neon. The decay time for the p-states is ~ 10 ns, that of the s-states 100 ns. The letters on the energy levels represent the angular momentum quantum number; the number in front of the letters gives the value for the principal quantum number; and the superscripts represent the multiplicity (singlet, doublet, etc.), see Appendix 3.

Table 13.1. Properties of Some Common Laser Materials.

Type of laser	Wave-length(s) (nm)	Beam divergence (mrad)	Peak power output (W)	Comments
Ruby (Cr^{3+}-doped Al_2O_3)	694.3	10 5 0.5	CW:[a] ~5 pulsed (1–3 ms): 10^6–10^8 Q-switched (10 ns): 10^9	Optically pumped three level laser. Lasing occurs between Cr^{3+} levels. Low efficiency (0.1%). Historic device.
Neodymium (Nd^{3+}-doped glass or YAG[b])	1,064	3–8	CW: 10^3 pulsed (0.1–1 μs): ~10^4	Optically pumped four level laser. High efficiency 2%.
HeNe	632.8 (1150; 3390)	1	10^{-3} – 10^{-2}	See Fig. 13.36 and text. Most widely used.
CO_2	10,600; 9,600	2	CW: 10–1.5 × 10^4 pulsed (10^2–10^3 ns): 10^5	High efficiency (20%). Lasing occurs between vibrational levels (Fig. 13.37).
Semiconductor GaAs GaAlAs[c]	 ~870 ~850	 250 500	Homojunction, pulsed: (10^2 ns) 10–30 Heterojunction, CW: 1–4 × 10^{-1}	Small size, direct conversion of electrical energy into optical energy. 10–55% efficiency. See Figs. 13.38 and 13.43.
Dye (organic dyes in solvents)	350–1000	3 10	CW: ~10^{-1} pulsed (6 ns) ~10^5	Lasing occurs between vibrational sublevels of molecules. Tunable by Littrow prism (Fig. 13.36(a)).

[a] CW: Continuous wave.
[b] Yttrium aluminum garnet ($Y_3Al_2O_{15}$).
[c] See Fig. 13.40.

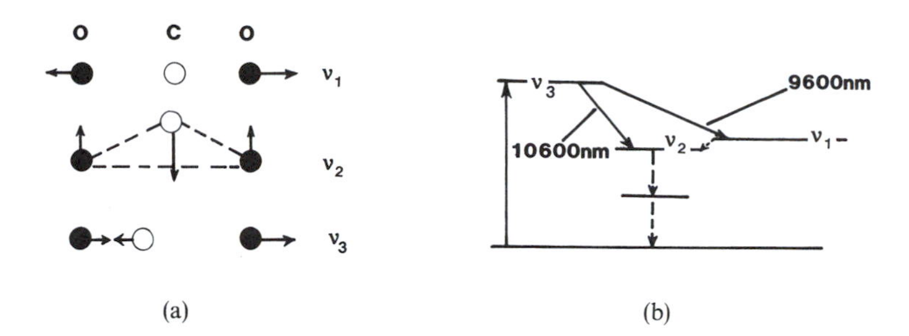

Figure 13.37. CO_2 laser. (a) Fundamental modes of vibration for a CO_2 molecule; ν_1: symmetric stretching mode; ν_2: bending mode; ν_3: asymmetric stretching mode. (b) Energy level diagram for various vibrational modes.

the neon 2*s*- and 3*s*-levels. (Direct electron–neon collisions also contribute to the pumping). Lasing occurs between the neon *s*- and *p*-levels and produces three characteristic wavelengths. Suppression of two of the wavelengths is accomplished by multilayer dielectric mirrors which provide a maximum reflectivity at the desired wavelength, or by a Littrow prism, as shown in Fig. 13.36(a).

13.8.3. Carbon Dioxide Laser

The CO_2 molecules possess three fundamental modes of vibration, as shown in Fig. 13.37(a). The lasing occurs between these levels as shown in Fig. 13.37(b). Pumping is accomplished by electron–atom collisions (see above) and by a resonant energy transfer from N_2 molecules which are added to the CO_2. Nitrogen (and helium) greatly improve the pumping efficiency of the CO_2 laser which is one of the most efficient and powerful lasers.

13.8.4. Semiconductor Laser

The "cavity" for this laser consists of heavily doped (10^{18} cm^{-3}) *n*- and *p*-type semiconductor materials such as GaAs. The energy band diagram for a *p*–*n* junction has been shown in Fig. 8.19 and is redrawn in Fig. 13.38(a) for the case of forward bias. We notice a population inversion of electrons in the depletion layer. Two opposite end faces of this *p*–*n* junction are made parallel and are polished or cleaved along crystal planes. The other faces are left untreated to suppress lasing in unwanted directions (Fig. 13.38(b)). A reflective coating of the window is usually not necessary since the reflectivity of the semiconductor is already 35%. The pumping occurs by direct injection of electrons and holes into the depletion region. Semiconductor lasers are small and can be quite efficient.

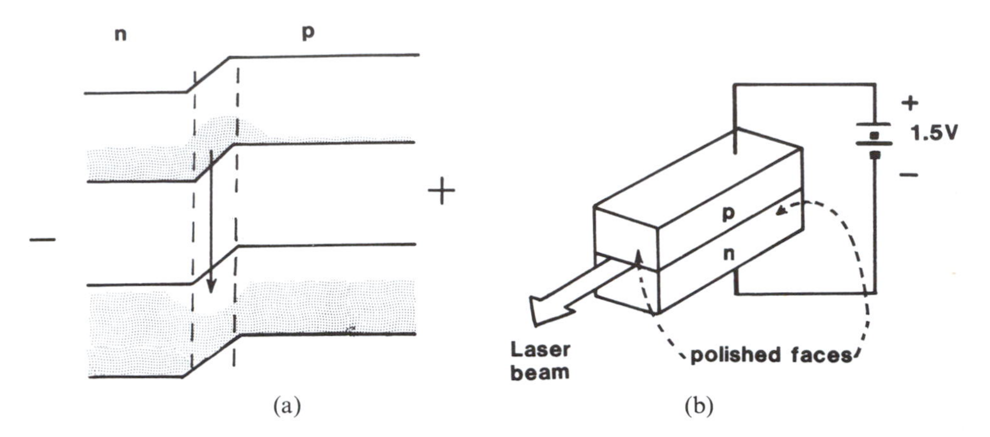

Figure 13.38. (a) Energy band diagram of a heavily doped, forward biased semiconductor. (b) Schematic setup of a semiconductor laser.

13.8.5. Direct Versus Indirect Band-Gap Semiconductor Lasers

We need to discuss now whether or not *all* semiconducting materials are equally well suited for a laser. Indeed, they are not. Direct band-gap materials, such as GaAs, have a much higher quantum efficiency for the emission of light than indirect band-gap materials, such as silicon. This needs some explanation. Let us assume that an electron at the top of the valence band in silicon has absorbed energy, and has thus been excited (pumped) by means of a direct interband transition into the conduction band, as shown in Fig. 13.39. This "*hot*" electron quickly *thermalizes*, i.e., it reverts down within 10^{-14} s to the bottom of the conduction band in a nonradiative process, involving a phonon (to conserve momentum, see Section 12.2). In order to recombine finally with the left-behind hole in the valence band (by means of an indirect transition)

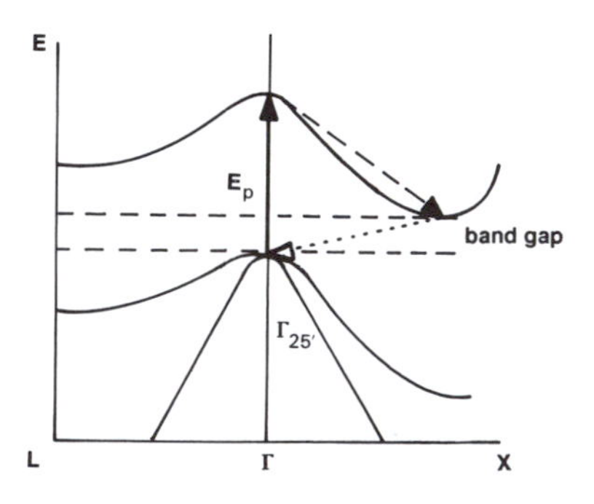

Figure 13.39. Direct interband transition pumping (E_p) and phonon-involved reversion of a *hot* electron by indirect transitions for an indirect band-gap semiconductor such as silicon. (Compare with Figs. 5.23 and 12.2.)

a second phonon-assisted process has to take place. This requirement substantially reduces the probability for emission. The time interval which elapses before such a recombination takes place may be as much as 0.25 s, which is substantially longer than it would take for a direct recombination in a pumped semiconductor. Before this quarter of a second has passed, the electron and also the hole have already recombined through some other nonradiative means involving impurity states, lattice defects, etc. Thus, the electron in question is lost before a radiative emission occurs. This does not mean that indirect emissive transitions would never take place. In fact, they do occur occasionally and have been observed, for example, in GaP, but with a very small quantum efficiency. Indirect band-gap semiconductors therefore seem to be not suited for lasers. It should be added, however, that silicon devices, when made porous by anodically HF etching, have been observed to emit visible light. It is speculated that the etching creates an array of columns which act as fine quantum lines, and thus alter the electronic band structure of silicon to render it direct.

13.8.6. Wavelength of Emitted Light

The wavelength of a binary GaAs laser is about 0.87 μm. This is, however, not the most advantageous wavelength for telecommunication purposes because glass attenuates light of this wavelength appreciably. By inspecting Fig. 13.30 we note that the optical absorption in glass is quite wavelength dependent having minima in absorption at 1.3 μm and 1.55 μm. Fortunately, the bandgap energy, i.e., the wavelength at which a laser emits light, can be adjusted to a certain degree by utilizing ternary or quaternary compound semiconductors (Fig. 13.40). Among them $In_{1-x}Ga_xAs_yP_{1-y}$ plays a consider-

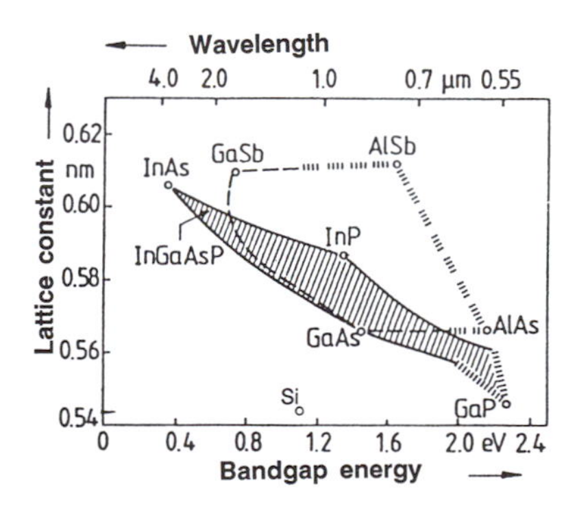

Figure 13.40. Lattice constants, energy gaps, and emission wavelengths of some ternary and quaternary compound semiconductors at 300 K. The lines between the binary compounds denote ternaries. The cross-hatched lines indicate indirect interband transitions. Pure silicon is also added for comparison.

able role for telecommunication purposes, because the useful emission wavelengths of these compounds can be varied between 0.886 μm and 1.55 μm (which corresponds to gap energies from 1.4 eV to 0.8 eV). In other words, the above-mentioned desirable wavelengths of 1.3 μm and 1.55 μm can be conveniently obtained by utilizing a properly designed indium–gallium–arsenide–phosphide laser.

A note on compound semiconductor fabrication needs to be inserted at this point. Semiconductor compounds are usually deposited out of the gaseous or liquid phase onto an existing semiconductor substrate, whereby a relatively close match of the lattice structure of substrate and layer has to be maintained. This process in which the lattice structure of the substrate is continued into the deposited layer is called "*epitaxial growth*." The important point is that, in order to obtain a strain-free epitaxial layer, the lattice constants of the involved components have to be nearly identical. Figure 13.40 shows, for example, that this condition is fulfilled for GaAs and AlAs. These compounds have virtually identical lattice constants. A near-perfect lattice match can also be obtained for ternary $In_{0.53}Ga_{0.47}As$ on an InP substrate, see Fig. 13.40. In short, the critical parameters for designing lasers from compound semiconductors include the band-gap energy, the similarities of the lattice constants of the substrate and active layer, the fact whether or not a direct band-gap material is involved, and the refractive indices of the core and cladding materials (see Section 13.8.8).

Finally, the emission wavelength depends on the temperature of operation because the band-gap decreases with increasing temperature (see Appendix 4) according to the empirical equation

$$E_{gT} = E_{g0} - \frac{\xi T^2}{T + \theta_D} \tag{13.19}$$

where E_{g0} is the band-gap energy at $T=0$ K, $\xi \approx 5 \times 10^{-4}$ eV/K, and $\theta \approx 200$ K is approximately the Debye temperature (see Table 19.2 and Section 19.4).

13.8.7. Threshold Current Density

A few more peculiarities of semiconductor lasers will be added to deepen our understanding. Each diode laser has a certain power output characteristic which depends on the input current density, as depicted in Fig. 13.41. Applying low pumping currents results in predominantly *spontaneous* emission of light. The light is in this case incoherent and is not strongly monochromatic, i.e., the spectral line width is spread over several hundred Ångströms. However, when the current density increases above a certain threshold, population inversion eventually occurs. At this point, the stimulated emission (lasing) dominates over spontaneous emission and the laser emits a single wavelength having a line width of about 1 Å. Above the threshold the laser operates about one hundred times more efficiently than below the threshold.

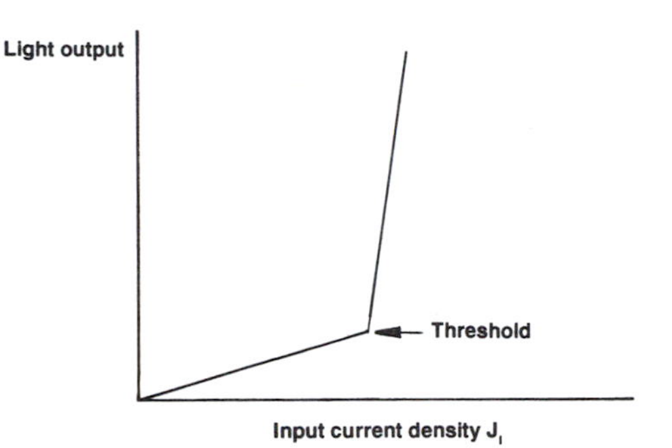

Figure 13.41. Schematic representation of the power output of a diode laser versus the pump current density. The threshold current density for a homojunction GaAs laser is on the order of 10^4 A/cm^2.

The electric vector vibrates perpendicular to the cavity, i.e., the emitted light is plane polarized. Additionally, standing waves are formed within the laser which avoids destructive interference of the radiation. The distance between the two cavity faces must therefore be an integer multiple of half a wavelength.

13.8.8. Homojunction Versus Heterojunction Lasers

Lasers for which the *p*-type and *n*-type base materials are alike (e.g., GaAs) are called *homojunction* lasers. In these devices the photon distribution extends considerably beyond the electrically active region (in which the lasing occurs) into the adjacent inactive regions, as shown in Fig. 13.42. The total light-emitting layer, *D*, for GaAs is about 10 μm wide whereas the depletion layer, *d*, i.e., the active region, might be as narrow as 1 μm. The photons which

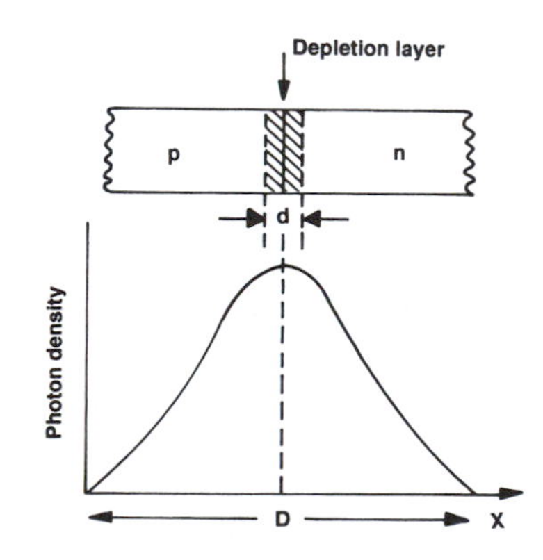

Figure 13.42. Schematic representation of the photon distribution in the vicinity of the depletion layer of a homojunction diode laser.

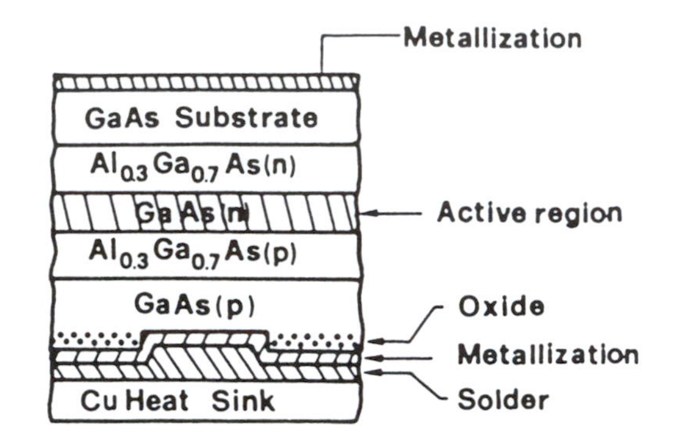

Figure 13.43. Schematic representation of a double heterojunction laser in which the active region consists of an *n*-doped GaAs layer.

penetrate into the nonactive region do not stimulate further emission and thus reduce the quantum efficiency (which in the present case is about 10%). In essence, some of these photons are eventually absorbed and thus increase the temperature of the laser. The homojunction laser has therefore to be cooled or operated in a pulsed mode employing bursts of 100 ns duration, allowing for intermittent cooling times as long as 10^{-2} s. This yields peak powers of about 10–30 W.

Cooling or pulsing is not necessary for heterojunction lasers in which, for example, *two* junctions are utilized as depicted in Fig. 13.43. If the refractive index of the active region is larger than that of the neighboring areas, an "*optical waveguide structure*" is effectively achieved which confines the photons within the GaAs layer (total reflection!). This way, virtually no energy is absorbed in the nonactive regions. The threshold current density can be reduced to 400 A/cm^2. The quantum efficiency can reach 55% and the output power in continuous mode may be as high as 390 mW. The disadvantage of a double heterojunction laser is, however, its larger angular divergence of the emerging beam which is between 20° and 40°.

13.8.9. Laser Modulation

For telecommunication purposes it is necessary to impress an a.c. signal on the output of a laser, i.e., to modulate directly the emerging light by, say, the speech. This can be accomplished, for example, by *amplitude modulation*, i.e., by biasing the laser initially above the threshold and then superimposing on this d.c. voltage an a.c. signal (Fig. 13.44). The amplitude of the emerging laser light depends on the slope of the power/current characteristic. Another possibility is *pulse modulation* i.e., the generation of subnanosecond pulses having nanosecond spacings between them. (For digitalization, see Section 13.10.) This high speed pulsing is possible because of the inherently short turn-on

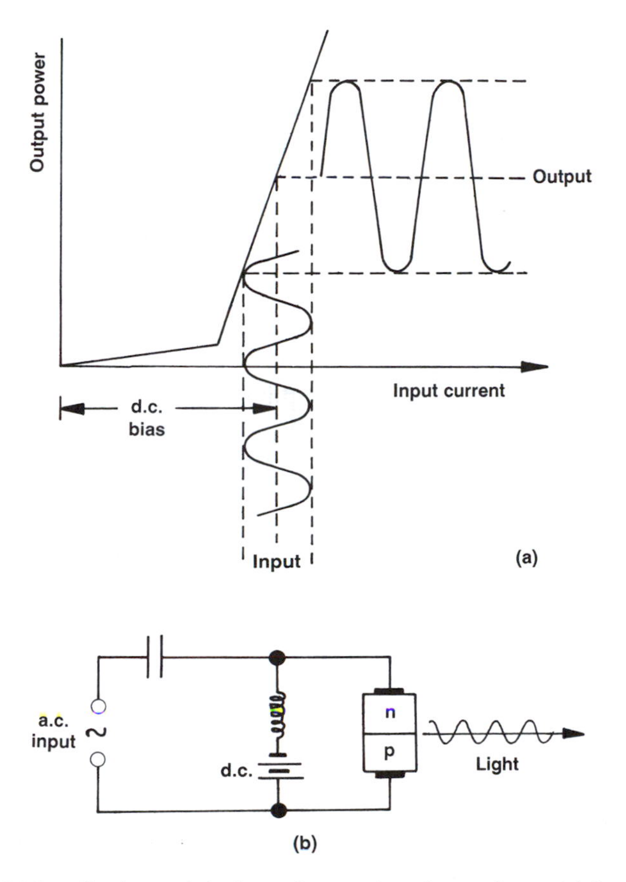

Figure 13.44. Amplitude modulation of a semiconductor laser: (a) input current/output power characteristic; (b) circuit diagram. (The d.c. power supply has to be electrically insulated from the a.c. source.)

and turn-off times (10^{-10} s) of semiconductor lasers when initially biased just below the threshold current density. Finally, *frequency modulation* can be achieved by applying, perpendicularly to the diode junction, a periodic-varying mechanical pressure (by means of a transducer), thus periodically altering the dielectric constant of the cavity. This way modulation rates of several hundred megahertz have been achieved.

13.8.10. Laser Amplifier

The laser can also function as an optical amplifier which is again used for telecommunication purposes. A weak optical signal enters a laser through one of its windows and there stimulates the emission of photons. The amplified signal leaves the other window after having passed the cavity only once. This "*traveling-wave*" laser is biased slightly below the threshold current in order to exclude spurious lasing not triggered by an incoming signal. Nevertheless, some photons are always spontaneously generated, which causes some background noise.

A new development is the *erbium-doped fiber amplifier* which works quite similar to the above-mentioned traveling-wave laser. Erbium atoms, contained in lengths of a coiled glass fiber, are pumped to higher energies by an indium–gallium–arsenide–phosphide laser at a wavelength of 0.98 μm or 1.48 μm. When a weakened signal enters one end of this erbium-doped fiber, the erbium atoms gradually transfer their energy to the incoming signal by stimulated emission, thus causing amplification. A mere 10 mW of laser power can thus achieve a gain of 30–40 dB. Networks which include fiber amplifiers, linked at certain distance intervals to cladded optical glass fibers (Fig. 13.30), have the potential of transmitting data at very high rates, e.g., 2.5 gigabits of information per second over more than 20,000 km. This is possible because fibers are able to support a large (but finite) number of channels. The advantage of erbium-doped optical fibers is that they do not interrupt the path of a light signal as conventional "*repeaters*" do (which convert light into an electric current, amplify the current, and then transform the electrical signal back into light).

13.8.11. Quantum Well Lasers

Quantum well lasers are the ultimate in miniaturization as already discussed in Section 8.7.10. We have explained there that some unique properties are observed when device dimensions become comparable to the wavelength of electrons. In essence, when a thin (20 nm wide) layer of a *small band-gap* material (such as GaAs) is sandwiched between two *large band-gap* materials (such as AlGaAs) a similar energy configuration is encountered as known for an *electron in a box* (Fig. 8.33). Specifically, the carriers are confined in this case to a potential well having "infinitely" high walls. Then, as we know from Section 4.2, the formerly continuous conduction or valence bands reduce to discrete energy levels, see Fig. 13.45.

The light emission in a *quantum well laser* occurs as a result of electron transitions from these conduction band levels into valence levels. It goes almost without saying that the line width of the emitted light is small in this case, because the transitions occur between *narrow* energy levels. Further, the threshold current density for lasing (Fig. 13.41) is reduced by one order of magnitude, and the number of carriers needed for population inversion is likewise smaller.

If a series of large and small band-gap materials are joined, thus forming a *multiple quantum well laser*, the gain is even further increased and the stability of the threshold current toward temperature fluctuations is improved. The

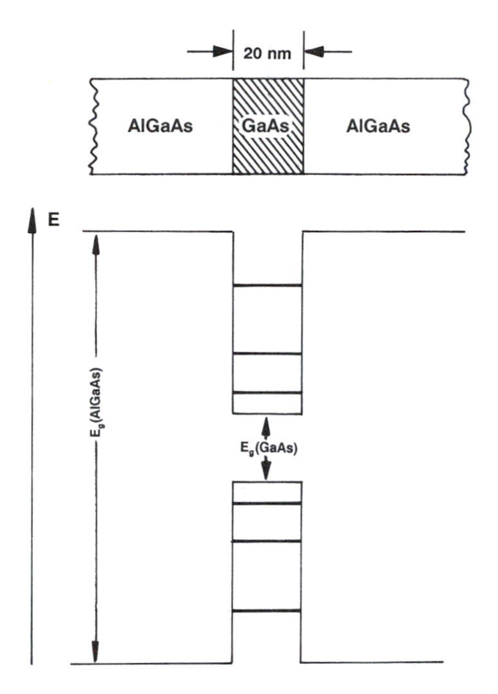

Figure 13.45. Band structure of a single quantum dot structure. See in this context Section 8.7.10 and Fig. 8.33(b).

GaAs/GaAlAs combination yields an emission wavelength somewhat below 0.87 μm, whereas InGaAsP quantum well lasers emit light near 1.3 μm or 1.5 μm (depending on their composition). It appears to be challenging to eventually fabricate quantum wire or quantum dot lasers (see Section 8.7.10) which are predicted to have even lower threshold current densities and higher modulation speeds.

13.8.12. Light-Emitting Diode (LED)

A brief description of light-emitting diodes (LED) will conclude this section. These devices are of great technical importance as an inexpensive, rugged, small, and efficient light source for display purposes. The LED consists, like the semiconductor laser, of a forward biased *p–n* junction. The above-mentioned special facing procedures are, however, omitted during the manufacturing process. Thus, the LED does not operate in the lasing mode. The emitted light is therefore neither phase coherent nor collimated. It is, of course, desirable that the light emission occurs in the visible spectrum. Certain III/V compound semiconductors, such as $Ga_xAs_{1-x}P$, GaP, and $Ga_xAl_{1-x}As$ fulfill this requirement. The light is emitted in the red, green, yellow, or orange part of the spectrum. Blue color is possible when using silicon carbide or zinc selenide. (Silicon carbide is, however, an indirect bandgap material and has therefore a low efficiency, see above.) The radiation may leave the device through a window which has been etched through the metallic contact (*surface emitter*).

13.9. Integrated Optoelectronics

Integrated optoelectronics deals with a family of optical components such as lasers, photodiodes, optical waveguides, optical modulators, optical storage devices, etc., which are integrated on a common substrate (if feasible) with the aim of fulfilling similar functions as electrical integrated circuits do. The main difference to electrical devices is that in optical integrated circuits (OICs) the signal is transmitted by light. Still, they need in most cases electrical energy to become functional which explains the name *optoelectronic*. Among the advantages of optical devices are reduced weight, the capability of light of different wavelengths to travel independently and simultaneously in the same wave guide (*multiplexing*), the immunity against receiving extraneous signals from surrounding devices by stray electromagnetic coupling (*crosstalk*), the difficulty in performing wire taps (because of the lack of electromagnetic fields which would extend beyond the optical fiber), high reliability, and notably, the *low-loss transmission* (<2 dB/km) of signals in optical fibers. We have discussed in previous chapters two major optoelectronic components, the laser (Section 13.8) and the photodetector (Sections 8.7.6 and 8.7.7). A few more building blocks need to be added to complete the picture. This will be done now.

13.9.1. Passive Waveguides

The interconnecting medium between various optical devices is called a *waveguide*. It generally consists of a thin, transparent layer whose index of refraction n_2 is *larger* than the refractive indices of the two surrounding media n_1 and n_3. If this condition is fulfilled and if the light impinges on the boundary between n_2 and n_1 (or n_3) at an angle which is larger than the angle of total reflection,[8] then the optical beam travels in zigzag paths between the internal boundaries of Region 2. In other words, by undergoing total reflection, the light wave is considered to remain in the center region. This statement needs, however, some refinement. As a rule, the light which travels in the center

---

[8] Total reflection occurs when $\sin \alpha_T > n_2/n_1$; see (10.1) for $\sin \beta = 1$ (since the sine cannot become larger than one).

medium extends, to a certain degree, into the neighboring media. The spatial distribution of the optical energy within all three media is called a *mode*. This spatial distribution can be calculated by solving the wave equation (10.4) while taking the appropriate boundary conditions into consideration. We have done this twice in earlier parts of this book (Section 10.3 and Section 4.3). We learned there that the electric field strength or, equivalently, the intensity of a wave decreases in the adjacent medium obeying an exponential function. If two boundaries need to be considered, as in the present case, and if the thickness, t, of the center region is comparable to the wavelength of light, then the solution of the wave equation yields an electric field strength distribution (as depicted in Fig. 13.46, lower curve). Now, we know from previous calculations (Section 4) that under certain conditions additional solutions, i.e., distribution functions, do exist (similarly, as a vibrating string can oscillate at higher harmonics). In the present case they are called first-order, second-order, etc., modes. They are likewise depicted in Fig. 13.46. The reader probably recognizes that this "optical tunnel effect" is equivalent to the quantum mechanical tunnel effect shown in Fig. 4.8.

We now consider the most common case in which n_1 is considerably smaller than n_3, e.g., $n_1 = 1$ for air and $n_3 = 3.6$ for GaAs (whereas n_2 is still made larger than n_3!). For this *asymmetric* case the condition for containing the light in the waveguide is

$$n_2 - n_3 \geq \frac{(2\kappa + 1)^2 \lambda_0^2}{32 n_2 t^2}, \tag{13.20}$$

where λ_0 is the wavelength of the light in vacuum, $\kappa = 0, 1, 2 \ldots$ is the mode number, and t is the thickness of the center layer. A calculation (see Problem 1) shows that the difference between n_2 and n_3 needs to be only about 1% in order to contain the light in the center medium.

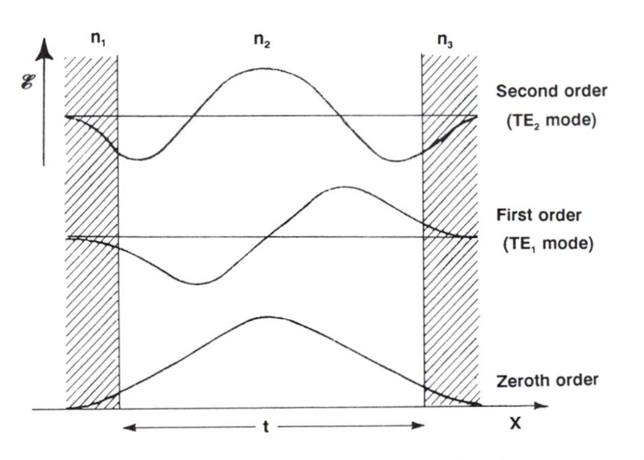

Figure 13.46. Electric field strength distribution (modes) in a waveguide assuming $n_1 = n_3$ (symmetric behavior). The zeroth order and higher order modes are shown. (Compare with Fig. 4.8.)

13.9.2. Electro-Optical Waveguides (EOW)

So far we tacitly implied that the various layers of a waveguide structure have been permanently manufactured by some type of deposition process out of the gaseous or liquid phase on a semiconducting substrate. This is indeed quite often done by employing, for example, molecular beam epitaxy or liquid phase epitaxy processes. However, a rather ingenious alternative method can be utilized instead. This technique involves a Schottky-barrier contact which, when reverse biased, forms (as we know from Section 8.7.2) a wide depletion layer (Fig. 13.47). We shall show in a short calculation that a depletion of charge carriers increases the index of refraction of a solid.

Recall that $n_2 > n_3$ (and n_1) is the prerequisite for a waveguide structure. We derived in Chapter 11 a relationship between the free carriers density (N_f) and the index of refraction

$$\hat{n}^2 = 1 - \frac{e^2 N_f}{\pi m^* \nu^2}, \tag{11.7}$$

where m^* is the effective mass of the electrons in the medium, ν is the frequency of light, e is the charge of the electrons, and $\hat{n} = n - ik$ is the complex index of refraction. We rewrite (11.7) twice for the substrate (Medium 3) and for the depletion layer from which some free carriers have been removed by the applied electric field (Medium 2), assuming that the damping constant k is negligibly small,

$$n_2^2 = 1 - \frac{e^2 N_{f2}}{\pi m^* \nu^2}, \tag{13.21}$$

$$n_3^2 = 1 - \frac{e^2 N_{f3}}{\pi m^* \nu^2}. \tag{13.22}$$

The difference in the indices of refraction is then

$$n_2^2 - n_3^2 = \frac{e^2}{\pi m^* \nu^2}(N_{f3} - N_{f2}), \tag{13.23}$$

which reduces with

$$n_2^2 - n_3^2 = (n_2 + n_3)(n_2 - n_3) \approx 2n_3(n_2 - n_3)^9 \tag{13.24}$$

and $c = \nu \cdot \lambda$ to

$$n_2 - n_3 = \frac{e^2 \lambda^2}{2n_3 \pi m^* c^2}(N_{f3} - N_{f2}). \tag{13.25}$$

Equation (13.25) demonstrates, as suggested above, that a reduction in the number of free carriers from N_{f3} to N_{f2} causes an increase in the index of

---

[9] n_2 can be assumed to be approximately equal to n_3 (see above and Problem 1) which yields $n_2 + n_3 \approx 2n_3$.

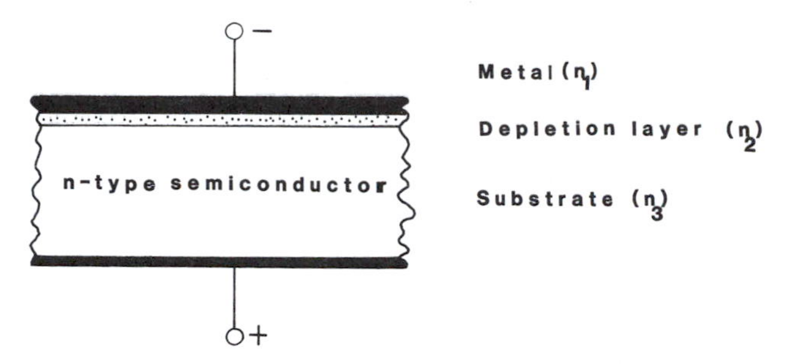

Figure 13.47. Electro-optical waveguide making use of a reverse biased Schottky-barrier contact. (See also Fig. 8.15.) The light travels in Medium 2 when a high enough voltage is applied to the device.

refraction in Medium 2. Then, the device becomes an optical waveguide. For this to happen, the doping of the substrate needs to be reasonably high in order that an appreciable change in the index of refraction is achieved (see Problem 4).

13.9.3. Optical Modulators and Switches

When discussing electronic devices in Section 8.7.12 we encountered a digital *switch* which is capable of turning the electric current *on* or *off* by applying a voltage to the gate of a MOSFET. An equivalent optical device is obtained by making use of the electro-optical waveguide (Fig. 13.47). In the present case, this device is biased initially just below the threshold, i.e., at a voltage which barely prevents the lowest mode optical wave to pass. Then, by an additional voltage between metal and substrate, the EOW becomes transparent. In analogy to its electrical equivalent (Fig. 8.30) this device may be called an *enhancement-type* or *normally-off* electro-optical waveguide. By varying the bias voltage periodically above the threshold the EOW can serve as an effective modulator of light.

A *depletion-type* or *normally-on* EOW can also be built. This device exploits the Franz–Keldysh effect, i.e., the shift of the absorption edge to lower energies when an electric field is applied to a semiconductor (Fig. 13.48). The photon energy of the light is chosen to be slightly smaller than the band-gap energy (dotted line in Fig. 13.48). Thus, the semiconductor is normally in the transparent mode. If, however, a large electric field (in the order of 10^5 V/cm) is applied to the device, then the band gap shifts to lower energies and the absorbance at that particular wavelength (photon energy) becomes several orders of magnitude larger, thus essentially blocking the light. (The Franz–Keldysh shift can be understood when inspecting Fig. 8.15, which shows a lowering of the conduction band and thus a reduction in the band-gap energy when a reverse bias is applied to a semiconductor.)

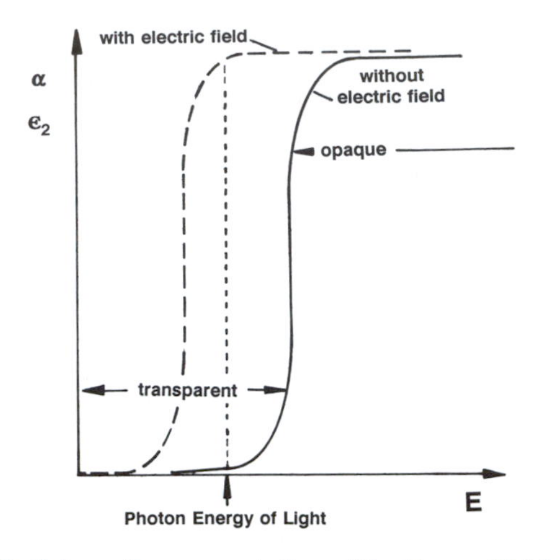

Figure 13.48. Schematic representation of the Franz–Keldysh effect.

Finally, if a piezoelectric transducer imparts some pressure on a waveguide, the index of refraction changes. This *photoelastic effect* can also be utilized for modulation and switching.

Electro-optical modulators can be switched rapidly. The range of frequencies over which the devices can operate is quite wide.

13.9.4. Coupling and Device Integration

We now need to discuss some procedures for transferring optical waves (i.e., information) from one optical (or optoelectronic) device to the next. Of course, *butting*, i.e., the end-on attachment of two devices, is always an option, particularly if their cross-sectional areas are comparable in size. This technique is indeed frequently utilized for connecting optical fibers (used in long-distance transmission) to other components. A special fluid or layer which matches the indices of refraction is inserted between the two faces in order to reduce reflection losses. Optical alignment and permanent mechanical attachment are nontrivial tasks. They can be mastered, however. In those cases where no end faces are exposed for butting, a *prism coupler* may be used. This device transfers the light through a *longitudinal surface*. In order to achieve low-loss coupling, the index of refraction of the prism must be larger than that of the underlying materials. This is quite possible for glass fibers ($n \approx 1.5$) in conjunction with prisms made out of strontium titanate ($n = 2.3$) or rutile ($n = 2.5$) but is difficult for semiconductors ($n \approx 3.6$).

Phase coherent energy transfer between two *parallel* waveguides (or an optical fiber and a waveguide) can be achieved by optical tunneling (Fig. 13.49). For this to occur, the indices of refraction of the two waveguides must

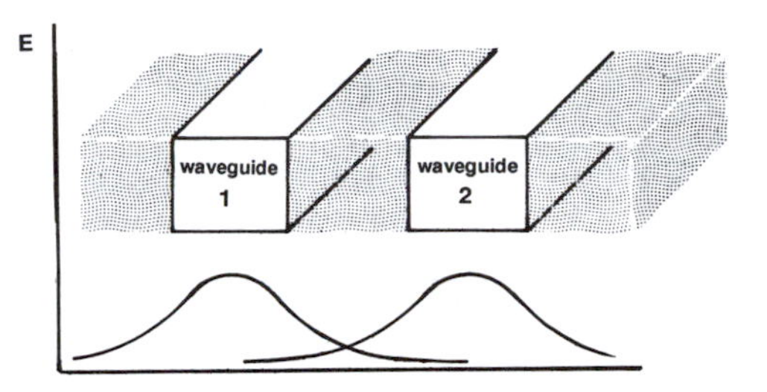

Figure 13.49. Schematic representation of energy transfer between two waveguides (or a waveguide and an optical fiber) by optical tunneling. Compare with Fig. 13.46. ($n_2 > n_1, n_3$.)

be larger than those of the adjacent substrates. Further, the width of the layer between the two waveguides must be small enough to allow the tails of the energy profiles to overlap.

The most elegant solution for efficient energy transfer is the monolithic integration of optical components on *one* chip. For example, a laser and a waveguide may be arranged in one building block, as schematically depicted in Fig. 13.50. Several points need to be observed however. First, the wavelength of the light, emitted by the laser, needs to be matched to a wavelength at which the absorption in the waveguide is minimal. Second, the end faces of the laser need to be properly coated (e.g., with SiO_2) to provide adequate feedback for stimulated emission.

Another useful integrated structure involves a transverse photodiode which is coupled to a waveguide, see Fig. 13.51. As explained in Section 8.7.6, this photodiode is reverse biased. The electron–hole pairs are created in or near a long and wide depletion layer by photon absorption. The losses are mini-

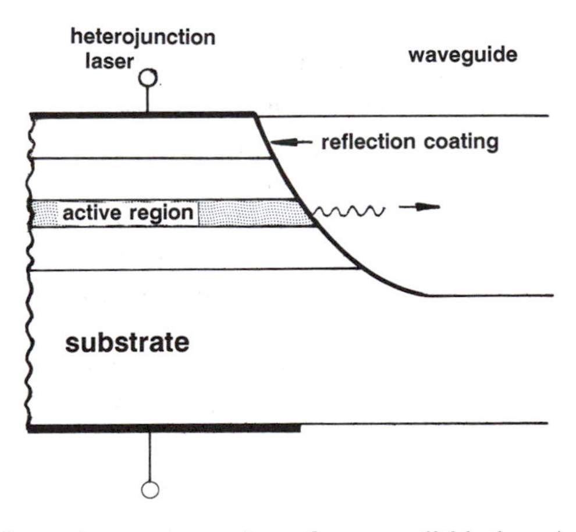

Figure 13.50. Schematic representation of a monolithic laser/waveguide structure. Compare with Fig. 13.43.

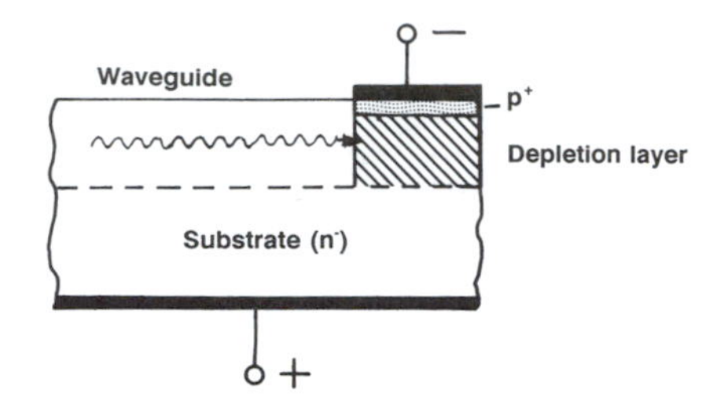

Figure 13.51. Schematic representation of a monolithic transverse photodiode/waveguide structure. A wide depletion layer (active region) is formed in the *n*-region by the reverse bias. For details, see Section 8.7.6.

mized owing to the fact that the light does not have to penetrate the (inactive) *p*-region as in flat-plate photovoltaics. The quantum efficiency of the transverse photodiode can be considerably enhanced by increasing the *length* of the depletion layer.

All taken, the apparently difficult task of connecting optical fibers, waveguides, lasers, or photodetectors and their integration on one chip have progressed considerably in the last decade and have found wide application in a multitude of commercial devices.

13.9.5. Energy Losses

Optical devices lose energy through absorption, radiation, or light scattering, similarly as the electrical resistance causes energy losses in wires, etc. The optical loss is expressed by the *attenuation* (or absorbance) α which was defined in Section 10.4. It is measured in cm^{-1} or, when multiplied by 4.3, in decibels per centimeter.

Scattering losses take place when the direction of the light is changed by multiple reflections on the "rough" surfaces in waveguides or glass fibers, or to a lesser extent by impurity elements and lattice defects.

Absorption losses occur when photons excite electrons from the valence band into the conduction band (interband transitions), as discussed in Chapter 12. They can be avoided by using light whose photon energy is smaller than the band-gap energy. Free carrier absorption losses take place when electrons *in the conduction band* (or in shallow donor states) are raised to higher energies by *intra*band transitions. These losses are therefore restricted to semiconductor waveguides, etc., and essentially do not occur in dielectric materials. We know from (10.21a) that the absorbance α is related to the imaginary part of the dielectric constant ε_2 through

$$\alpha = \frac{2\pi}{\lambda n}\varepsilon_2. \tag{13.26}$$

On the other hand, the free electron theory provides us with an expression for ε_2 (11.27) which is, for $\nu^2 \gg \nu_2^2$,

$$\varepsilon_2 = \frac{\nu_2 \nu_1^2}{\nu^3}, \tag{13.27}$$

where

$$\nu_1^2 = \frac{e^2 N_f}{\pi m^*} \tag{13.28}$$

(see (11.8)) and

$$\nu_2 = \frac{\nu_1^2}{2\sigma_0} \tag{13.29}$$

(see (11.23)) and

$$\sigma_0 = N_f e \mu \tag{13.30}$$

(see (8.13)). Combining equations (13.26) through (13.30) yields

$$\alpha = \frac{e^3 N_f \lambda^2}{\pi n (m^*)^2 c^3 \mu}. \tag{13.31}$$

We note in (13.31) that the free carrier absorbance is a linear function of N_f and is inversely proportional to the mobility of the carriers. The absorbance is also a function of the square of the wavelength.

Radiation losses are, in essence, only significant for curved channel waveguides in which case photons are emitted into the surrounding media. A detailed calculation reveals that the radiation loss depends exponentially on the radius of the curvature. The minimal tolerable radius differs considerably in different materials and ranges between a few micrometers to a few centimeters. The energy loss is particularly large when the difference in the indices of refraction between the waveguide and the surrounding medium is small.

13.10. Optical Storage Devices

Optical techniques have been used for thousands of years to retrieve stored information. Examples are ancient papyrus scrolls or stone carvings. The book you are presently reading likewise belongs in this category. It is of the *random access* type, because a particular page can be viewed immediately without first exposing all previous pages. Other examples of optical storage devices are the conventional photographic movie film (with or without optical sound track) or the microfilm used in libraries. The latter are *sequential* storage media because all previous material has to be scanned before the information of interest can be accessed. They are also called *read only memories* (ROM) because the information content *cannot* be changed by the user. All examples given so far are *analog* storage devices.

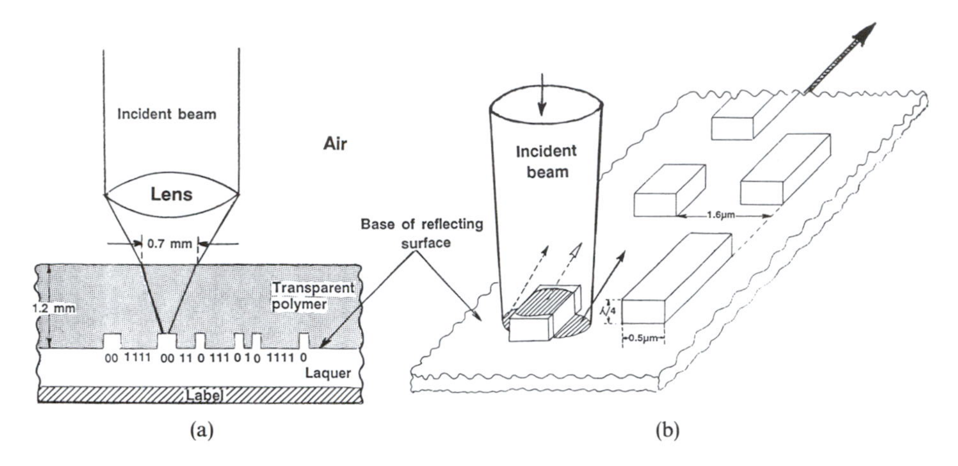

Figure 13.52. Schematic of a compact disc optical storage device. Readout mode. (Not drawn to scale.) The reflected beams in Fig. 13.52(b) are drawn under an angle for clarity. The base and bump areas covered by the probing light have to be of equal size in order that destructive interference can occur (see the hatched areas covered by the incident beam in Fig. 13.52(b)).

Another form of storage utilizes the optical disc which has recently gained wide-spread popularity. Here, the information is generally stored in *digital* form. The most common application, the compact disc (CD), is a *random access, read only memory* device. However, "*write once, read many*" (WORM) and erasable magneto-optic discs (Section 17.5) are also available for special applications. The main advantage of optical techniques is that the readout involves a noncontact process (in contrast to magnetic tape or mechanical systems). Thus, no wear is encountered.

Let us now discuss the optical compact disc. Here, the information is stored below a transparent, polymeric medium in the form of bumps, as shown in Fig. 13.52. The height of these bumps is one-quarter of a wavelength ($\lambda/4$) of the probing light. Thus, the light which is reflected from the base of these bumps travels half a wavelength farther than the light reflected from the bumps. If a bump is encountered, the combined light reflected from bump and base is extinguished by destructive interference. *No light* may be interpreted as a *zero* in binary code, whereas *full intensity* of the reflected beam would then constitute a *one*. For audio purposes, the initial analog signal is sampled at a frequency of 44.1 kHz (about twice the audible frequency) to digitize the information into a series of ones and zeros (similarly as known from computers, Section 8.7.12). Quantization of the signal into 16 digit binary numbers gives a scale of 2^{16} or 65,536 different values. This information is transferred to a disc (see below) in the form of bumps and absences of bumps. For readout from the disc the probing light is pulsed with the same frequency so that it is synchronized with the digitized storage content.

The spiral path on the useful area of a 120 mm diameter CD is 5.7 km long and contains 22,188 tracks which are 1.6 μm apart. (As a comparison, 30 tracks can be accommodated on a human hair.) The spot diameter of the readout beam near the bumps is about 1.2 μm. The information density on a CD is 800 kbits/mm^2, i.e., a standard CD can hold about 7×10^9 kbits. This provides a playback time of about one hour. A disc of the same diameter can also be digitally encoded with 600 megabytes of computer data which is equivalent to three times the text of a standard 24 volume encyclopedia.

The manufacturing process of CDs requires an optically flat glass plate which has been covered with a light sensitive layer (photoresist) about $\lambda/4$ in thickness. Then a helium–neon laser whose intensity is modulated (pulsed) by the digitized information is directed onto this surface while the disc is rotated. Developing of the photoresist causes a hardening of the unexposed areas. Subsequent etching removes the exposed areas and thus creates pits in the photoresist. The pitted surface is then coated with silver (to facilitate electrical conduction) and then electroplated with nickel. The nickel mould thus created (or a copy of it) is used to transfer the pit structure to a transparent polymeric material by injection moulding. The disc is then coated by a reflective aluminum film and finally covered by a protective lacquer and a label.

The CD is read from the back side, i.e., the information is now contained in the form of bumps (see Fig. 13.52). In order to facilitate focusing onto a narrow spot, monochromatic light, as provided by a laser, is essential. A GaAlAs heterojunction laser having a wavelength in air of 780 nm is utilized. The beam size at the surface of the disc is relatively large (0.7 mm in diameter) to minimize possible light obstruction by small dust particles. However, the beam converges as it traverses through the polymer disc to reach the reflecting surface which contains the information. Small scratches on the polymer surface are also tolerated quite well. The aligning of the laser beam on the extremely narrow tracks is a nontrivial task but can be managed.

13.11. The Optical Computer

We have learned in Section 8.7 that transistors are used as switching devices. We know that a small voltage applied to the base terminal of a transistor triggers a large electron flow from emitter to collector. The question arises whether or not a purely optical switching device can be built for which a light beam, having a small intensity, is capable of triggering the emission of a light beam that has a large intensity. Such an *optical transistor* (called *transphasor*) has indeed been constructed which may switch as much as 1000 times faster (picoseconds) than an electronic switch (based on a transistor).

The main element of a transphasor is a small (a few millimeters long) piece of nonlinear optical material (see below) which has, similar to a laser (or a

Fabry–Perot interferometer), two exactly parallel surfaces at its longitudinal ends. These surfaces are coated with a suitable thin film in order to render them semitransparent. Once monochromatic light, stemming from a laser, has entered this "cavity" through one of its semitransparent windows, some of the light is reflected back and forth between the interior windows, whereas another part of the light eventually escapes *through* the windows (Fig. 13.53). If the length between the two windows just happens to be an integer multiple

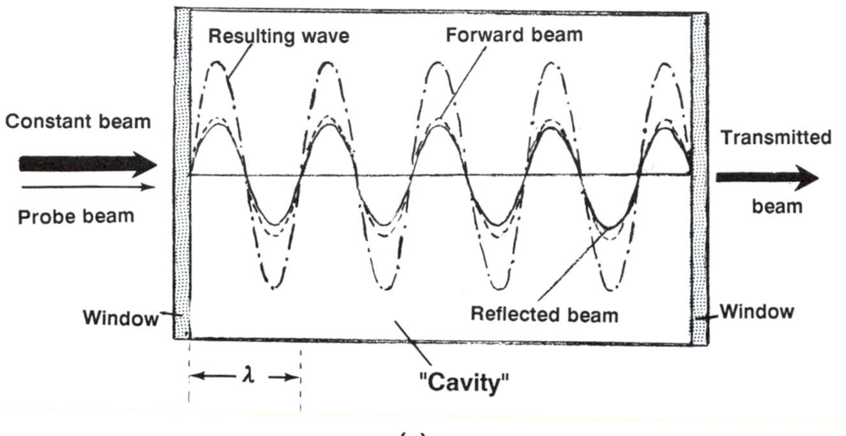

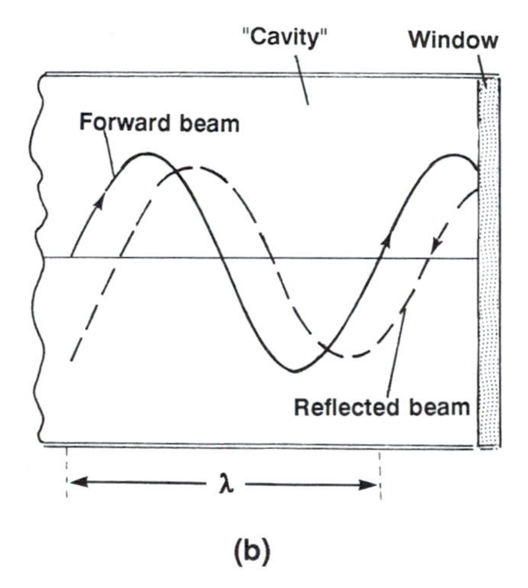

Figure 13.53. Schematic representation of some light waves in a *transphasor*. The reflectivity of the windows is about 90%. (a) Constructive interference. The length of the "cavity" equals an integer multiple of $\lambda/2$. (b) Condition (a) above is not fulfilled. The sum of many forward and reflected beams decreases the total intensity of the light. (*Note*: No phase shift occurs on the boundaries inside the "cavity" because $n_{\text{cavity}} > n_{\text{air}}$.)

of half a wavelength of the light, *constructive interference* occurs and the amplitude (or the intensity) of the light in the "cavity" increases rapidly (Fig. 13.53(a)). As a consequence, the intensity of the transmitted light is also strong. In contrast to this, if the distance between the two windows is not an integer multiple of half a wavelength of the light, the many forward and reflected beams in the "cavity" weaken each other mutually, with the result that the intensity of the transmitted light is rather small (Fig. 13.53(b)). In other words, all conditions which do not lead to constructive (or near constructive) interference produce rather small transmitted intensities (particularly if the reflectivity of the windows is made large).

The key ingredient of a transphasor is a specific substance, namely, the above-mentioned nonlinear optical material which changes its index of refraction as a function of the intensity of light. As we know from Section 10.2 the index of refraction is

$$n_{\text{med}} = \frac{c_{\text{vac}}}{c_{\text{med}}} = \frac{\lambda_{\text{vac}}}{\lambda_{\text{med}}}. \tag{13.32}$$

Thus, we have at our disposal a material which, as a result of high light intensity, changes its index of refraction which, in turn, changes λ_{med} until an integer multiple of $\lambda_{\text{med}}/2$ equals the cavity length and constructive interference may take place. Moreover, just shortly before this condition has been reached, a positive feedback mechanism mutually reinforces the parameters involved and brings the beams rapidly closer into the constructive interference state.

An optical switch involves a "constant laser beam" whose intensity is not yet strong enough to trigger constructive interference (Fig. 13.53(a)). This light intensity is supplemented by a second laser beam, the "*probe beam*," which is directed onto the same spot of the window of the transphasor and which provides the extra light energy to trigger a large change in n and thus constructive interference (Fig. 13.54). All taken, a small intensity change

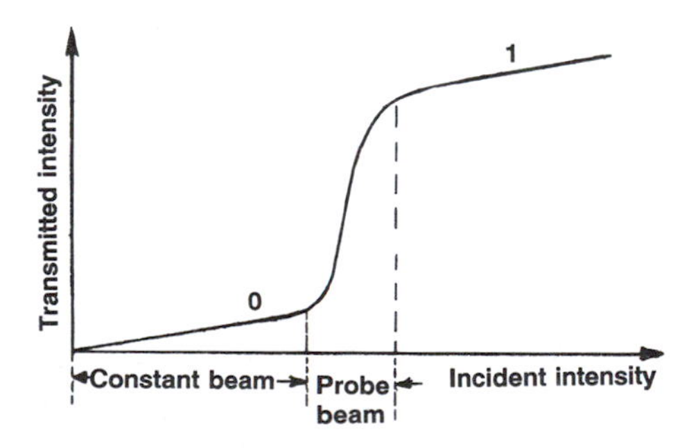

Figure 13.54. Schematic representation of an optical AND gate as obtained from an optical transistor (transphasor) constructed from a material with nonlinear refractive index. The low transmission state may represent a "zero" in binary logic whereas the high transmission of light may stand for a "one."

caused by the probe beam invokes a large intensity of the transmitted beam. This combination of two signals which interact with a switching device can be utilized as an "AND" logic circuit, as described in Section 8.7.12. Likewise, "OR" gates (either of the two beams is already strong enough to trigger critically a change in n) or "NOT" gates (which involve the reflected light) can be constructed.

One important question still remains to be answered. It pertains to the mechanisms involved in a nonlinear optical material. Such a material consists, for example, of indium antimonide, a narrow-band-gap semiconductor, having a gap energy of only 0.2 eV. (It therefore needs to be cooled to 77 K in order to suppress thermally-induced conduction band electrons.) Now, we know from Chapter 12 that when photons of sufficiently high energy interact with the valence electrons of semiconductors, some of these electrons are excited across the gap into the conduction band. The number of excited electrons is, of course, larger the smaller the gap energy (see Chapter 12), and the larger the number of impinging photons. On the other hand, the index of refraction, n, depends on the number of free electrons, N_f (in the conduction band, for example), as we know from (11.7)

$$\hat{n}^2 = 1 - \frac{e^2 N_f}{\pi m \nu^2}. \tag{13.33}$$

Thus, a high light intensity substantially changes N_f and therefore n as stated above.

A crude photonic computer was introduced in 1990 by Bell Laboratories. However, new nonlinear materials need to be found before optical computers become competitive with their electronic counterparts.

Problems

1. Calculate the difference in the refractive indices which is necessary in order that an asymmetric waveguide operates in the zeroth mode. Take $\lambda_0 = 840$ nm, $t = 800$ nm, and $n_2 = 3.61$.

2. How thick is the depletion layer for an electro-optical waveguide when the index of refraction ($n_3 = 3.6$) increases in Medium 2 by 0.1%. Take $n_1 = 1$, $\lambda_0 = 1.3$ μm, and zeroth-order mode.

3. Calculate the angle of total reflection in (a) a GaAs waveguide ($n = 3.6$), and (b) a glass waveguide ($n = 1.5$) against air.

4. Of which order of magnitude does the doping of an electro-optical waveguide need to be in order that the index of refraction changes by one-tenth of one percent? Take $n_3 = 3.6$, $m^* = 0.067$ m$_0$, and $\lambda = 1.3$ μm.

5. Calculate the free carrier absorption loss in a semiconductor assuming $n = 3.4$, $m^* = 0.08$ m$_0$, $\lambda = 1.15$ μm, $N_f = 10^{18}$ cm^{-3}, and $\mu = 2 \times 10^3$ cm^2/Vs.

6. Show that the energy loss in an optical device, expressed in decibels per centimeter, indeed equals 4.3α.
7. Calculate the necessary step height of a "bump" on a compact disc in order that destructive interference can occur. (Laser wavelength in air, 780 nm; index of refraction of transparent polymeric materials, 1.55.)
8. Calculate the gap energy and the emitting wavelength of a GaAs laser which is operated at 100° C. Take the necessary data from the tables in Appendix 4 and Section 19.2.

Suggestions for Further Reading (Part III)

F. Abelès, ed., Optical properties and electronic structure of metals and alloys, *Proceedings of the International Conference*, Paris, 13–16 Sept. 1965, North-Holland, Amsterdam (1966).
F. Abelès, ed., *Optical Properties of Solids*, North-Holland, Amsterdam (1972).
M. Born and E. Wolf, *Principles of Optics*, 3rd ed., Pergamon Press, Oxford (1965).
G. Bouwhuis, ed., *Principles of Optical Disc Systems*, Adam Hilger, Bristol (1985).
M. Cardona, *Modulation Spectroscopy, Solid State Physics*, Suppl. 11, Academic Press, New York, (1969).
W. W. Duley, *Laser Processing and Analysis of Materials*, Plenum Press, New York (1983).
K.J. Ebeling, *Integrierte Optoelektronik*, Springer-Verlag, Berlin (1989).
M.P. Givens, Optical properties of metals, in *Solid State Physics*, Vol. 6, Academic Press, New York (1958).
O.S. Heavens, *Optical Properties of Thin Solid Films*, Academic Press, New York (1955).
L.L. Hench and J.K. West, *Principles of Electronic Ceramics*, Wiley, New York (1990).
R.E. Hummel, *Optische Eigenschaften von Metallen und Legierungen*, Springer-Verlag, Berlin (1971).
R.G. Hunsperger, *Integrated Optics, Theory and Technology*, 3rd ed., Springer-Verlag, New York (1991).
T.S. Moss, *Optical Properties of Semiconductors*, Butterworth, London (1959).
P.O. Nilsson, Optical properties of metals and alloys, *Solid State Physics*, Vol. 29, Academic Press, New York (1974).
B.O. Seraphin, ed., *Optical Properties of Solids—New Developments*, North-Holland/American Elsevier, Amsterdam, New York (1976).
A.V. Sokolov, *Optical Properties of Metals*, American Elsevier, New York (1967).
O. Svelto, *Principles of Lasers*, 2nd ed., Plenum Press, New York (1982).
A Vašíček, *Optics of Thin Films*, North-Holland, Amsterdam (1960).
F. Wooten, *Optical Properties of Solids*, Academic Press, New York (1972).

PART IV

MAGNETIC PROPERTIES OF MATERIALS

CHAPTER 14

Foundations of Magnetism

14.1. Introduction

The phenomenon of magnetism, i.e., the mutual attraction of two pieces of iron or iron ore, was surely known to the antique world. The ancient Greeks have been reported to experiment with this "mysterious" force. The designation *magnetism* is said to be derived from a region in Turkey which was known by the name of *Magnesia* and which had plenty of iron ore.

Interestingly enough, a piece of magnetic material such as iron ore does not immediately attract other pieces of the same material. For this, at least one of the pieces has to be *magnetized*. Simply said, its internal "elementary magnets" need alignment in order for it to become a permanent magnet. Magnetizing causes no problem in modern days. One merely places iron into an electric coil through which a direct current passes for a short time. (This was discovered by Oersted at the turn of the last century.) But how did the ancients do it? There may have been at least two possibilities. First, a bolt of lightning could have caused a magnetic field large enough to magnetize a piece of iron ore. Once one magnet had been produced and identified, more magnets could have been obtained by rubbing virgin pieces of iron ore with the first magnet. There could have been another possibility. It is known that if a piece of iron is repeatedly hit very hard, the "elementary magnets" will be "shaken loose" and align in the direction of the earth's magnetic field. An iron hammer, for example, is *north-magnetic* on its face of impact in the northern hemisphere. Could it have been that a piece of iron ore was used as a hammer and thus it became a magnet?

Magnetic materials made an important contribution to the development of the consciousness of mankind because they paved the way to discoveries of new continents once the compass had been invented. (A compass needle is a pivoted bar magnet which positions itself approximately in the north–south direction. We call the tip that points to geographic north, the north-seeking pole, or simply the north pole, and the opposite end the south pole.) The British coined the word *lodestone* for the iron ore Fe_3O_4 which meant to say that this mineral points the way. Our modern technology would be unthinkable without magnetic materials and magnetic properties. Magnetic tapes or discs (computers), television, motors, generators, telephones, and transformers are only a few examples of their applications.

Thus far, we have used the word magnetism very loosely when implying the mutual magnetic attraction of pieces of iron. There are, however, several other classes of magnetic materials that differ in kind and degree in their mutual interaction. We shall distinguish in the following between ferromagnetism (which term we restrict to the classical magnetism in iron and a few other metals and alloys) and para-, dia-, antiferro-, and ferrimagnetism. The oldest known magnetic ore, the magnetite, or lodestone, Fe_3O_4, is actually a *ferri*magnet $(FeO) \cdot Fe_2O_3$ called *iron ferrite.*

In the sections to come, we will first define the magnetic constants and then remind the reader of some fundamental equations in magnetism before discussing magnetism by classical and quantum theory. Practical applications of magnetic materials are presented in the final chapter.

14.2. Basic Concepts in Magnetism

The goal of this chapter is to characterize the magnetic properties of materials. At least five different types of magnetic materials exist, as mentioned in the Introduction. A qualitative, as well as a quantitative, distinction between these different types can be achieved in a relatively simple way by utilizing a method proposed by Faraday. The magnetic material to be investigated is suspended from one of the arms of a sensitive balance and is allowed to reach into an inhomogeneous magnetic field (Fig. 14.1). Diamagnetic materials are expelled from this field, whereas para-, ferro-, antiferro-, and ferrimagnetic

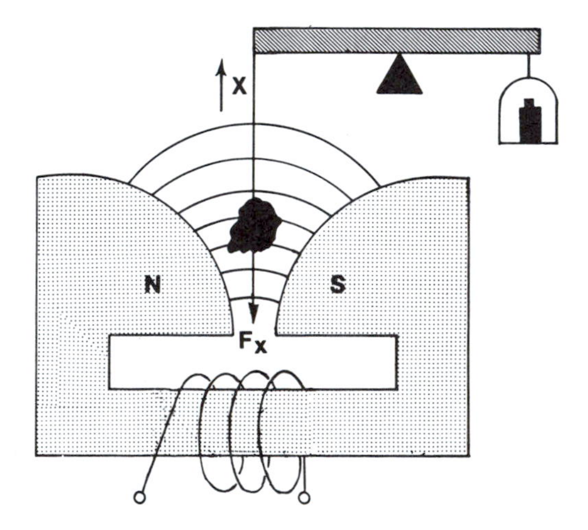

Figure 14.1. Measurement of the magnetic susceptibility in an inhomogeneous magnetic field.

materials are attracted in different degrees. It has been found empirically that the apparent loss or gain in mass, i.e., the force F on the sample exerted by the magnetic field, is

$$F = V\chi H \frac{dH}{dx}, \tag{14.1}$$

where dH/dx is the change of the "*magnetic field strength*" $|\mathbf{H}|$ in the x-direction and V is the volume of the sample. The magnetic material can be characterized by a material constant, χ, called the *susceptibility*, which expresses how responsive this material is to an applied magnetic field (see (14.1)). Characteristic values for χ are given in Table 14.1. Frequently, a second material constant, the *permeability* μ, is used. This constant is related to the susceptibility by the definition

$$\mu = 1 + 4\pi\chi. \tag{14.2}$$

For empty space, and for all practical purposes, also for air, χ is zero and thus $\mu = 1$ (see (14.2)). For diamagnetic materials one finds χ to be small and negative and thus μ slightly less than 1 (see Table 14.1). For para- and antiferromagnetic materials χ is again small, but positive. Thus, μ is slightly larger than one. Finally, χ and μ are large and positive for ferro- and ferri-

Table 14.1. Magnetic Constants of Some Materials at Room Temperature. (Source: Landolt–Börnstein, *Zahlenwerte der Physik*, Vol. 11/9, 6th ed., Springer-Verlag, Berlin (1962).)

Material	χ_m [cm³/g]	χ_v unitless	μ unitless	Type of magnetism
Bi	-1.34×10^{-6}	-13.13×10^{-6}	0.99983	
Be	-1.0×10^{-6}	-1.85×10^{-6}	0.99998	
Ag	-0.192×10^{-6}	-2.016×10^{-6}	0.99997	Diamagnetic
Au	-0.142×10^{-6}	-2.74×10^{-6}	0.99996	
Ge	-0.106×10^{-6}	-0.564×10^{-6}	0.99999	
Cu	-0.086×10^{-6}	-0.77×10^{-6}	0.99999	
Superconductor[a]		$\sim -8 \times 10^{-2}$		
Snβ	$+0.026 \times 10^{-6}$	$+0.19 \times 10^{-6}$	1	
W	$+0.32 \times 10^{-6}$	$+6.18 \times 10^{-6}$	1.00008	
Al	$+0.61 \times 10^{-6}$	$+1.65 \times 10^{-6}$	1.00002	Paramagnetic
Pt	$+0.983 \times 10^{-6}$	$+21.04 \times 10^{-6}$	1.00026	
Low carbon steel			5×10^3	
Fe–3% Si (grain oriented)			4×10^4	Ferromagnetic
Ni–Fe–Mo			10^6	

[a] See Section 7.6.1.

Note: The table lists the mass susceptibility (or specific susceptibility), χ_m, whose unit is cm³/g and the unitless volume susceptibility, χ_v, which was calculated from χ_m by multiplying χ_m by the density. Other sources may provide mass, atomic, molar, or gram equivalent susceptibilities in cgs or mks units.

magnetic materials. The magnetic constants are temperature-dependent, except for diamagnetic materials, as we will see later.

The magnetic field parameters at a given point in space are defined to be the magnetic field strength **H**, which we introduced above, and the magnetic flux density or magnetic induction **B**. In free (empty) space, **B** and **H** are identical. Inside a magnetic material, the induction **B** consists of the free-space component (**H**), plus a contribution to the magnetic field which is due to the presence of matter (Fig. 14.2(a))

$$\mathbf{B} = \mathbf{H} + 4\pi\mathbf{M}, \tag{14.3}$$

where **M** is called the "magnetization" of the material. For a material in which the magnetization is though to be proportional to the applied field strength we define

$$\mathbf{M} = \chi\mathbf{H}. \tag{14.4}$$

Combining (14.2), (14.3), and (14.4) yields

$$\mathbf{B} = \mathbf{H}(1 + 4\pi\chi) = \mu\mathbf{H}. \tag{14.5}$$

Above, we define **B** to be the magnetic flux density in a material, i.e., the magnetic flux per unit area. The magnetic flux ϕ is then defined as the scalar product of **B** and area (represented by vector **A**), i.e.,

$$\phi = \mathbf{B} \cdot \mathbf{A}. \tag{14.6}$$

In free space for which $\mathbf{M} = 0$ we obtain, by using (14.3),

$$\phi = \mathbf{H} \cdot \mathbf{A}. \tag{14.7}$$

Finally, we need to define the magnetic moment $\boldsymbol{\mu}_m$ through the following equation:

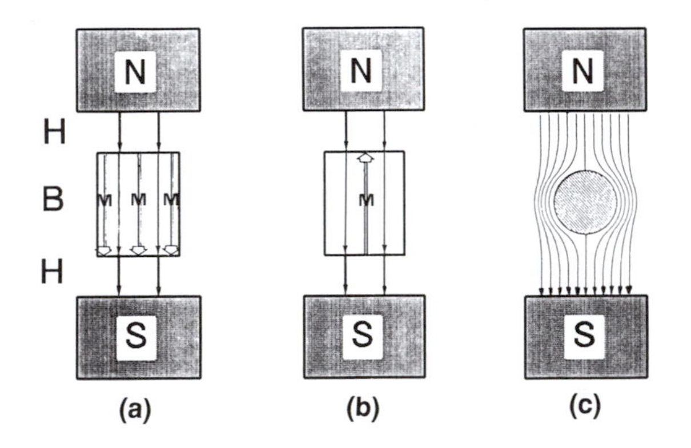

Figure 14.2. Schematic representation of magnetic field lines in and around different types of materials. (a) *Para- or ferromagnetics*. The magnetic induction (B) inside the material consists of the free-space component (H) plus a contribution by the material (M); see (14.3). (b) In *diamagnetics*, the response of the material counteracts the external magnetic field. (c) In a thin surface layer of a *superconductor*, a supercurrent is created (below its transition temperature) which causes a magnetic field that opposes the external field. As a consequence, the magnetic flux lines are expelled from the interior of the material.

$$\mathbf{M} = \frac{\boldsymbol{\mu}_m}{V}, \tag{14.8}$$

i.e., the magnetization is the magnetic moment per unit volume. **B**, **H**, **M**, and $\boldsymbol{\mu}_m$ have been written above in vector form. However, in the following sections, we will mostly utilize their moduli.

*14.3. Units

It needs to be noted that in magnetic theory several unit systems are commonly in use. The scientific and technical literature on magnetism is still widely written in electromagnetic cgs (emu) units. Thus, we shall use this system here. In some European countries, the SI units are mandatory. Conversion factors from emu into SI units are given in Appendix 4. The magnetic field strength in cgs units is measured in Oersted and the magnetic induction in Gauss. In SI units H is measured in A/m and B is given in Tesla (T). Equation (14.3) reads in SI units

$$\mathbf{B} = \mu_0(\mathbf{H} + \mathbf{M}), \tag{14.9}$$

where $\mu_0 = 4\pi \times 10^{-7}$ (kg·m/C^2) is called the permeability of free space. Writing (14.5) in SI units yields

$$\mathbf{B} = \mu_0(1 + \chi)\mathbf{H} = \mu_0\mu_r\mathbf{H}, \tag{14.10}$$

where

$$\mu_r = (1 + \chi) \tag{14.11}$$

is now called the relative permeability. Comparison between (14.11) and (14.2) explains the difference of 4π in the permeabilities mentioned above. The (relative) permeability in the cgs and SI system is unitless and has the same value. The volume susceptibility is also unitless in both systems but differs by a factor 4π (see Appendix 4).

Problems

1. Show that the unit for 1 Oe is equivalent to [g$^{1/2}$/cm$^{1/2}$·s] by making use of (14.1).
2. An electromagnet is a helical winding of wire through which an electric current flows. The magnetic field strength of such a solenoid is (in cgs units)

$$H = \frac{4\pi}{10}\frac{In}{l}\,\text{(Oe)}, \tag{14.12}$$

   where I is the current (in A), n is the number of turns, and l is the length of the solenoid (in cm). In SI units the field strength is $H = In/l$ (A-turns/m or A/m). A solenoid of 1000 turns is 10 cm long and is passed through by a current of 2A. What is the field strength in Oe and A/m?
3. Familiarize yourself with the units of **H**, **B**, and **M** in the different unit systems. Convert (14.9) into (14.3) by making use of the conversion tables in Appendix 4.

CHAPTER 15

Magnetic Phenomena and Their Interpretation—Classical Approach

15.1. Overview

We stated in the last chapter that different types of magnetism exist, and that they are characterized by the magnitude and the sign of the susceptibility (see Table 14.1).

Since various materials respond so differently in a magnetic field, we suspect that several fundamentally different mechanisms must be responsible for the magnetic properties. In the first part of this chapter we shall attempt to unfold the multiplicity of the magnetic behavior of materials by describing some pertinent experimental findings and giving some brief interpretations. In the sections to follow, we shall treat the atomistic theory of magnetism in more detail.

15.1.1. Diamagnetism

Ampère postulated more than one hundred years ago that *molecular currents* are responsible for the magnetism in a solid. He compared the molecular currents to an electric current in a loop-shaped piece of wire which is known to cause a magnetic moment. Today, we replace Ampère's molecular currents by *orbiting valence electrons.*

For the understanding of diamagnetism, a second aspect needs to be considered. It was found by Lenz that a current is induced in a wire loop whenever a bar magnet is moved toward (or from) this loop. The current thus induced causes, in turn, a magnetic moment which is opposite to the one of the bar magnet (Fig. 15.1). (This has to be so in order for mechanical work to be expended in producing the current; otherwise, a perpetual motion would

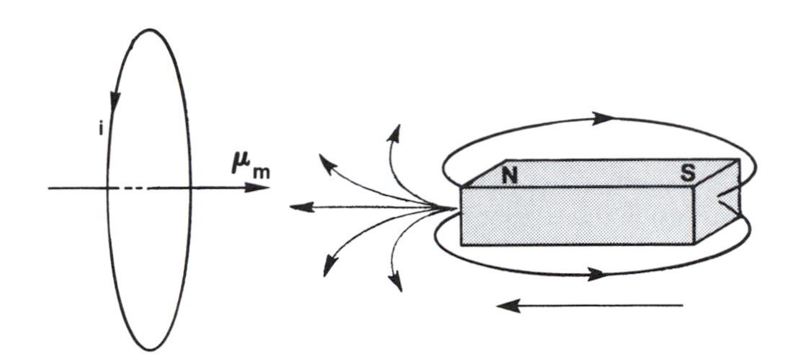

Figure 15.1. Induction of a current in a loop-shaped piece of wire by moving a bar magnet toward the wire loop. The current in the loop causes a magnetic field which is directed opposite to the magnetic field of the bar magnet (Lenz's law).

be created!) Diamagnetism may then be explained by postulating that the external magnetic field induces a change in the magnitude of inner-atomic currents, i.e., *the external field accelerates or decelerates the orbiting electrons*, in order that their magnetic moment is in the opposite direction from the external magnetic field. In other words, the responses of the orbiting valence electrons counteract the external field (Fig. 14.2(b)) and thus shield the inner electrons from an external magnetic field. A more accurate and quantitative explanation of diamagnetism replaces the induced currents by precessions of the electron orbits about the magnetic field direction (Larmor precession, see Fig. 15.2).

So far, we implicitly considered only electrons which are *bound* to their respective nuclei. Now, metals are known also to have *free electrons*. They are forced to move in a magnetic field in a circular path. This leads to a second

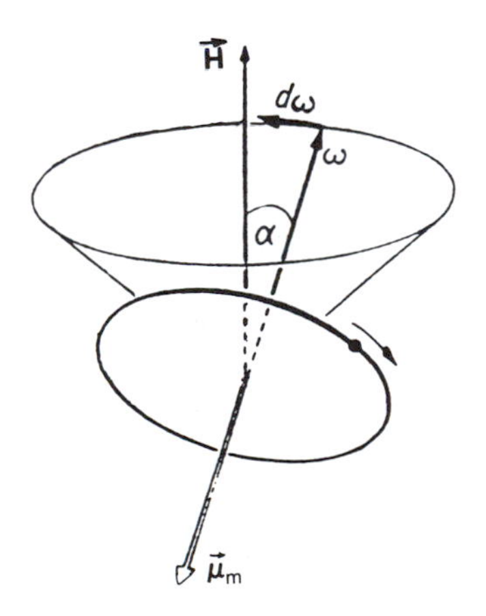

Figure 15.2. Precession of an orbiting electron in an external magnetic field. Precession is the motion which arises as a result of external torque acting on a spinning body (such as a spinning top) or, as here, on an orbiting electron.

contribution to the diamagnetic moment; specifically, the circulating *free* electrons cause a magnetic moment, similarly as described above.

It has been observed that superconducting materials (Section 7.6) expel the magnetic flux lines when in the superconducting state (Meissner effect). In other words, a superconductor behaves in an external magnetic field as if **B** would be zero inside the superconductor (Fig. 14.2(c)). Thus, with (14.3), we obtain

$$\mathbf{H} = -4\pi\mathbf{M},$$

which means that the magnetization is equal and opposite to the external magnetic field strength. The result is a perfect diamagnet. The susceptibility (14.4)

$$\chi = \frac{\mathbf{M}}{\mathbf{H}}$$

in superconductors is $-1/4\pi$ compared to -10^{-6} in the normal state (see Table 14.1). This strong diamagnetism can be used for frictionless bearings, i.e., for support of loads by a repelling magnetic field. The levitation effect in which a magnet hovers above a superconducting material, and the suspension effect where a chip of superconducting material hangs beneath a magnet can be explained with the strong diamagnetic properties of superconductors. (See also Problem 12.)

15.1.2. Paramagnetism

Paramagnetism in *solids* is attributed, to a large extent, to a magnetic moment that results from electrons which spin around their own axis. We have already introduced the *electron spin* in Section 6.4 and mentioned there that, because of the Pauli principle, no two electrons having the same energy can have the same value and sign for the spin moment. In other words, each electron state can be occupied by two electrons only; one with positive spin and one with negative spin, or as is often said, one with spin up and one with spin down. An external magnetic field tries to turn the unfavorably oriented spin moments in the direction of the external field. We will talk about the quantum mechanical aspect of spin paramagnetism in more detail in Chapter 16. Spin paramagnetism is slightly temperature-dependent. It is in general very weak and is observed in some metals and in salts of the transition elements.

Free atoms (dilute gases) as well as rare earth elements and their salts and oxides possess an additional source of paramagnetism. It stems from the magnetic moment of the orbiting electrons. Without an external magnetic field, these magnetic moments are randomly oriented and thus they mutually cancel one another. As a result, the net magnetization is zero. However, when an external field is applied, the individual magnetic vectors tend to turn into the field direction. Thermal agitation counteracts the alignment. Thus, electron-orbit paramagnetism is temperature-dependent.

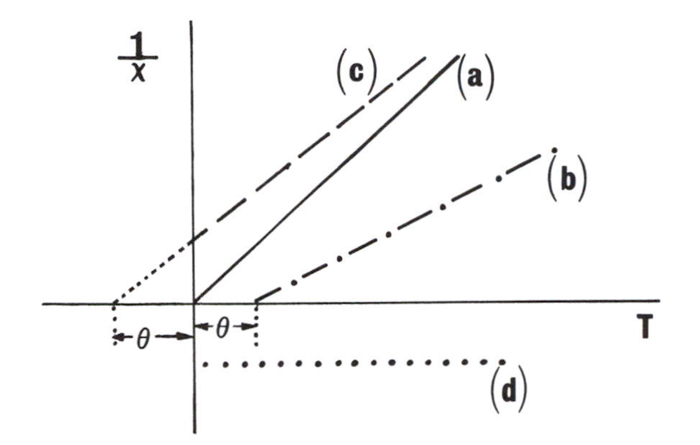

Figure 15.3. Schematic representation of (a) the Curie law and (b) and (c) the Curie–Weiss law. (d) The diamagnetic behavior is also shown for comparison.

The temperature dependence of many paramagnetic materials is governed by the experimentally found Curie law, which states that the susceptibility χ is inversely proportional to the absolute temperature T,

$$\chi = \frac{C}{T}, \tag{15.1}$$

where C is called the Curie constant. For many other substances, a more general relationship is observed, which is known as the Curie–Weiss law

$$\chi = \frac{C}{T - \theta}, \tag{15.2}$$

where θ is another constant which has the same unit as the temperature and may have positive as well as negative values (see Fig. 15.3). We will explain the meaning of the constants C and θ in Section 15.3.

Metals, with a few exceptions, do *not* obey the Curie–Weiss law, as we shall see in Chapter 16. However, Ni (above the Curie temperature, see below) (and in a limited temperature interval also Fe and β-Co), the rare earth elements, and salts of the transition elements (e.g., the carbonates, chlorides, or sulfates of Fe, Co, Cr, Mn) obey the Curie–Weiss law quite well.

We have just mentioned that in most solids only spin paramagnetism is observed. This is believed to be due to the fact that in crystals the electron orbits are essentially coupled to the lattice, which prevents the orbital magnetic moments from turning into the field direction. One says in this case that the orbital moments are "quenched." Exceptions are the rare earth elements and their derivatives, which have "deep-lying" 4*f*-electrons.[1] The latter ones are shielded by the outer electrons from the crystalline field of the neighboring ions. Thus, the orbital magnetic moments of the *f*-electrons may turn into the

[1] See Appendix 3.

external field direction and contribute to electron-orbit paramagnetism. The fraction of the total magnetic moment contributed by orbital motion versus by spin is defined as the "g-factor."

It is now possible to make some general statements about whether para- or diamagnetism might be expected in certain materials. For paramagnetic materials, the magnetic moment of the electrons is thought to point *in* the direction of the external field, i.e., the magnetic moment enhances the external field. Diamagnetism counteracts an external field, as we have seen in Section 15.1.1. Thus, para- and diamagnetism oppose each other. Solids which have both orbital as well as spin paramagnetism are clearly paramagnetic. Rare earth metals with unfilled 4f-electron bands are an example of this. In most other solids, however, the orbital paramagnetism is "quenched," as we said above. Yet, they still might have spin paramagnetism. The possible presence of a net spin-paramagnetic moment depends upon whether or not the magnetic moments of the individual spins cancel each other. More specifically, if a solid has *completely filled electron bands*, we anticipate (because of the Pauli principle) the same number of electrons with spins up as well as with spins down. For example, a completely filled d-band contains $5N$ electrons with spins up and $5N$ electrons with spins down. This results in a cancellation of the spin moments and no net spin paramagnetism is expected. These materials are thus diamagnetic (no orbital and no spin paramagnetic moment). We mention as examples for filled bands intrinsic semiconductors, insulators, and ionic crystals such as NaCl. (In the latter case, an electron transfer occurs between cations and anions which causes closed electron shells, i.e., filled bands.)

In materials with partially filled bands, the electron spins are arranged, according to "Hund's rule," in such a manner that the total spin moment is maximized. This condition is energetically more favorable, as quantum mechanics shows. For example, in an atom with eight valence d-electrons, five of the spins align, say, up, and three spins point down which results in a net total of two spins up (Fig. 15.4). The atom is then expected to have two units of (para-)magnetism.

The smallest unit (or quantum) of the magnetic moment is called one Bohr magneton

$$\mu_B = \frac{eh}{4\pi m} = 0.927 \times 10^{-20} \left(\frac{\text{erg}}{\text{Oe}}\right). \tag{15.3}$$

(The symbols have the usual meaning.) We shall derive equation (15.3) in Chapter 16. In the above example, the metal is said to have two Bohr magnetons per atom.

Figure 15.4. Schematic representation of the spin alignment in a d-band which is partially filled with eight electrons (Hund's rule).

One word of caution should be added about applying the above general principles too rigidly. Some important exceptions do exist. They must be explained by considering additional information (see Chapter 16). For example, copper, which has one *s*-electron in its valence band, should be paramagnetic according to our considerations brought forward so far. In reality, copper is diamagnetic. Other examples are superconductors which are perfect diamagnetics below a transition temperature; they repel the magnetic flux lines from their interior, as we explained in Section 15.1.1.

15.1.3. Ferromagnetism

We turn now to ferromagnetics and commence with the experimentally found magnetization curve for these materials. A newly cast piece of iron (often called *virgin iron*) is inserted into a ring-shaped solenoid (Fig. 15.5). (The ring shape is used to contain the magnetic field within the coil.) If the external field

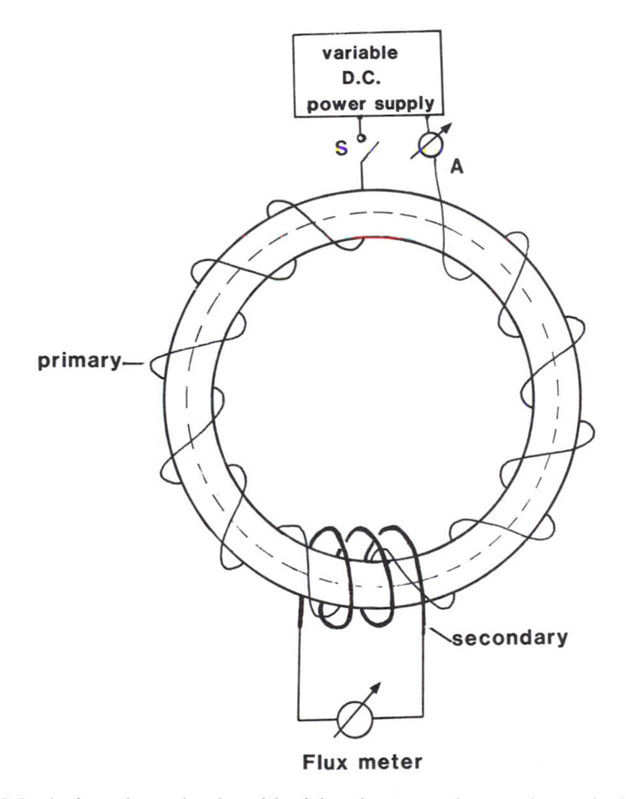

Figure 15.5. A ring-shaped solenoid with primary and secondary windings. The magnetic flux lines are indicated by a dashed circle. Note, that a current can flow in the secondary circuit only if the current (and therefore the magnetic flux) in the primary winding changes with time. An on–off switch in the primary circuit may serve for this purpose.

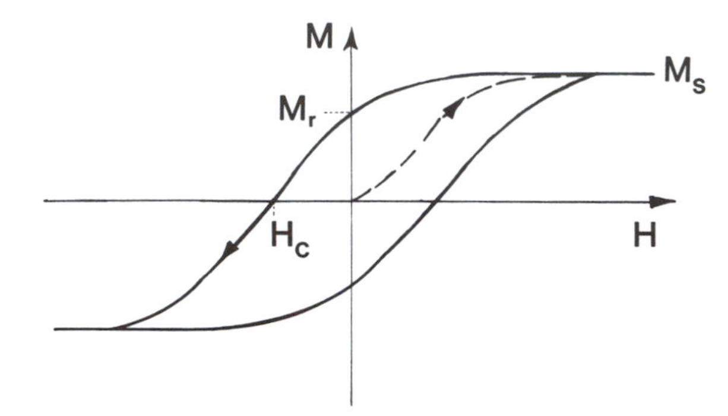

Figure 15.6. Schematic representation of a hysteresis loop of a ferromagnetic material. The dashed curve is for virgin material.

strength is increased (by increasing the current in the primary winding), then the magnetization (measured in a secondary winding with a flux meter) rises at first slowly and then more rapidly (Fig. 15.6). Finally, M levels off and reaches a constant value, called the saturation magnetization M_s. When H is reduced to zero, the magnetization retains a positive value, called the remanent magnetization, or remanence, M_r. It is this retained magnetization which is utilized in permanent magnets. The remanent magnetization can be removed by reversing the magnetic field strength to a value H_c, called the coercive field. Solids having a large combination of M_r and H_c are called *hard magnetic materials* (in contrast to *soft* magnetic materials for which the area inside the loop of Fig. 15.6 is very small). A complete cycle through positive and negative H-values, as shown in Fig. 15.6, is called a *hysteresis loop*. It should be noted that a second type of hysteresis curve is often used, in which B (instead of M) is plotted versus H. No saturation value for B is observed. (The residual induction B_r at $H = 0$ is called the retentivity. Removal of B_r requires a field which is called *coercivity*. However, remanence and retentivity, as well as coercive field, coercive force, and coercivity are often used interchangeably.)

The saturation magnetization is temperature-dependent (Fig. 15.7(a)). Above the *Curie temperature*, T_C, ferromagnetics become paramagnetic. Table 15.1 lists saturation magnetizations and Curie temperatures of some elements. For ferromagnetics the Curie temperature, T_C, and the constant θ in the Curie–Weiss law are nearly identical. A small difference exists, however, because the transition from ferromagnetism to paramagnetism is gradual, as can be seen in Fig. 15.7(b).

The magnetization of ferromagnetics is also stress-dependent (Fig. 15.8). As an example, a compressive stress increases the magnetization for nickel, while a tensile stress reduces M and therefore μ. This effect is just the opposite in certain nickel–iron alloys (permalloy, see Table 17.1) where a tensile stress *increases* M or μ. In polycrystalline iron the situation is more complex. At low

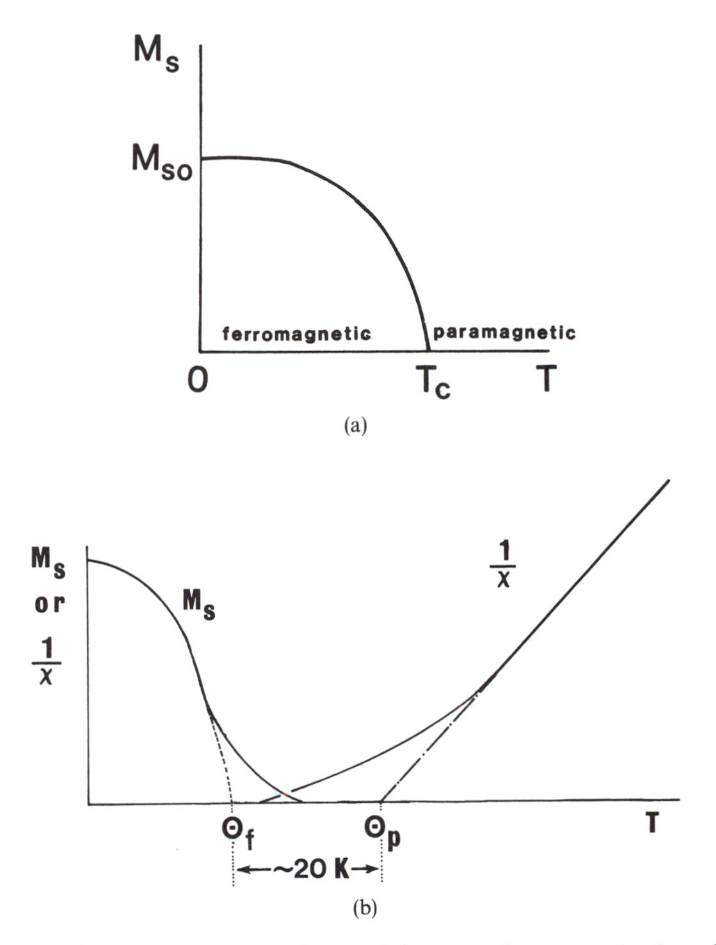

Figure 15.7. (a) Temperature dependence of the saturation magnetization of ferromagnetic materials. (b) Enlarged area near the Curie temperature showing the paramagnetic Curie point θ_p (which is identical to the θ in the Curie–Weiss law) and the ferromagnetic Curie point θ_f (which is identical to T_C).

fields, iron behaves like permalloy, whereas at high fields it behaves similar to nickel. These observations can be traced back to *magnetostriction*, an effect which describes a change in dimensions when a ferromagnetic substance is exposed to a magnetic field. (Incidentally, the periodic dimensional change by an alternating magnetic field causes the humming noise in transformers, see also Section 17.2.)

A few preliminary words should be said to explain the above-mentioned observations. In ferromagnetic materials, the spins of unfilled *d*-bands spontaneously align parallel to each other below T_C, i.e., they align within small domains without the presence of an external magnetic field (Fig. 15.9). The individual domains are magnetized to saturation. The spin direction in each

Table 15.1. Saturation Magnetization at 0 K and Curie Temperature (T_C) for Some Ferromagnetic Materials.

Metal	$4\pi M_{s0}$ (Maxwells/cm^2)	M_{s0} (A/m)	T_C (K)
Fe	2.20×10^4	1.75×10^6	1043
Co	1.82×10^4	1.45×10^6	1404
Ni	0.64×10^4	0.51×10^6	631
Gd	7.11×10^4	5.66×10^6	289

domain is, however, different, so that the individual magnetic moments for the material as a whole cancel each other and the net magnetization is zero. An external magnetic field causes those domains whose spins are parallel or nearly parallel to the external field to grow at the expense of the unfavorably aligned domains. (See the transition from Fig. 15.9(c) to Fig. 15.9(d).) When the entire crystal finally contains *one* single domain, having spins aligned parallel to the external field direction, the material is said to have reached *technical saturation magnetization*, M_s. Nevertheless, if the external magnetic field is further increased a small, additional rise in M is observed. This is caused by the forced alignment of those spins which precess about the field direction due to thermal activation. Eventually M_{s0} is, however, reached. An increase in temperature progressively destroys the spontaneous alignment. The gradual transition from ferromagnetism to paramagnetism is believed to be due to the fact that, slightly above T_C, small clusters of spins are still aligned parallel to each other, a phenomenon which is called magnetic short-range order.

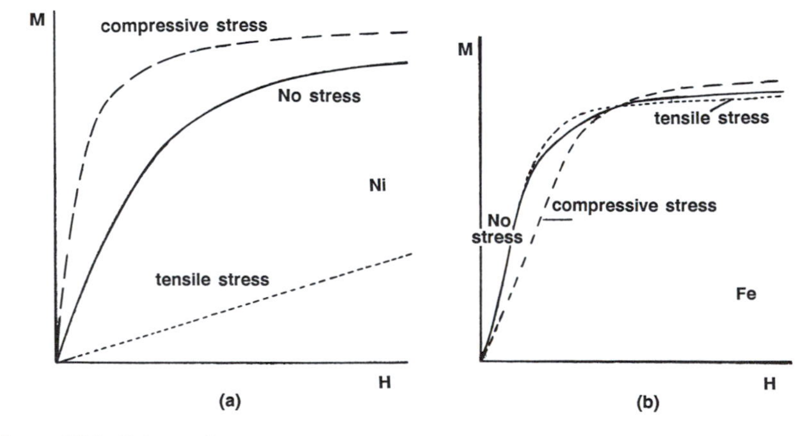

Figure 15.8. Schematic representation of the effect of tensile and compressive stresses on the magnetization behavior of (a) nickel and (b) iron.

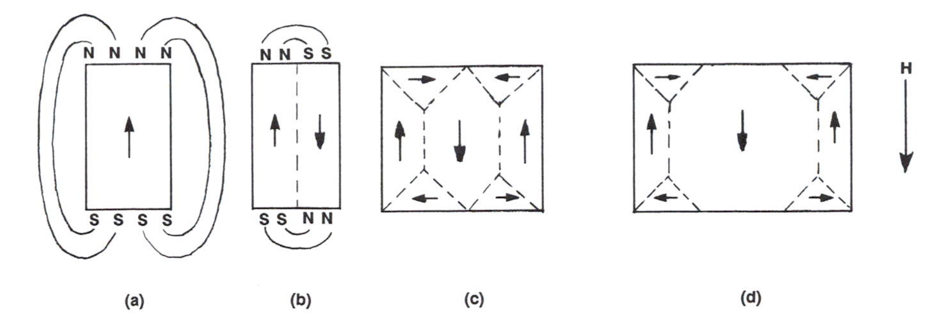

Figure 15.9. (a) Spontaneous alignment of all spins in a single direction. (b) Division into two magnetic domains having opposite spin directions. (c) Closure domains in a cubic crystal. (d) Growth of a domain whose spins are parallel to an external magnetic field. (The domain walls are *not* identical with the grain boundaries.)

There are a number of fundamental questions which come immediately to mind; e.g., in the virgin state, why is the spontaneous division into many individual domains apparently preferred to one single domain? To answer this, let us assume for a moment that all electron spins in a *crystal* are indeed aligned in parallel, Fig. 15.9(a). As a consequence, north and south poles would be created on opposite ends of the solid. This would be energetically unfavorable because it would be the source of a large external magnetic field. The magnetostatic energy of this field can be approximately halved if the crystal contains *two* domains which are magnetized in opposite directions. This way north and south poles are closer together and the external magnetic field is confined to a smaller area (Fig. 15.9(b)). Further divisions into still smaller and smaller domains with concomitant reductions in magnetostatic energies lead, however, eventually to an optimal domain size. Apparently, an opposing mechanism must be active. The energy involved for the latter has been found to be the quantum mechanical *exchange energy*. As we will learn in Section 16.2, this exchange energy causes adjacent spins to align parallel to each other. It is this interplay between exchange energy which demands parallel spin alignment and magnetostatic energy, which supports antiparallel spins that leads eventually to an energetically most favorable domain size (which is about 1–100 μm).

A further reduction in magnetostatic energy can be obtained if the magnetic flux follows a completely closed path within a crystal so that no exterior poles are formed. Indeed, "closure" domain structures, as shown in Fig. 15.9(c), are observed in cubic crystals.

Another question which needs to be answered pertains to whether the flip from one spin direction into the other occurs in one step, i.e., between two adjacent atoms or instead over an extended range of atoms. Again, the above-mentioned exchange energy which supports a parallel spin alignment hinders a spontaneous flip-over. Instead, a gradual rotation over several hundred atomic distances is energetically more favorable. The region between individual domains in which the spins rotate from one direction into the next is called a domain wall or a *Bloch wall.*

Finally, we may ask the question whether and how those domain walls can be made visible. The most common method, devised by Bitter in 1931, utilizes an aqueous suspension of very fine-dispersed Fe_3O_4 particles which is applied to the polished surface of a test material. These particles are attracted to the domain wall endings and can then be observed as fine lines under an optical microscope. Another method exploits the rotation of the plane of polarization of reflected light from differently magnetized areas (Kerr effect).

We mentioned above that an external magnetic field causes a movement of the domain walls. The movement is, as a rule, not continuous, but occurs most of the time in distinct jumps. This is known as the Barkhausen effect which utilizes an induction coil wound around a ferromagnetic rod. The former is connected to an amplifier and a loudspeaker. Audible clicks are heard when a permanent magnet approaches the iron rod. The wall motions may be impeded by imperfections in the crystal such as by particles of a second phase, oxides, holes, or cracks. A second type of impediment to free domain wall motion stems from dislocations, i.e., from residual stresses in the crystal caused by grinding, polishing, or plastic deformation.

Cold work enlarges the coercivity and the area within the hysteresis loop. Further, cold work decreases the permeability and causes a clockwise rotation of the hysteresis curve. In short, mechanical hardness and magnetic hardness parallel each other in many cases. (There exist exceptions, however, such as in the case of silicon additions to iron which makes the material magnetically softer and mechanically harder, see Section 17.2.3.) Recrystallization and grain growth by annealing at suitable temperatures relieve the stresses and restore the soft-magnetic properties.

We shall return to ferromagnetism in Section 15.4 and Chapter 16.

15.1.4. Antiferromagnetism

Antiferromagnetic materials exhibit, just as ferromagnetics, a spontaneous alignment of moments below a critical temperature. However, the responsible neighboring atoms in antiferromagnetics are aligned in an antiparallel fashion (Fig. 15.10). Actually, one may consider an antiferromagnetic crystal to be divided into two interpenetrating sublattices A and B, each of which has a spontaneous parallel alignment of spins. Figure 15.10 depicts the spin alignments for two manganese compounds. (Only the spins of the manganese ions contribute to the antiferromagnetic behavior.) Figure 15.10(a) implies that the ions in a given (111) plane possess parallel spin alignment, whereas ions in the adjacent plane have antiparallel spins with respect to the first plane. Thus, the magnetic moments of the solid cancel each other and the material as a whole has no net magnetic moment.

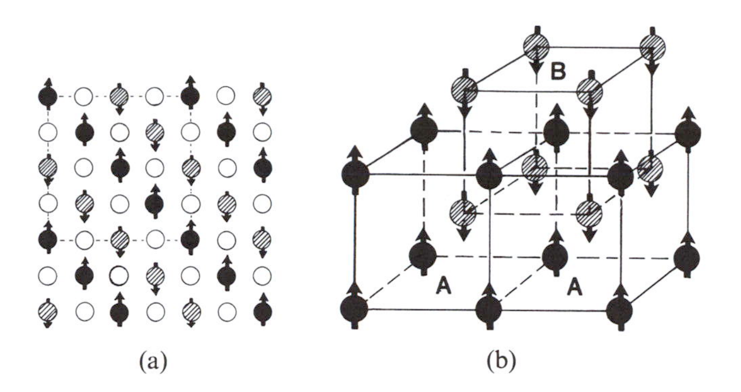

Figure 15.10. Schematic representation of spin alignments for antiferromagnetics at 0 K. (a) Display of a (100) plane of MnO. The gray (spin down) and black (spin up) circles represent the Mn-ions. The oxygen ions (open circles) do not contribute to the antiferromagnetic behavior. MnO has a NaCl structure. (b) Three-dimensional representation of the spin alignment of manganese ions in MnF_2. (The fluorine ions are not shown.) This figure demonstrates the interpenetration of two manganese sublattices A and B having antiparallel aligned moments.

Antiferromagnetic materials are paramagnetic above the *Néel temperature* T_N, i.e., they obey there a linear $T = f(1/\chi)$ law (see Fig. 15.11). Below T_N, however, the inverse susceptibility may rise with decreasing temperature. The extrapolation of the paramagnetic line to $1/\chi = 0$ yields a negative θ. Thus, the Curie–Weiss law (15.2) needs to be modified for antiferromagnetics to read

$$\chi = \frac{C}{T - (-\theta)} = \frac{C}{T + \theta}. \tag{15.4}$$

The Néel temperature is often below room temperature (Table 15.2). Most antiferromagnetics are found among ionic compounds. They are insulators or semiconductors. No practical application for antiferromagnetism is known at this time. (See, however, the use of "canted" antiferromagnetics, described in Section 17.5, which are materials in which the magnetic moments of the two

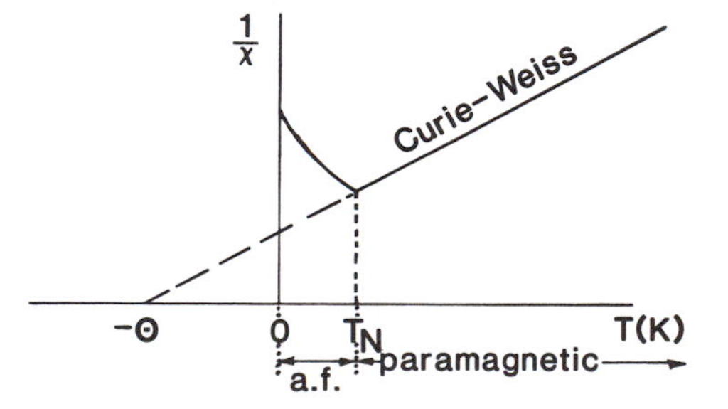

Figure 15.11. Schematic representation of the temperature dependence of a polycrystalline antiferromagnetic (a.f.) material.

Table 15.2. Characteristic Data for Some Antiferromagnetic Materials.

Substance	T_N (K)	$-\theta$ (K)
MnO	116	610
MnF_2	67	82
α-Mn	100	?
FeO	198	570
NiO	523	~2000
CoO	293	330
Cr	310	?

sublattices are not completely antiparallel. This results in a small net magnetization.)

15.1.5. Ferrimagnetism

Ferrimagnetic materials are of great technical importance. They exhibit a spontaneous magnetic moment (Fig. 15.9) and hysteresis (Fig. 15.6) below a Curie temperature, just as iron, cobalt, or nickel. In other words, ferrimagnetic materials possess, similarly as ferromagnetics, small domains in which the electron spins are spontaneously aligned in parallel. The main difference to ferromagnetics is, however, that ferrimagnetics are ceramic materials (oxides) and that they are poor electrical conductors. A large resistivity is often desired for high frequency applications (e.g., to prevent eddy currents in cores of coils, see Chapter 17).

To explain the spontaneous magnetization in ferrimagnetics, Néel proposed that two sublattices should exist in these materials (just as in antiferromagnetics) each of which contains ions whose spins are aligned parallel to each other. The crucial point is that each of the sublattices contain a *different* amount of magnetic ions. This causes some of the magnetic moments to remain uncancelled. As a consequence, a net magnetic moment results. Ferrimagnetic materials can thus be described as *imperfect antiferromagnetics*. The crystallography of ferrites is rather complex. We defer its discussion until later. For the time being it suffices to know that there are two types of lattice sites which are available to be occupied by the metal ions. They are called A sites and B sites. (As before, oxygen ions do not contribute to the magnetic moments).

We will now discuss as an example nickel ferrite, $NiO \cdot Fe_2O_3$. The Fe^{3+} ions are equally distributed between A and B sites (Fig. 15.12) and since ions on A and B sites exhibit spontaneous magnetization in opposite directions, we expect overall cancellation of spins for these ions. Specifically, atomic iron

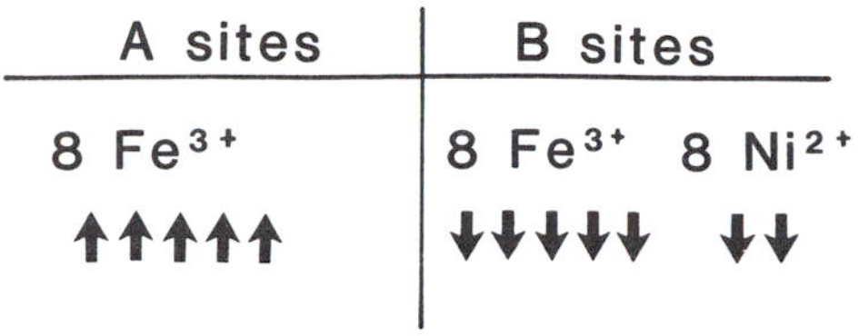

A sites	B sites	
8 Fe^{3+}	8 Fe^{3+}	8 Ni^{2+}
↑↑↑↑↑	↓↓↓↓↓	↓↓

Figure 15.12. Distribution of spins upon A and B sites for the inverse spinel $NiO \cdot Fe_2O_3$. The spins within one site are arranged considering Hund's rule (Fig. 15.4). The iron ions are equally distributed among the A and B sites. The nickel ions are only situated on B sites. The relevance of the number of ions per unit cell is explained later on in the text.

possesses six $3d$-electrons and two $4s$-electrons ($3d^6 4s^2$, see Appendix 3). The Fe^{3+} ions are deprived of three electrons, so that five d-electrons, or five spin moments per atom remain in its outermost shell. This is indicated in Fig. 15.12.

The electron configuration of nickel in its atomic state is $3d^8 4s^2$. Two electrons are stripped in the Ni^{2+}-ion so that eight d-electrons per atom remain. They are arranged, according to Hund's rule (Fig. 15.4), to yield two net magnetic moments (Fig. 15.12). All nickel ions are accommodated on the B sites. Nickel ferrite is thus expected to have two uncancelled spins, i.e., two Bohr magnetons (per formula unit), which is essentially observed (see Table 15.3).

The small discrepancy between experiment and calculation is believed to be caused by some contributions of orbital effects to the overall magnetic moment, and by a slight deviation of the distribution of metal ions on the A and B sites from that shown in Fig. 15.12.

The unit cell of cubic ferrites contains a total of 56 ions. Some of the metal ions are situated inside a *tetrahedron formed by the oxygen ions*. These are the above-mentioned A sites (Fig. 15.13(a)). Other metal ions are arranged in the center of an *octahedron* and are said to be on the B sites (Fig. 15.13(b)). The A and B sites are nestled inside a unit cell (Fig. 15.13(c)). Now, only 8 tetrahedral sites and 16 octahedral sites are occupied by metal ions. In $NiO \cdot Fe_2O_3$ twice as many iron ions as nickel ions are present. Eight of the Fe^{3+} ions per unit cell occupy the A sites, eight of them occupy some of the B sites and the eight Ni^{2+} ions fill the remaining B sites (Fig. 15.12). This

Table 15.3. Calculated and Measured Number of Bohr Magnetons for Some Ferrites.

Ferrite	Mn	Fe	Co	Ni	Cu
Calculated μ_B	5	4	3	2	1
Measured μ_B	4.6	4.1	3.7	2.3	1.3

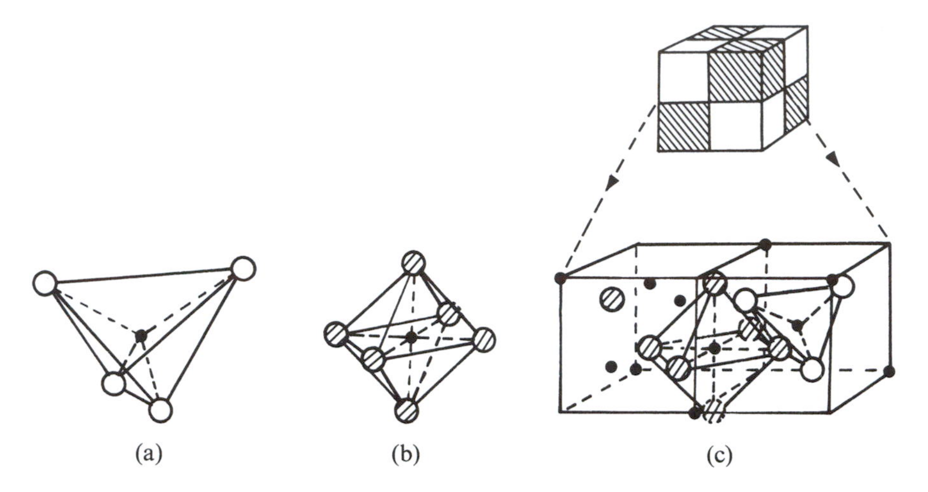

Figure 15.13. Crystal structure of cubic ferrites. The small filled circles represent metal ions, the large open or shaded circles represent oxygen ions: (a) tetrahedral or A sites; (b) octahedral or B sites; and (c) one-fourth of the unit cell of a cubic ferrite. A tetrahedron and an octahedron are marked. Adapted from J. Smit, and H.P.J. Wijn, *Ferrites*, Wiley, New York (1959).

distribution is called an *inverse spinel structure* (in contrast to a *normal spinel* such as for $ZnO \cdot Fe_2O_3$ in which *all* Fe^{3+} ions occupy the B sites).

The temperature dependence of most ferrimagnetics is very similar to ferromagnetics (Fig. 15.14): The saturation magnetization decreases with increasing temperature until it vanishes at a Curie temperature T_C. Above T_C, ferrimagnetics behave paramagnetically, having a nonlinear $1/\chi$ versus T relationship.

In conclusion, this section described, in a mostly qualitative way, the difference between dia-, para-, ferro-, antiferro-, and ferrimagnetism. In the

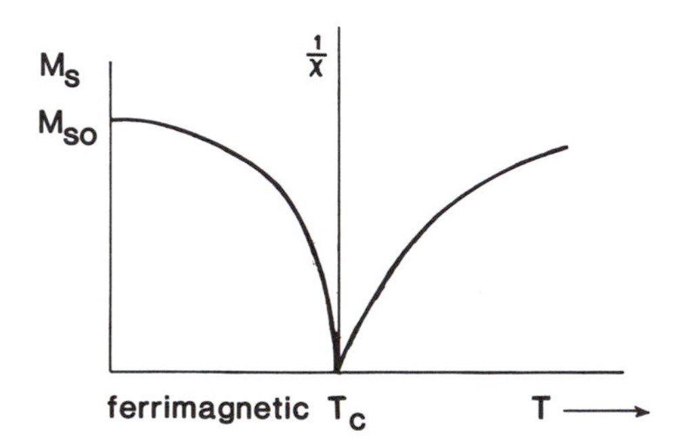

Figure 15.14. Schematic representation of the temperature dependence of the saturation magnetization M_s and the reciprocal susceptibility for ferrites.

sections to come, we shall again pick up the different forms of magnetism and deepen our understanding of these phenomena by following essentially the train of thought brought forward by Langevin, Weiss, and Néel.

15.2. Langevin Theory of Diamagnetism

We shall now develop the classical theory of diamagnetism in a quantitative way as put forward by Langevin at the turn of this century.

We stated before that the orbital motion of an electron about its nucleus induces a magnetic moment μ_m. We compared the latter with a magnetic moment which is created by a current passing through a loop-shaped wire. This magnetic moment is naturally larger, the larger the current I, and the larger the area A, of the orbit or loop:

$$\mu_m = I \cdot A = \frac{e}{t} A = \frac{e}{s/v} A = \frac{ev\pi r^2}{2\pi r} = \frac{evr}{2} \tag{15.5}$$

(e is the electron charge, and r is the radius of the orbit).

We know that an external magnetic field accelerates (or decelerates) the orbiting electrons, which results in a change in magnetic moment. We shall now calculate this change in μ_m.

The external magnetic field induces an electric field (Section 15.1.1) which, in turn, exerts an electrostatic force $|\mathbf{F}|$ on the orbiting electron which is

$$F = ma = \mathscr{E}e, \tag{15.6}$$

where $|\mathscr{E}|$ is the electric field strength and m is the mass of the electron. From this equation we obtain the acceleration of the electron

$$a = \frac{dv}{dt} = \frac{\mathscr{E}e}{m}. \tag{15.7}$$

To calculate the acceleration we need to know the electric field strength $\mathscr{E}$. It is defined as the ratio of the induced voltage (or emf) V_e per orbit length L (see Section 7.1), i.e.,

$$\mathscr{E} = \frac{V_e}{L}. \tag{15.8}$$

As we said earlier, a change in an external magnetic flux ϕ induces in a loop-shaped wire an emf which opposes, according to Lenz's law, the change in flux

$$V_e = -\frac{d\phi}{dt} = -\frac{d(HA)}{dt} \tag{15.9}$$

(see (14.7)). Thus, the acceleration of the electron becomes, by combining

(15.7)–(15.9),

$$\frac{dv}{dt} = \frac{\mathscr{E}e}{m} = \frac{V_e e}{Lm} = -\frac{eA}{Lm}\frac{dH}{dt} = -\frac{e\pi r^2}{2\pi r m}\frac{dH}{dt} = -\frac{er}{2m}\frac{dH}{dt}. \tag{15.10}$$

A change in the magnetic field strength from 0 to H yields a change in the velocity of the electrons

$$\int_{v_1}^{v_2} dv = -\frac{er}{2m}\int_0^H dH \tag{15.11}$$

or

$$\Delta v = -\frac{erH}{2m}. \tag{15.12}$$

This change in electron velocity yields in turn a change in magnetic moment, as we see by combining (15.5) with (15.12)

$$\Delta\mu_m = \frac{e\Delta v r}{2} = -\frac{e^2 r^2 H}{4m}. \tag{15.13}$$

So far we tacitly assumed that the magnetic field is perpendicular to the plane of the orbiting electron. In reality, however, the orbit plane varies constantly in direction with respect to the external field. Thus, we have to find an average value for $\Delta\mu_m$ which we expect to be slightly smaller than that given in (15.13) since $\Delta\mu_m$ approaches zero when the field direction and the orbit plane become parallel. A simple calculation (see Problem 2) yields

$$\overline{\Delta\mu_m} = -\frac{e^2\bar{r}^2 H}{6m}. \tag{15.14}$$

One further consideration needs to be made. Up to now, we treated only *one* electron. If we take all Z electrons in the outermost orbit into account (Section 15.1.1), then the *average change in magnetic moment per atom* is

$$\overline{\Delta\mu_m} = -\frac{e^2 Z\bar{r}^2 H}{6m}. \tag{15.15}$$

The *magnetization, caused by this change of magnetic moment* is, according to (14.8),

$$M = \frac{\mu_m}{V} \equiv -\frac{e^2 Z\bar{r}^2 H}{6mV}. \tag{15.16}$$

This, finally, yields together with (14.4) the diamagnetic susceptibility

$$\boxed{\chi_{\text{dia}} = \frac{M}{H} = -\frac{e^2 Z\bar{r}^2}{6mV} = -\frac{e^2 Z\bar{r}^2}{6m}\frac{N_0\delta}{W},} \tag{15.17}$$

where $N_0\delta/W$ is the number of atoms per unit volume (with N_0 = Avogadro constant, δ = density, and W = atomic mass. Since we are working in electromagnetic cgs units, H is given in Oe, μ_m in erg/Oe, and e in abcoul.). Inserting specific numbers into (15.17) yields susceptibilities between -10^{-5} and -10^{-7}, quite in agreement with the experimental values listed in Table 14.1 (see Problem 1).

The quantities in (15.17) are essentially temperature-independent which is in agreement with the experimental observation that χ does not vary much with temperature for diamagnetic materials.

*15.3. Langevin Theory of (Electron Orbit) Paramagnetism

We turn now to the atomistic theory of paramagnetism as brought forward by Langevin. This theory should explain the observations made by Curie and Weiss, i.e., it should explain the temperature dependence of the susceptibility, as shown in Fig. 15.3. The Langevin theory does *not* treat spin paramagnetism which is, as we said before, responsible for the paramagnetic behavior of many metals and which is only *slightly* temperature-dependent.

Langevin postulated that the magnetic moments of the orbiting electrons are responsible for paramagnetism. The magnetic moments of these electrons are thought to point in random directions. An external magnetic field tries to align the individual magnetic moments, $\boldsymbol{\mu}_m$, parallel to the field direction. Once aligned, the magnetic moments have a potential energy E_p which is naturally greater the larger the field strength $\mathbf{H}$ and the larger $\boldsymbol{\mu}_m$. As a matter of fact, the maximum potential energy is reached when the magnetic moments are completely aligned, i.e., when $\mathbf{H}$ is parallel to $\boldsymbol{\mu}_m$. In general, the potential energy is

$$E_p = -\mu_m H \cos\alpha, \tag{15.18}$$

where α is the angle between field direction and $\boldsymbol{\mu}_m$ (see Fig. 15.15). The sign in (15.18) defines the direction in which $\boldsymbol{\mu}_m$ points to with respect to $\mathbf{H}$.

As we explained earlier, thermal agitation tends to counteract the alignment caused by the external magnetic field. The randomizing effect obeys, as usual, the laws of Boltzmann statistics. The probability of an electron to have the energy E_p is thus proportional to $\exp(-E_p/k_B T)$, where k_B is the Boltzmann constant and T is the absolute temperature.

Let us assume the electrons to be situated at the center of a sphere. The vectors, representing their magnetic moments, may point in all possible directions. Let us consider at present a small number, dn, of these vectors only. They are thought to point in the direction interval $d\alpha$ and thus penetrate an area, dA, situated at the surface of the unit sphere; see Fig. 15.16. This infinitesimal number dn of magnetic moments which have the energy E_p is

$$dn = \text{const.}\ dA \exp(-E_p/k_B T). \tag{15.19}$$

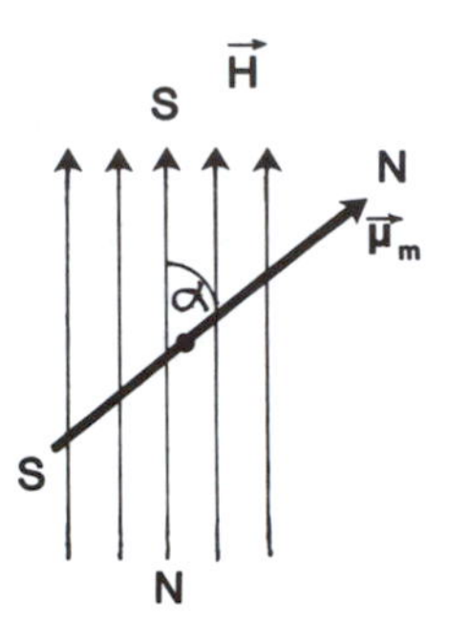

Figure 15.15. Schematic representation of the magnetic moment of an electron which has been partially aligned by an external magnetic field.

We relate the area dA to the angle interval $d\alpha$ which yields, because of trigonometric considerations (see Problem 2),

$$dA = 2\pi R^2 \sin \alpha \, d\alpha, \tag{15.20}$$

where $R = 1$ is the radius of the unit sphere. Combining (15.18)–(15.20) gives

$$dn = \text{const.}\ 2\pi \sin \alpha \, d\alpha \exp\left(\frac{\mu_m H}{k_B T} \cos \alpha\right). \tag{15.21}$$

We use for abbreviation

$$\xi = \frac{\mu_m H}{k_B T}. \tag{15.22}$$

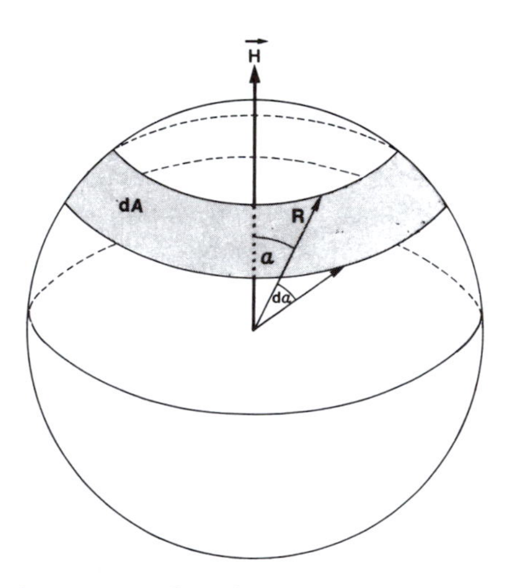

Figure 15.16. Schematic representation of a unit sphere in whose center the electrons are thought to be located.

Integrating (15.21) provides

$$n = 2\pi \text{ const.} \int_0^\pi \sin\alpha \exp(\xi \cos\alpha)\, d\alpha \tag{15.23}$$

which yields

$$\text{const.} = \frac{n}{2\pi \displaystyle\int_0^\pi \sin\alpha \exp(\xi \cos\alpha)\, d\alpha}. \tag{15.24}$$

Now, the magnetization M is, according to (14.8), the magnetic moment μ_m per unit volume. In our case, the total magnetization must be the sum of all individual magnetic moments. And if we consider the magnetic moments in the field direction, then the magnetization per unit volume is

$$M = \int_0^n \mu_m \cos\alpha\, dn \tag{15.25}$$

which yields with (15.21)

$$M = \text{const. } 2\pi\mu_m \int_0^\pi \cos\alpha \sin\alpha \exp(\xi \cos\alpha)\, d\alpha \tag{15.26}$$

and with (15.24)

$$M = \frac{n\mu_m \displaystyle\int_0^\pi \cos\alpha \sin\alpha \exp(\xi \cos\alpha)\, d\alpha}{\displaystyle\int_0^\pi \sin\alpha \exp(\xi \cos\alpha)\, d\alpha}. \tag{15.27}$$

This function can be brought into a standard form by setting $x = \cos\alpha$ and $dx = -\sin\alpha\, d\alpha$ (see Problem 5) which yields

$$M = n\mu_m\left(\coth\zeta - \frac{1}{\zeta}\right) = n\mu_m\left(\frac{\zeta}{3} - \frac{\zeta^3}{45} + \frac{2\zeta^5}{945} - \cdots\right). \tag{15.28}$$

The term $\zeta = \mu_m H/k_B T$ is usually much smaller than one (Problem 6), so that (15.28) reduces to

$$M = n\mu_m \frac{\zeta}{3} = \frac{n\mu_m^2 H}{3k_B T}, \tag{15.29}$$

which yields, for the susceptibility (14.4) at not too high field strengths

$$\boxed{\chi_{\text{para}}^{\text{orbit}} = \frac{M}{H} = \frac{n\mu_m^2}{3k_B}\frac{1}{T} \equiv C \cdot \frac{1}{T}.} \tag{15.30}$$

This is Curie's law (15.1) which expresses that the susceptibility is inversely

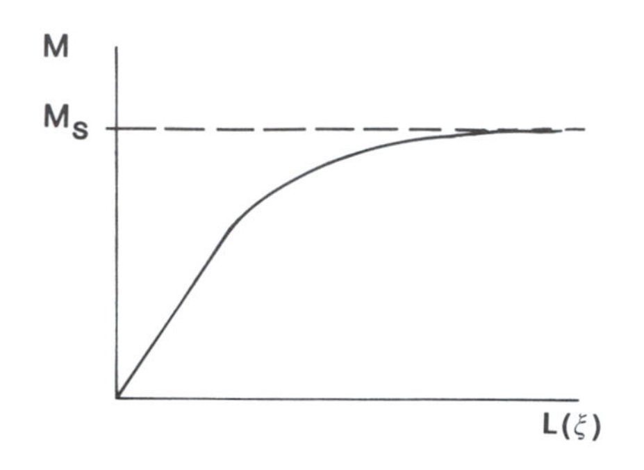

Figure 15.17. Schematic representation of the Langevin function $L(\zeta) = \coth \zeta - 1/\zeta$ where $\xi = \mu_m H/k_B T$.

proportional to the temperature. The Curie constant is thus

$$C = \frac{n\mu_m^2}{3k_B}. \tag{15.31}$$

Let us now discuss the results of the Langevin theory for electron-orbit paramagnetism. If we insert actual values in (15.30), we obtain susceptibilities which are small and positive, which is quite in agreement with experimental findings (see Table 14.1 and Problem 7).

The Langevin theory for paramagnetism yields that for a given temperature and for small values of the field strength the magnetization is a linear function of H (Fig. 15.17 and equation (15.29)). For large field strengths the magnetization eventually reaches a saturation value M_s. (This behavior is quite similar to the one observed for virgin iron or other ferromagnetics.) It indicates that eventually a limit is reached at which all magnetic moments are aligned to their maximum value. The Langevin model yields a temperature dependence of the susceptibility as found experimentally by Curie for many substances. The $1/\chi$ dependence of T is characteristic for electron-orbital paramagnetism.

One can refine the Langevin result by applying quantum theory. This was done by Brillouin who took into account that not all values for the magnetic moment (or the angular moment) are allowed, i.e., that the angular moments are quantized in an external magnetic field (Appendix 3). This restriction is termed *space quantization*. The calculation leads to the *Brillouin function* which improves the quantitative agreement between theory and experiment.

Finally, we know from Section 15.1.2 that the temperature dependence of the susceptibility for many solids does not always obey the Curie (or the Curie–Weiss) law. Actually, the susceptibility for most metals and alloys varies only very little with temperature. We have learned that in these solids the spin paramagnetism is predominant, which is not considered in the atomistic Langevin model. Quantum theory can explain the relative temperature insensitivity of spin paramagnetism, as we shall see in Section 16.1.

*15.4. Molecular Field Theory

So far, we implied that the magnetic field, which tries to align the magnetic moments, stems from an external source only. This assumption seems to be not always correct. Weiss observed that some materials obey a somewhat modified Curie law, as shown in Fig. 15.3(b) and (c). He postulated, therefore, that the magnetic moments of the individual electrons (or atoms) interact with each other. In this case, the total magnetic field, H_t, acting on a magnetic moment, is thought to be composed of two parts, namely, the external field H_e and the *molecular field* H_m

$$H_t = H_e + H_m, \tag{15.32}$$

where

$$H_m = \gamma M \tag{15.33}$$

contains the molecular field constant γ. The susceptibility is calculated by using (15.30), (15.32), and (15.33)

$$\chi = \frac{M}{H_t} = \frac{M}{H_e + \gamma M} = \frac{C}{T}. \tag{15.34}$$

Solving (15.34) for M yields

$$M = \frac{H_e C}{T - \gamma C}. \tag{15.35}$$

Finally, we obtain

$$\chi = \frac{M}{H_e} = \frac{C}{T - \gamma C} = \frac{C}{T - \theta}, \tag{15.36}$$

which is the experimentally observed Curie–Weiss law (15.2). If θ is found to be positive then the interactions of the individual magnetic moments reinforce each other, i.e., the magnetic moments align parallel. In this case the susceptibility becomes larger, as can be deduced from (15.36).

We now attempt to interpret *ferromagnetism* by making use of the molecular field theory. We already know from Section 15.1.3 that, in ferromagnetic materials, the neighboring magnetic moments interact with each other, which leads to a spontaneous magnetization in small domains below T_C. Weiss postulated that the above-introduced internal or molecular field is responsible for this parallel alignment of spins, and considered ferromagnetics to be essentially paramagnetics having a very large molecular field. In essence, he applied the Langevin theory to ferromagnetics. In the light of quantum theory

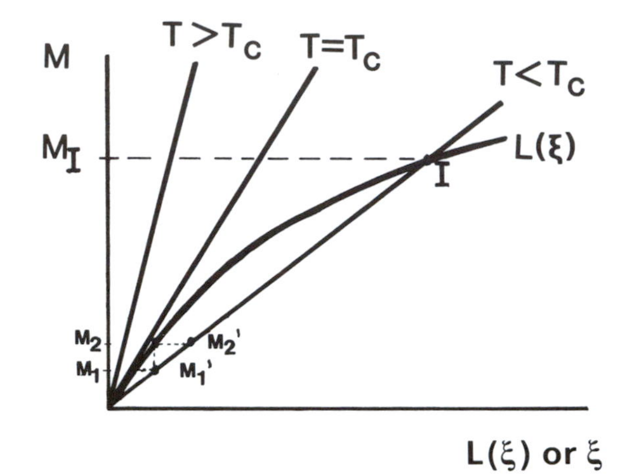

Figure 15.18. Langevin function $L(\zeta)$, i.e., (15.28) and plot of (15.38) for three temperatures.

the molecular field is essentially the *exchange force*, as we shall see in Section 16.2.

We follow the train of thought put forward by Weiss. Let us consider the case for no external magnetic field. Then the spins are only subjected to the molecular field H_m. This yields for the Langevin variable ζ (see (15.22)) with (15.33)

$$\zeta = \frac{\mu_m H_m}{k_B T} = \frac{\mu_m \gamma M}{k_B T} \tag{15.37}$$

and provides for the magnetization by rearranging (15.37)

$$M = \frac{k_B T}{\mu_m \gamma} \zeta. \tag{15.38}$$

We note from (15.38) that for the present case the magnetization is a linear function of ζ with the temperature as a proportionality factor (see Fig. 15.18). The intersection I of a given temperature line with the Langevin function $L(\zeta)$ represents the finite spontaneous magnetization, M_I, at this temperature.[2] With increasing temperature, the straight lines in Fig. 15.18 increase in slope, thus decreasing the point of intercept I and therefore the value for the sponta-

[2] The intersection at the origin is an unstable state, as can easily be seen: If the ferromagnetic material is exposed to, say, the magnetic field of the earth, its magnetization will be, say, M_1. This causes a molecular field of the same value (M_1') which in turn magnetizes the material to the value M_2, and so on until the point I is reached.

neous magnetization. Finally, at the Curie temperature T_C, no intercept, i.e., no spontaneous magnetization is present anymore. The slope $k_B T/\mu_m \gamma$ in (15.38) is then identical to the slope of the Langevin function near the origin, which is $n\mu_m/3 = M/3$ according to (15.29) and (14.8). This yields, for T_C,

$$\frac{k_B T_C}{\mu_m \gamma} = \frac{M}{3}. \tag{15.39}$$

A value for the molecular field constant γ can then be calculated by measuring the Curie temperature and inserting T_C into the rearranged equation (15.39)

$$\gamma = \frac{3k_B T_C}{\mu_m M}. \tag{15.40}$$

This yields, for the molecular magnetic field strength (15.33),

$$H_m = \gamma M = \frac{3k_B T_C}{\mu_m}. \tag{15.41}$$

Numerical values for the molecular field are around 10^7 Oe (see Problem 10). This hypothetical field is several orders of magnitude larger than any steady magnetic field which can be produced in a laboratory. We should note that even though the molecular field theory gives some explanation of ferromagnetism, it cannot predict which solids are ferromagnetic. Quantum theory extends considerably ouf understanding of this matter.

We mention in closing that the molecular field theory can also be applied to antiferromagnetics and to ferrimagnetic materials. As we know from Section 15.1.4 we need to consider in this case two interpenetrating sublattices A and B, each having mutually antiparallel aligned spins. This means that we now have to consider a molecular field H_{mA} acting on the A ions which stems from the magnetization M_B of the B ions. Since the magnetization of A and B ions point in opposite directions, the molecular field from an adjacent ion is now negative. The calculations, which follow similar lines as shown above, yield equation (15.4), i.e., the Curie–Weiss law for antiferromagnetics.

Problems

1. Calculate the diamagnetic susceptibility of germanium. Take for r the atomic radius (1.37 Å). (*Note*: Check your units! Does χ come out unitless? Have you inserted e in abcoul? Compare your result with that listed in Table 14.1.)

2. In the text, we introduced an average value for the magnetic moment, which we said is somewhat smaller than the maximal value for $\boldsymbol{\mu}_m \| \mathbf{H}$. Calculate this $\Delta\mu_m$. (*Hint*: Consider all orbits projected on a plane perpendicular to the field direction and calculate thus an average value for the square of the orbit radius. Refer to the figure on the next page. Show at first that $dA = 2\pi R^2 \sin\alpha \, d\alpha$.)

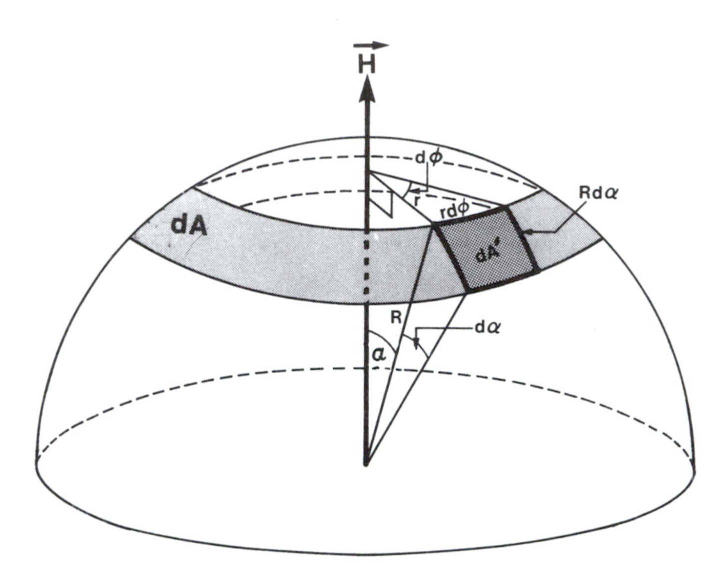

3. Convince yourself that the units in (15.15), (15.16), (15.17), and (15.5) are consistent with the electromagnetic cgs system.

4. Confirm the numerical value of the Bohr magneton listed in (15.3) and confirm the unit given there.

5. Evaluate the function

$$\frac{\int_0^\pi \cos\alpha \sin\alpha \exp(\xi \cos\alpha)\, d\alpha}{\int_0^\pi \sin\alpha \exp(\xi \cos\alpha)\, d\alpha}$$

by substituting $x = \cos\alpha$ and $dx = -\sin\alpha\, d\alpha$. Compare your result with (15.28).

6. Calculate a value for ξ in the Langevin function assuming $\mu_m = 3\mu_B$, $H = 10$ kOe, and room temperature.

7. Calculate the susceptibility for a paramagnetic substance at room temperature, assuming $\mu_m = \mu_B$ and 10^{23} magnetic moments per cubic centimeter. Compare your result with Table 14.1. What is the implication of $n = 10^{23}$ magnetic moments per cubic centimeter?

8. Estimate the number of Bohr magnetons for iron and cobalt ferrite from their electron configuration, as done in the text. Compare your results with those listed in Table 15.3. Explain the discrepancy between experiment and calculation. Give the chemical formula for these ferrites.

9. Explain the term "mixed ferrites." Explain also why the *lodestone* Fe_3O_4 is a ferrimagnetic material. Give its chemical formula.

10. Calculate the molecular field for iron ($\mu_m = 2.22\ \mu_B$, $T_C = 1043$ K).

11. You are given two identical rectangular iron rods. One of the rods is a (ferro) magnet, the other is not. The rods are now placed on a wooden table. Using only the two rods and nothing else, you are asked to determine which is which. Can this be done?

12. Explain the "suspension effect" of superconductors mentioned in Section 15.1.1. (*Hint*: Refer to Fig. 15.1 and keep in mind that if the bar magnet is moved in the opposite direction from that shown, the current direction in the loop is reversed.)

13. *Computer problem.* Plot the Langevin function using various parameters. For which values of H does one obtain saturation magnetization?

CHAPTER 16

Quantum Mechanical Considerations

We have seen in the previous chapter that the classical electromagnetic theory is quite capable of explaining the essentials of the magnetic properties of materials. Some discrepancies between theory and experiment have come to light, however, which need to be explained. Therefore, we now refine and deepen our understanding by considering the contributions which quantum mechanics provides to magnetism. We will see in the following that quantum mechanics yields answers to some basic questions. We will discuss why certain metals that we expect to be paramagnetic are in reality diamagnetic; why the paramagnetic susceptibility is relatively small for most metals; and why most metals do not obey the Curie–Weiss law. We will also see that ferromagnetism can be better understood by applying elements of quantum mechanics.

16.1. Paramagnetism and Diamagnetism

We mentioned at the beginning of the previous chapter that, for most solids, the dominant contribution to paramagnetism stems from the magnetic moment of the *spinning electrons.* We recall from Chapter 6 that each electron state may be occupied by a maximum of two electrons, one having positive spin and the other having negative spin (called spin up and spin down). To visualize the distribution of spins, we consider an electron band to be divided into two halves, each of which is thought to be occupied under normal conditions by an identical amount of electrons of opposite spin, as shown in Fig. 16.1(a). Now, if we apply an external magnetic field to a free electron solid, some of the electrons having unfavorably oriented spins tend to change their field direction. This can only be achieved, however, when the affected

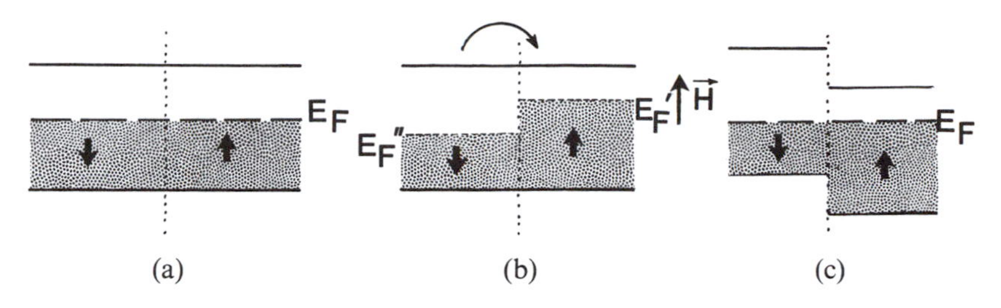

Figure 16.1. Schematic representation of the effect of an external magnetic field on the electron distribution in a partially filled electron band, (a) without magnetic field, (b) and (c) with magnetic field.

electrons assume an energy which is higher than the Fermi energy, E_F, since all lower electron states of opposite spin direction are already occupied (Fig. 16.1(b)). Thus, theoretically, the transfer of electrons from one half-band into the other would cause two individual Fermi energies (E'_F and E''_F) to occur. Of course, this is not possible. In reality the two band halves shift relative to each other until equilibrium, i.e., a common Fermi energy is reached (Fig. 16.1(c)).

Now, we recall from Chapter 6 that the electron distribution within a band is not uniform. We rather observe a parabolic distribution of energy states, as shown in Fig. 6.4. Thus, we refine our treatment by replacing Fig. 16.1(c) by Fig. 16.2, which depicts the density of states of the two half-bands. We observe a relatively large $Z(E)$ near E_F. Thus, a small change in energy (provided by the external magnetic field) may cause a large number of electrons to switch to the opposite spin direction.

We calculate now the susceptibility from this change in energy ΔE. It is evident that ΔE is larger, the larger the external magnetic field strength $|\mathbf{H}|$, and the larger the magnetic moment of the spinning electrons $|\boldsymbol{\mu}_{ms}|$, i.e.,

$$\Delta E = H\mu_{ms}. \tag{16.1}$$

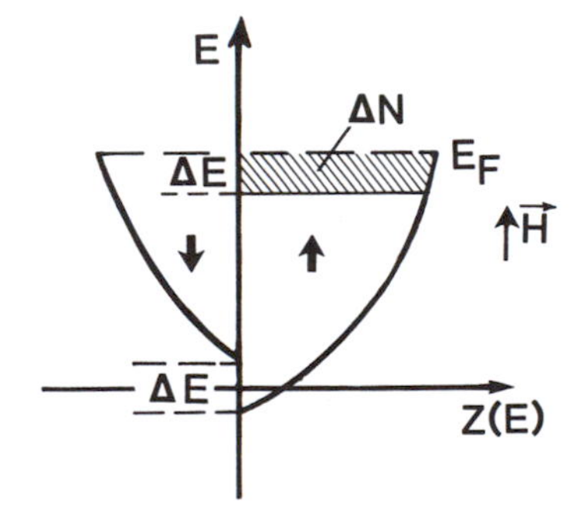

Figure 16.2. Schematic representation of the density of states $Z(E)$ in two half-bands. The shift of the two half-bands occurs as a result of an external magnetic field. Free electron case. (See also Fig. 16.1(c).) The area ΔN equals $\Delta E \cdot Z(E)$.

As mentioned already, the number of electrons ΔN transferred from the spin down into the spin up direction depends on the density of states at the Fermi energy, $Z(E_F)$, and the energy difference ΔE (Fig. 16.2), i.e.,

$$\Delta N = \Delta E Z(E_F) = H\mu_{ms} Z(E_F). \tag{16.2}$$

The magnetization $|\mathbf{M}|$ of a solid, caused by an external magnetic field is, according to (14.8),

$$M = \frac{\mu_m}{V}. \tag{16.3}$$

The magnetization is, of course, larger, the more electrons are transferred from spin down into spin up states. We thus obtain, for the present case,

$$M = \frac{\mu_{ms}}{V} \Delta N = \frac{\mu_{ms}^2 H Z(E_F)}{V}, \tag{16.4}$$

which yields for the susceptibility

$$\chi = \frac{M}{H} = \frac{\mu_{ms}^2 Z(E_F)}{V}. \tag{16.5}$$

The spin magnetic moment of one electron equals one Bohr magneton μ_B (see below). Thus, (16.5) finally becomes

$$\boxed{\chi_{para}^{spin} = \frac{\mu_B^2 Z(E_F)}{V}.} \tag{16.6}$$

The susceptibilities for paramagnetic metals calculated with this equation agree fairly well with those listed in Table 14.1 (see Problem 1). Thus, (16.6) substantiates, in essence, that only the electrons close to the Fermi energy are capable of realigning in the magnetic field direction. If we postulate instead that *all* valence electrons contribute to χ_{para} we would wrongfully calculate a susceptibility which is two or even three orders of magnitude larger than that obtained by (16.6).

It is important to realize that the ever-present diamagnetism makes a sizable contribution to the overall susceptibility, so that χ for metals might be positive or negative depending on which of the two components predominates. This will be elucidated now in a few examples.

To begin with, we discuss beryllium which is a bivalent metal having a filled 2*s*-shell in its atomic state (see Appendix 3). However, in the crystalline state, we observe band overlapping (see Chapter 6) which causes some of the 2*s*-electrons to spill over into the 2*p*-band. They populate the very bottom of this band (see Fig. 16.3). Thus, the density of states at the Fermi level, and consequently, χ_{para}, is very small. In effect, the diamagnetic susceptibility predominates which makes Be diamagnetic.

In order to understand why copper is diamagnetic, we need to remember

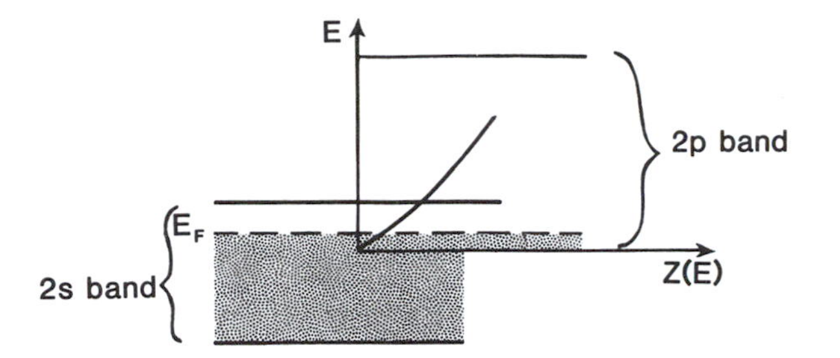

Figure 16.3. Overlapping of $2s$- and $2p$-bands in Be and the density of states curve for the $2p$-band.

that for this metal the Fermi energy is close to the band edge (Fig. 5.22). Thus, the density of states near E_F and the paramagnetic susceptibility (16.6) are relatively small. Furthermore, we have to recall that the diamagnetic susceptibility (15.17)

$$\boxed{\chi_{\text{dia}} = -\frac{e^2 Z \bar{r}^2}{6mV}} \tag{16.7}$$

is proportional to the square of the average orbit radius $\bar{r}$, and proportional to the total number of electrons, Z, in a given orbit. Copper has about ten $3d$-electrons which makes $Z \approx 10$. Further, the radius of d-shells is fairly large. Thus, for copper, χ_{dia} is large because of two contributions. The diamagnetic contribution predominates over the paramagnetic one. As a result, copper is diamagnetic. The same is true for silver and gold and the elements which follow copper in the periodic chart such as zinc and gallium.

Intrinsic semiconductors which have filled valence bands and whose density of states at the top of the valence band is zero (Fig. 6.6) have, according to (16.6), no paramagnetic susceptibility and are therefore diamagnetic. However, a small paramagnetic contribution might be expected for highly doped *extrinsic* semiconductors which have, at high enough temperatures, a considerable amount of electrons in the conduction band (see Chapter 8).

We turn now to the temperature dependence of the susceptibility of metals. The relevant terms in both (16.6) as well as (16.7) do not vary much with temperature. Thus, it is conceivable that the susceptibility of diamagnetic metals is not temperature-dependent, and that the susceptibility of paramagnetic metals often does *not* obey the Curie–Weiss law. In fact, the temperature dependence of the susceptibility for different paramagnetic *metals* has been observed to decrease, to increase, or to remain essentially constant (Fig. 16.4). However, nickel (above T_C) and rare earth metals obey the Curie–Weiss law reasonably well.

At the end of this section we remind the reader that in dilute gases (and also in rare earth metals and their salts) a second component contributes to

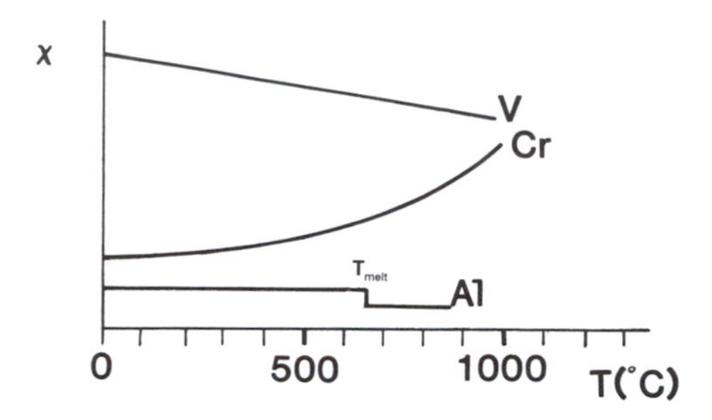

Figure 16.4. Temperature dependence of the paramagnetic susceptibility for vanadium, chromium, and aluminum in arbitrary units. From Landolt–Börnstein, *Zahlenwerte der Physik*, 6th ed., Vol. II/9, Springer-Verlag, Berlin (1962).

paramagnetism. It stems from a magnetic moment which is caused by the angular momentum of the orbiting electrons. We mentioned already in Section 15.1 that this contribution is said to be "quenched" (nonexistent) in most solids.

Finally, we want to find a numerical value for the magnetic moment of an orbiting electron from a quantum-mechanical point of view. We recall from (15.5):

$$\mu_m = \frac{evr}{2}. \tag{16.8}$$

Now, quantum theory postulates that the angular momentum mvr of an electron is not continuously variable but that it rather changes in discrete amounts of integer multiples of $\hbar$ only, i.e.,

$$mvr = n\hbar = \frac{nh}{2\pi}. \tag{16.9}$$

If one combines (16.8) with (16.9) one obtains

$$\mu_m = \frac{enh}{4\pi m}. \tag{16.10}$$

Using $n = 1$ for the first electron orbit (ground state) yields, for the magnetic moment of an orbiting electron,

$$\mu_m = \frac{eh}{4\pi m}. \tag{16.11}$$

It was found experimentally and theoretically that the magnetic moment of an electron due to orbital motion, as well as the magnetic moment of the spinning electron, are identical. This smallest unit of the magnetic moment is given by (16.11) and is called the Bohr magneton

$$\mu_B = \frac{eh}{4\pi m} = 0.927 \times 10^{-20} \text{ (erg/Oe)} \tag{16.12}$$

which we already introduced without further explanation in (15.3).

16.2. Ferromagnetism and Antiferromagnetism

The ferromagnetic metals, iron, cobalt, and nickel are characterized by unfilled d-bands (see Appendix 3). These d-bands overlap the next higher s-band in a similar manner as shown in the band structure of Fig. 5.22. The density of states for a d-band is relatively large because of its potential to accommodate up to ten electrons. This is schematically shown in Fig. 16.5, along with the Fermi energies for iron, cobalt, nickel, and copper. Since the density of states for, say, nickel is comparatively large at the Fermi energy, one needs only a relatively small amount of energy to transfer a considerable amount of electrons from spin down into spin up configurations, i.e., from one half-band into the other. We have already discussed in the previous section this transfer of electrons under the influence of an external magnetic field (Fig. 16.1). Now, there is an important difference between paramagnetics and ferromagnetics. In the former case, an external energy (i.e., the magnetic field) is needed to accomplish the flip in spin alignment, whereas for ferromagnetic materials the parallel alignment of spins occurs spontaneously in small domains of about 1–100 μm diameter. Any theory of ferromagnetism must be capable of satisfactorily explaining the origin of this energy which transfers electrons into a higher energy state.

The energy in question was found to be the *exchange energy*. It is "set free" when equal atomic systems are closely coupled, and in this way exchange their energy. This needs some further explanation.

We digress for a moment and compare two ferromagnetic atoms with two identical pendula which are interconnected by a spring. (The spring represents the interactions of the electrical and magnetic fields.) If one of the pendula is deflected, its amplitude slowly decreases until all energy has been

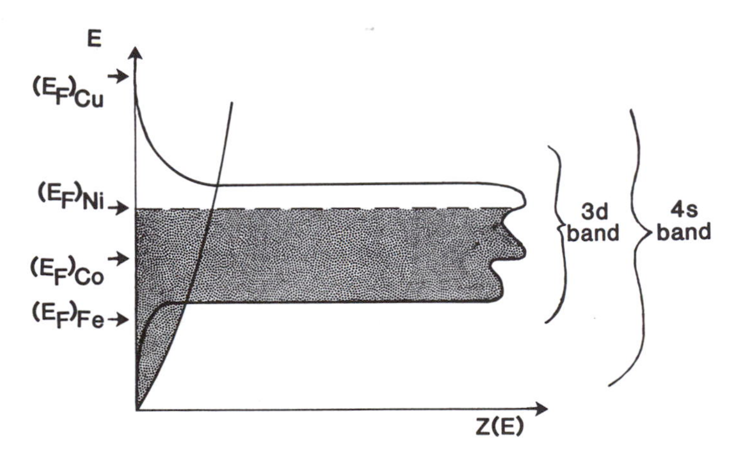

Figure 16.5. Schematic representation of the density of states for $4s$- and $3d$-bands and the Fermi energies for iron, cobalt, nickel, and copper. The population of the bands by the ten nickel ($3d + 4s$)-electrons is indicated by the shaded area.

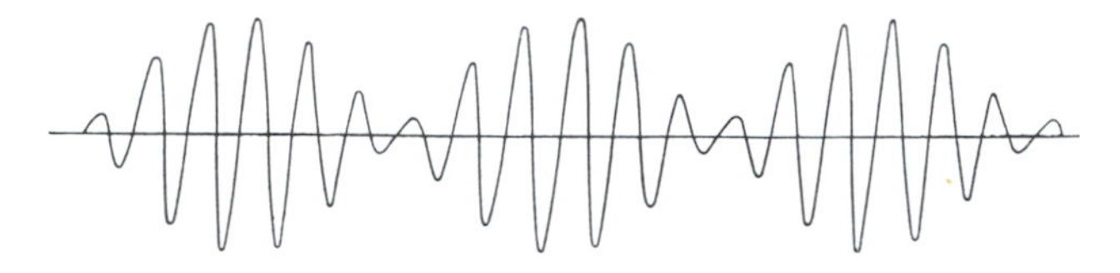

Figure 16.6. Amplitude modulation resulting from the coupling of two pendula. The vibrational pattern shows *beats*, similarly as known from two oscillators which have almost identical pitch. Compare with Fig. 2.1.

transferred to the second pendulum, which then in turn transfers its energy back to the first one and so on. Thus, the amplitudes decrease and increase periodically with time, as shown in Fig. 16.6. The resulting vibrational pattern is similar to that of two violin strings tuned at almost equal pitch. A mathematical expression for this pattern is obtained by adding the equations for two oscillators having similar frequencies ω_1 and ω_2

$$X_1 = b \sin \omega_1 t, \tag{16.13}$$

$$X_2 = b \sin \omega_2 t, \tag{16.14}$$

which yields

$$X_1 + X_2 = X = 2b \cos \frac{\omega_1 - \omega_2}{2} t \cdot \sin \frac{\omega_1 + \omega_2}{2} t. \tag{16.15}$$

Equation (16.15) provides two frequencies, $(\omega_1 - \omega_2)/2$ and $(\omega_1 + \omega_2)/2$, which can be identified in Fig. 16.6. The difference between the resulting frequencies is larger, the stronger the coupling. If the two pendula vibrate in a parallel fashion, the "pull" on the spring, i.e., the restoring force κ, is small. As a consequence, the frequency

$$\nu_0 = \frac{1}{2\pi} \sqrt{\frac{\kappa}{m}} \tag{16.16}$$

(see Appendix 1) is likewise small and is smaller than for independent vibrations. (On the other hand, antisymmetric vibrations cause large values of κ and ν_0.) This classical example demonstrates that two coupled and symmetrically vibrating systems may have a lower energy than two individually vibrating systems would have.

Quantum mechanics treats ferromagnetism in a similar way. The exact calculation involving many atoms is, however, not a trivial task. Thus, one simplifies the problem by solving the appropriate Schrödinger equation for two atoms only. The potential energy in the Schrödinger equation then contains the exchange forces between the nuclei a and b, the forces between two electrons 1 and 2, and the interactions between the nuclei and their neighboring electrons. This simplification seems to be justified because the exchange forces decrease rapidly with distance.

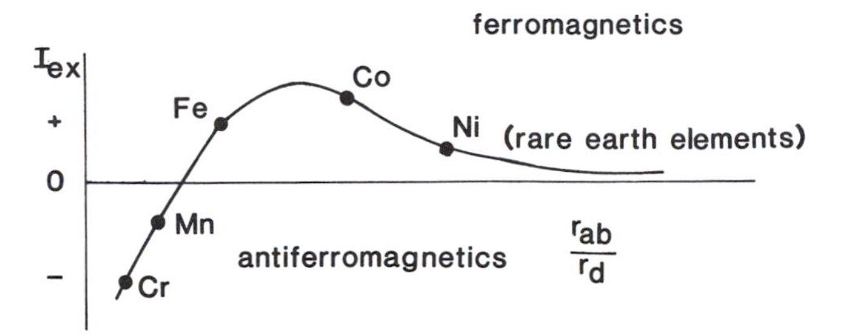

Figure 16.7. Exchange integral I_{ex} versus the ratio of interatomic distance r_{ab} and the radius of an unfilled d-shell. The position of the rare earth elements (which have unfilled f-shells) are also shown for completeness.

The calculation, first performed by Slater and Bethe, leads to an *exchange integral*

$$I_{ex} = \int \psi_a(1)\psi_b(2)\psi_a(2)\psi_b(1)\left[\frac{1}{r_{ab}} - \frac{1}{r_{a2}} - \frac{1}{r_{b1}} + \frac{1}{r_{12}}\right] d\tau. \qquad (16.17)$$

A positive value for I_{ex} means that parallel spins are energetically more favorable than antiparallel spins (and vice versa). We see immediately from (16.17) that I_{ex} becomes positive for a small distance r_{12} between the electrons i.e., a small radius of the d-orbit, r_d. Similarly, I_{ex} becomes positive for a large distance between the nuclei and neighboring electrons r_{a2} and r_{b1}.

I_{ex} is plotted in Fig. 16.7 versus the ratio r_{ab}/r_d. The curve correctly separates the ferromagnetics from manganese, which is not ferromagnetic. Figure 16.7 suggests that if the interatomic distance r_{ab} in manganese is increased (e.g., by inserting nitrogen atoms into the manganese lattice), the crystal thus obtained should become ferromagnetic. This is indeed observed. The ferromagnetic alloys named after Heusler, such as Cu_2MnAl or Cu_2MnSn, are particularly interesting in this context because they contain constituents which are not ferromagnetic, but all contain manganese.

The Bethe–Slater curve (Fig. 16.7) suggests that cobalt should have the highest, and nickel (and the rare earth elements) the lowest, Curie temperature among the ferromagnetics because of the magnitude of their I_{ex} values. This is indeed observed (Table 15.1). Overall, quantum theory is capable of explaining some ferromagnetic properties which cannot be understood with classical electromagnetic theory.

We turn now to a discussion on the number of Bohr magnetons in ferromagnetic metals as listed in Table 16.1. Let us consider nickel as an example and reinspect, in this context, Fig. 16.5. We notice that because of band overlapping the combined ten ($3d$ + $4s$)-electrons occupy the lower s-band and fill, almost completely, the $3d$-band. It thus comes as no surprise that nickel behaves experimentally as if the $3d$-band is filled by 9.4 electrons. To estimate μ_B we need to apply Hund's rule (Fig. 15.4), which states that the electrons in a solid occupy the available electron states in a manner which

Table 16.1. Magnetic Moment μ_m at 0 K for Ferromagnetic Metals.

Metal	μ_m
Fe	2.22 μ_B
Co	1.72 μ_B
Ni	0.60 μ_B
Gd	7.12 μ_B

maximizes the imbalance of spin moments. For the present case, this rule would suggest five electrons with, say, spin up, and an average of 4.4 electrons with spin down, i.e., we obtain a spin imbalance of 0.6 spin moments or 0.6 Bohr magnetons per atom (see Table 16.1). The average number of Bohr magnetons may also be calculated from experimental values of the saturation magnetization M_{s0} (see Table 15.1). Similar considerations can be made for the remaining ferromagnetics as listed in Table 16.1.

We now proceed one step further and discuss the magnetic behavior of certain nickel-based alloys. We use nickel–copper alloys as an example. Copper has one valence electron more than nickel. If copper is alloyed to nickel the extra copper electrons progressively fill the *d*-band and therefore compensate some of the unsaturated spins of nickel. Thus, the magnetic moment per atom of this alloy (and also its Curie temperature) is reduced. Nickel lacks about 0.6 electrons per atom for complete spin saturation because the 3*d*-band of nickel is filled by only 9.4 electrons (see above). Thus, about 60% copper atoms are needed until the magnetic moment (and μ_B) of nickel has reached a zero value (Fig. 16.8). Nickel–copper alloys, having a copper concentration of more than about 60% are consequently no longer

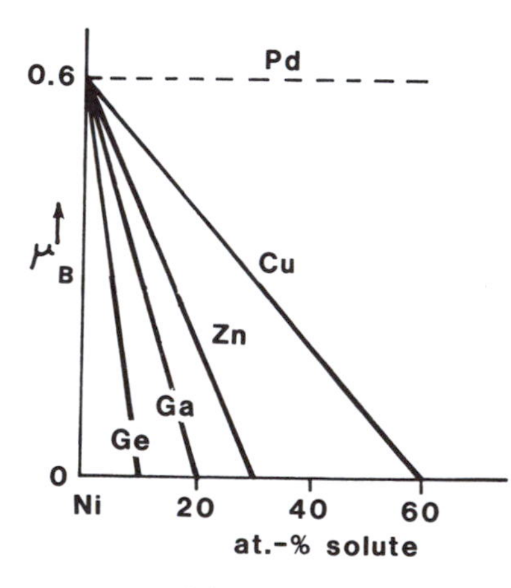

Figure 16.8. Magnetic moment per nickel atom as a function of solute concentration.

ferromagnetic; one would expect them to be diamagnetic. (In reality, however, they are strongly paramagnetic, probably owing to small traces of undissolved nickel.)

Zinc contributes about two extra valence electrons to the electron gas when alloyed to nickel. Thus, we expect a zero magnetic moment at about 30 at.% Zn, etc. Palladium, on the other hand, has the same number of valence electrons as nickel and thus does not change the magnetic moment of the nickel atoms when alloyed to nickel. The total magnetization (14.8) of the alloy is, of course, diluted by the nonferromagnetic palladium. The same is also true for the other alloys.

We conclude our discussion by adding a few interesting details. The rare earth elements are weakly ferromagnetic. They are characterized by unfilled *f*-shells. Thus, their electronic structure and their density of states have several features in common with iron, cobalt and nickel. They have a positive I_{ex} (see Fig. 16.7).

Copper has one more valence electron than nickel, which locates its Fermi energy slightly above the *d*-band (Fig. 16.5). Thus, the condition for ferromagnetism, i.e., an unfilled *d*- or *f*-band is not fulfilled for copper. The same is true for the following elements such as zinc or gallium.

We noted already that manganese is characterized by a negative value of the exchange integral. The distance between the manganese atoms is so small that their electron spins assume an antiparallel alignment. Thus, manganese and many manganese compounds are antiferromagnetic (see Fig. 15.10). Chromium has also a negative I_{ex} and thus is likewise antiferromagnetic (see Table 15.2).

Problems

1. The density of states near the Fermi surface of 1 cm^3 of a paramagnetic metal at $T = 0$ K is approximately 5×10^{41} energy states per Joule. Calculate the volume susceptibility. Compare your value with those of Table 14.1. What metal could this value represent? Explain possible discrepancies between experiment and calculation.

2. Derive (16.15) by adding (16.13) and (16.14).

3. Compare the experimental saturation magnetization M_{s0} (Table 15.1) with the magnetic moment μ_m at 0 K for ferromagnetic metals (Table 16.1). What do you notice? Estimate the degree of *d*-band filling for iron and cobalt.

4. From the results obtained in Problem 3 above, calculate the number of Bohr magnetons for crystalline (solid) iron and cobalt and compare your results with those listed in Table 16.1. What is the number of Bohr magnetons for an iron *atom* and a cobalt atom? What is μ_B for iron and cobalt ferrite?

5. Refer to Figure 16.1(b). Why are two different Fermi energies not possible within the same metal?

CHAPTER 17

Applications

17.1. Introduction

The production of ferro- and ferrimagnetic materials is a large-scale operation, measured in quantity as well as in currency. (This is in contrast to the products of the computer industry, where the price of the *material* which goes into a chip is a minute fraction of the device fabrication cost.) As an example, the annual sales of so-called electrical steel, used for electromotors and similar devices, reach the millions of tons and their market values are in the hundreds of millions of dollars. Other large-scale production items are permanent magnets for loudspeakers, etc., or magnetic recording materials. The following sections will give some impression about the technology (i.e., mostly materials science) which has been developed to improve the properties of magnetic materials.

17.2. Electrical Steels (Soft Magnetic Materials)

Electrical steel is used to multiply the magnetic flux in the cores of electromagnetic coils. These materials are therefore widely incorporated in many electrical machines in daily use. Among their applications are cores of transformers, electromotors, generators, or electromagnets.

In order to make these devices most energy efficient and economical, one needs to find magnetic materials which have the highest possible permeability (at the lowest possible price). Furthermore, magnetic core materials should be capable of being easily magnetized or demagnetized. In other words, the area within the hysteresis loop (or the coercive force, H_c) should be as small as possible (Fig. 15.6). We remember that materials whose hysteresis loops are narrow are called *soft* magnetic materials.

Electrical steels are classified by some of their properties, for example, by the amount of their core losses, by their composition, by their susceptibility, and whether or not they are *grain oriented*. We shall discuss these different properties in detail.

The energy losses which are encountered in electromotors (efficiency between 50% and 90%) or transformers (efficiency 95–99.5%) are estimated to be, in the United States, as high as 3×10^{10} kWh per year, which is equivalent to the energy consumption by about 3 million households, and which wastes about $\$2 \times 10^9$ per year. If by means of improved design of the magnetic cores, the energy losses would be reduced by only 5%, one could save about $\$10^8$ per year and several electric power stations. Thus, there is a clear incentive for improving the properties of magnetic materials.

17.2.1. Core Losses

The core loss is the energy which is dissipated in the form of heat within the core of electromagnetic devices when the core is subjected to an alternating magnetic field. Several types of losses are known, among which the eddy current loss and the hysteresis loss contribute the most. Typical core losses are between 0.3 and 3 watts per kilogram of core material (Table 17.1).

Let us first discuss the eddy current. Consider a transformer whose primary and secondary coils are wound around the legs of a rectangular iron yoke (Fig. 17.1(a)). An alternating electric current in the primary coil causes an alternating magnetic flux in the core which, in turn, induces in the secondary coil an alternating electromotive force V_e proportional to $d\phi/dt$, see (14.6) and (15.9),

$$V_e \propto -\frac{d\phi}{dt} = -A\frac{dB}{dt}. \tag{17.1}$$

Concurrently, an alternating emf is induced within the core itself, as shown in Fig. 17.1(a). This emf gives rise to the eddy current I_e. The eddy current is larger, the larger the permeability μ (because $B = \mu \cdot H$), the larger the conductivity σ of the core material, the higher the applied frequency, and the larger the cross-sectional area A of the core. (A is perpendicular to the magnetic flux ϕ, see Fig. 17.1(a).) We note in passing that, particularly at high frequencies, the eddy current shields the interior of the core from the magnetic field, so that only a thin exterior layer of the core contributes to the flux multiplication (skin effect).

In order to decrease the eddy current, several remedies are possible. First, the core can be made of an insulator in order to decrease σ. Ferrites are thus effective but also expensive materials to build magnetic cores (see Section 15.1.5). They are indeed used for high frequency applications. Second, the core

Table 17.1. Properties of Some Soft Magnetic Materials.

Name	Composition (mass %)	Permeability μ_{max} (unitless)	Coercivity H_c		Saturation induction[a] B_s		Resistivity $\rho(\mu\Omega \cdot cm)$	Core loss at 15 kG and 60 Hz (W/kg)
			(Oe)	(A/m)	(kG)	(T)		
Low carbon steel	Fe–0.05% C	5×10^3	1.0	80	21.5	2.1	10	2.8
Nonoriented silicon iron	Fe–3% Si, 0.005% C, 0.15% Mn	7×10^3	0.5	40	19.7	2	60	0.9
Grain-oriented silicon iron	Fe–3% Si, 0.003% C, 0.07% Mn	4×10^4	0.1	8	20	2	47	0.3
78 Permalloy	Ni–22% Fe	10^5	0.05	4	10.8	1.1	16	≈2
Mumetal	77% Ni; 16% Fe, 5% Cu, 2% Cr	10^5	0.05	4	6.5	0.6	62	
Supermalloy	79% Ni; 16% Fe, 5% Mo	10^6	0.002	0.1	7.9	0.8	60	
Supermendur	49% Fe, 49% Co, 2% V	6×10^4	0.2	16	24	2.4	27	
Metglas #2605 annealed	$Fe_{80}B_{20}$	3×10^5	0.04	3.2	15	1.5	≈200	0.3

[a] Above B_s the magnetization is constant and dB/dH is unity.

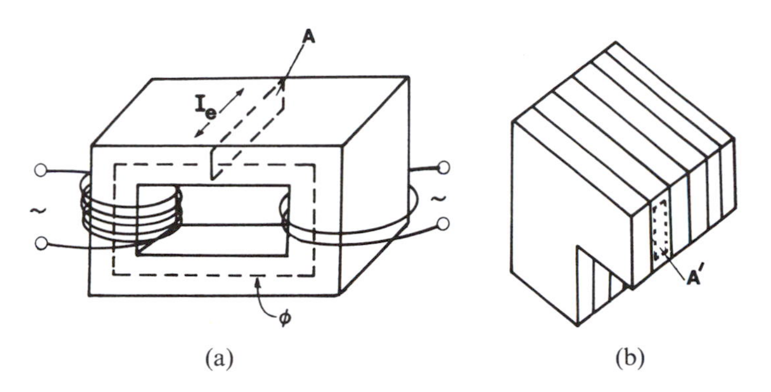

Figure 17.1. (a) Solid transformer core with eddy current I_e in a cross-sectional area A. Note the magnetic flux lines ϕ. (b) Cross section of a laminated transformer core. The area A' is smaller than area A in (a).

can be manufactured from pressed iron powder whereby each particle (which is about 50–100 μm in diameter) is covered by an insulating coating. However, the decrease in σ, in this case, is at the expense of a large decrease in μ. Third, the most widely applied method to reduce eddy currents is the utilization of cores made out of thin sheets which are electrically insulated from each other (Fig. 17.1(b)). This way the cross-sectional area A is reduced, which in turn decreases V_e (17.1), and additionally reduces losses due to the skin effect. Despite the lamination, a residual eddy current loss still exists which is caused by current losses within the individual laminations, and due to interlaminar losses that may arise if laminations are not sufficiently insulated from each other. These losses are, however, less than 1% of the total energy transferred.

Hysteresis losses are encountered when the magnetic core is subjected to a complete hysteresis cycle (Fig. 15.6). The work thus dissipated into heat is proportional to the area enclosed by a B/H loop. Proper materials selection and rolling of the materials with subsequent heat treatment greatly reduces the area of a hysteresis loop (see below).

17.2.2. Grain Orientation

The permeability of electrical steel can be substantially increased and the hysteresis losses can be decreased by making use of favorable grain orientations in the material. This needs some explanation. The magnetic properties of crystalline ferromagnetic materials depend on the crystallographic direction in which an external field is applied, an effect which is called *magnetic anisotropy*. Let us use iron as an example. Figure 17.2(a) shows magnetization curves of single crystals for three crystallographic directions. We observe that if the external field is applied in the $\langle 100 \rangle$ direction, saturation is achieved with the smallest possible field strength. The $\langle 100 \rangle$ direction is thus called the

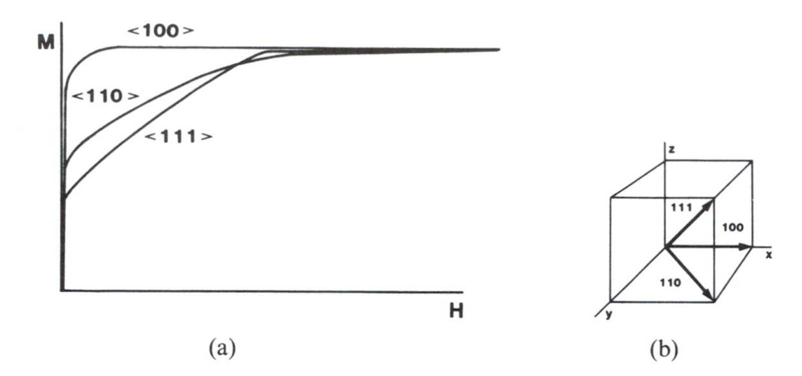

Figure 17.2. (a) Schematic magnetization curves for rod-shaped *iron single crystals* having different orientations (virgin curves). The magnetic field was applied in three different crystallographic directions. (Compare with Fig. 15.6 which refers to polycrystalline material). (b) Reminder of the Miller indices.

"easy direction." (In nickel, on the other hand, the $\langle 111 \rangle$ direction is the easy direction and $\langle 100 \rangle$ is the hard direction.)

This experimental finding gives us, incidentally, some clues about the spontaneous orientation of the spin magnetic moments in the demagnetized state. They are aligned in the easy direction. As an example, in virgin iron the spins are aligned along the $\langle 100 \rangle$ directions. Now, suppose that an external field is applied parallel to an easy direction. Then, the domains already having favorable alignment grow without effort at the expense of other domains until eventually the crystal contains one single domain (Fig. 15.9). The energy which is consumed during this process (which is proportional to the area within the hysteresis loop) is used to move the domain walls through the crystal.

A second piece of information needs to be considered too. Metal sheets, which have been manufactured by rolling and heating, often possess a "*texture*," i.e., they have a preferred orientation of the grains. It just happens that in α-iron and α-iron alloys the $\langle 100 \rangle$ direction is parallel to the rolling direction. This property is exploited when utilizing electrical steel.

Grain-oriented electrical steel is produced by initially hot-rolling the alloy followed by two stages of cold reduction with intervening anneals. During the rolling, the grains are elongated and their orientation is altered as described above. Finally, the sheets are recrystallized whereby some crystals grow in size at the expense of others (occupying the entire sheet thickness).

In summary, the magnetic properties of grain-oriented steels are best in the direction parallel to the direction of rolling. Electrical machines having core material of grain-oriented steel need less iron and are therefore smaller; the price increase due to the more elaborate fabrication procedure is often compensated by the savings in material. For details, see Table 17.1.

17.2.3. Composition of Core Materials

The least expensive core material is commercial low carbon steel (0.05% C). It possesses a relatively small permeability and has about ten times higher core losses than grain-oriented silicon iron (Table 17.1). Low carbon steel is used where low cost is more important than the efficient operation of a device. Purification of iron increases the permeability but also increases conductivity (eddy current!) and price.

Iron–silicon alloys containing between 1.4 and 3.5% Si and very little carbon have a higher permeability and a lower conductivity than low carbon steel (see Table 17.1). Furthermore, because of special features in the phase diagram ("γ-loop"), heat treatments of these alloys can be performed at much higher temperatures without interference from phase changes during cooling. The core losses decrease with increasing silicon content. However, for silicon concentrations above 4 or 5 weight % the material becomes too brittle to allow rolling. Grain orientation in iron–silicon alloys (see above) further increases the permeability and decreases the hysteresis losses. Other constituents in iron–silicon alloys are aluminum and manganese in amounts less than 1%. They are added mainly for metallurgical reasons because of their favorable influence on the grain structure and their tendency to reduce hysteresis losses. Grain-oriented silicon "steel" is the favored commercial product for highly efficient–high flux multiplying core applications.

The highest permeability is achieved for certain multicomponent nickel-based alloys such as Permalloy, Supermalloy, or Mumetal (Table 17.1). The latter can be rolled into thin sheets and is used to shield electronic equipment from stray magnetic fields.

17.2.4. Amorphous Ferromagnetics

The electrical properties of amorphous metals (metallic glasses) and their methods of production have already been discussed in Section 9.4. In the present context, we are interested only in their magnetic properties, in particular, as flux multipliers in transformers, motors, etc. Some amorphous metals (consisting of iron, nickel, or cobalt with boron, silicon, or phosphorus) have, when properly annealed below the crystallization temperature (for strain relaxation), a considerably higher permeability and a lower coercivity than the commonly used grain-oriented silicon–iron, see Table 17.1. Further, the electrical resistivity of amorphous alloys is generally larger than their crystalline counterparts, which results in smaller eddy-current losses. However, amorphous ferromagnetics possess a somewhat lower saturation induction (Table 17.1) (which sharply decreases even further at elevated temperatures) and their core losses increase rapidly at higher flux densities (e.g., above 1.4 T). Thus, the application of metallic glasses for flux multiplication purposes is, at the present, limited to devices with small flux densities, i.e., low currents, such as for transformers (e.g., for communication equipment), magnetic sensors, or magnetostrictive transducers.

17.3. Permanent Magnets (Hard Magnetic Materials)

Permanent magnets are devices which retain their magnetic field indefinitely. They are characterized by a large remanence B_r (or M_r), a relatively large coercivity H_c, and a large area within the hysteresis loop. They are called *hard magnetic materials* (see Section 15.1.3).

The best means to visualize the properties of permanent magnets is to inspect their *demagnetization curve* (Fig. 17.3) which is a part of a hysteresis loop, as shown in Fig. 15.6. Another parameter which is used to characterize hard magnetic materials is the *maximum energy product*, $(BH)_{max}$, which is related to the area within the hysteresis loop. We see immediately from Fig. 17.3 that B times H is zero at the intercepts of the demagnetization curve with the coordinate axes, and that the energy product peaks somewhere between these extreme values depending on the shape and size of the hysteresis curve. The values of B_r, H_c, and $(BH)_{max}$ for some materials which are used as permanent magnets are listed in Table 17.2.

The remanence B_r shown in Fig. 17.3 or listed in Table 17.2 is the maximal residual induction which can be obtained in a circular, close-loop magnet inserted in a coil. However, all permanent magnets need to have exposed poles in order to be useful. The necessary *air gap* between the north and south poles reduces the remanence, because the exposed poles create a *demagnetizing field* H_d which acts in the opposite direction to the B lines. We understand intuitively that the demagnetizing field depends on the shape, size, and gap

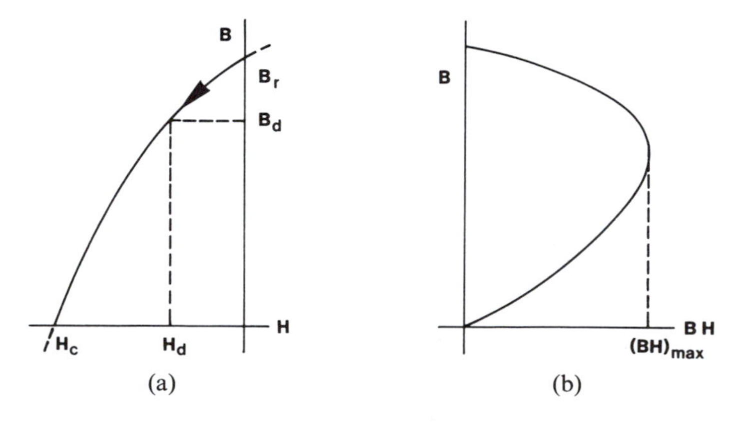

Figure 17.3. (a) Demagnetization curve for a ferromagnetic material. (Second quadrant in a B–H diagram.) (b) Energy product BH as a function of induction B.

Table 17.2. Properties of Materials Used for Permanent Magnets.

Material	Composition (mass %)	Remanence B_r (kG)	Remanence B_r (T)	Coercivity H_c (Oe)	Coercivity H_c (A/m)	Maximum energy product $(BH)_{max}$ per Vol. (MGOe)	Maximum energy product $(BH)_{max}$ per Vol. (kJ/m^3)
Steel	Fe–1% C	9	0.9	51	4×10^3	0.2	1.6
36 Co steel	36 Co, 3.75 W, 5.75 Cr, 0.8 C	9.6	0.96	228	1.8×10^4	0.93	7.4
Alnico 2	12 Al, 26 Ni, 3 Cu, 63 Fe	7	0.7	650	5.2×10^4	1.7	13
Alnico 5	8 Al, 15 Ni, 24 Co, 3 Cu, 50 Fe	12	1.2	720	5.7×10^4	5.0	40
Alnico 5 DG	same as above	13.1	1.3	700	5.6×10^4	6.5	52
Ba-ferrite (Ceramic 5)	$BaO \cdot 6\,Fe_2O_3$	3.95	0.4	2,400	1.9×10^5	3.5	28
PtCo	77 Pt, 24 Co	6.45	0.6	4,300	3.4×10^5	9.5	76
Remalloy	12 Co, 17 Mo, 71 Fe	10	1	230	1.8×10^4	1.1	8.7
Vicalloy 2	13 V, 52 Co, 35 Fe	10	1	450	3.6×10^4	3.0	24
Cobalt–Samarium	Co_5Sm	9	0.9	8,700	6.9×10^5	20	159
Iron–Neodymium–Boron	$Fe_{14}Nd_2B_1$	13	1.3	14,000	1.1×10^6	40	318

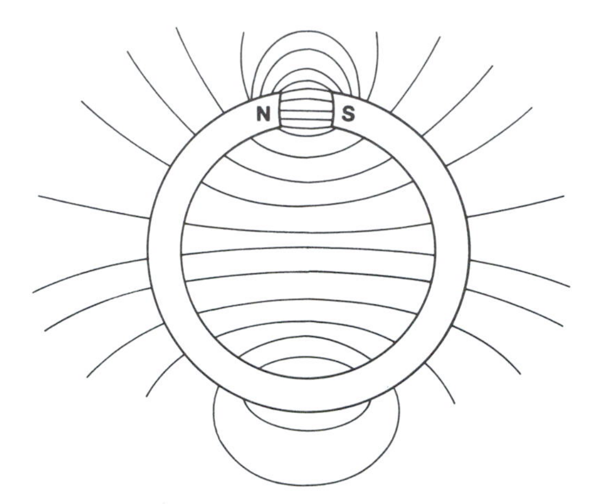

Figure 17.4. Fringing and leakage of a permanent magnet.

length of a magnet. Thus, a reduced value for the residual induction, termed B_d, is obtained as shown in Fig. 17.3. Another effect which reduces the useful magnetic field is *fringing* near the air gap and *leakage* from the sides of a magnet (Fig. 17.4).

We now turn to the properties of some common hard magnetic materials. Today, many permanent magnets are made of Alnico alloys which contain various amounts of aluminum, nickel, cobalt, and iron, along with some minor constituents such as copper and titanium (Table 17.2). Their properties are improved by heat treatments (homogenization at 1250° C, fast cooling, and tempering at 600° C, Alnico 2). Further improvement is accomplished by cooling the alloys in a magnetic field (Alnico 5). The best properties are achieved when the grains are made to have a preferred orientation. This is obtained by cooling the bottom of the crucible after melting, thus forming long columnar grains with a preferred $\langle 100 \rangle$ axis in the direction of heat flow. A magnetic field parallel to the $\langle 100 \rangle$ axis yields Alnico 5-DG (directional grain).

The superior properties of heat-treated Alnico stems from the fact that during cooling and tempering of these alloys, rod-shaped iron and cobalt-rich α-precipitates are formed which are parallel to the $\langle 100 \rangle$ directions (*shape anisotropy*). These strongly magnetic precipitates are single domain particles and are imbedded in a weakly magnetic nickel and aluminum matrix (α). Alnico alloys are mechanically hard and brittle and can, therefore, only be shaped by casting or by pressing and sintering of metal powders.

The newest hard magnetic materials are made of neodymium–boron–iron, see Table 17.2. They possess a superior coercivity and thus a larger $(BH)_{max}$. The disadvantage is a relatively low Curie temperature of about 300° C.

Ceramic ferrite magnets, such as barium or strontium ferrite ($BaO \cdot 6Fe_2O_3$ or $SrO \cdot 6Fe_2O_3$), are brittle and relatively inexpensive. They crystallize in the form of plates with the hexagonal c-axis (which is the easy axis) perpendicular to the plates. Some preferred orientation is observed because the flat plates arrange parallel to each other during pressing and sintering. Ferrite powder is often imbedded in plastic materials which yields flexible magnets. They are used, for example, in the gaskets of refrigerator doors.

High carbon steel magnets with or without cobalt, tungsten, or chromium are only of historic interest. Their properties are inferior to other magnets. It is believed that the permanent magnetization of quenched steel stems from the martensite-induced internal stress which impedes the domain walls from moving through the crystal.

Research on permanent magnetic materials still proceeds with unbroken intensity. The goal is to improve corrosion resistance, price, remanence, coercivity, magnetic ordering temperature, and processing procedures. A few examples are given here. Carbon and nitrogen are increasingly used as the metalloid in iron/rare earth magnets such as in Fe–Nd–C or in $Fe_{17}Sm_2N_x$. Nitrogen treatment of sintered $Fe_{14}Nd_2B$ raises the Curie temperature by more than 100 K. Nitriding of $F_{17}Sm_2$ (at 400° to 500° C) yields a room temperature coercivity as high as 30 kOe (2.4×10^6 A/m), a remanence of 15.4 kG (1.5 T) and a T_c of 470° C. Corrosion of the Fe–Nd–B sintered magnets is a serious problem. The principal corrosion product is $Nd(OH)_3$. The corrosion resistance can be improved by utilizing intermetallic compounds such as Fe–Nd–Al or Fe–Nd–Ga, or by applying a moisture-impervious coating. Other approaches are the rare-earth-free Co–Zr–B alloys (with or without silicon) which have a Curie temperature around 500° C and a coercivity of 6.7 kOe (5.3×10^5 A/m).

17.4. Magnetic Recording

Magnetic recording tapes, discs, drums, or magnetic strips on credit cards consist of small, needlelike oxide particles about 0.1×0.5 μm in size which are imbedded in a nonmagnetic binder. The particles are too small to sustain a domain wall. They consist therefore of a single magnetic domain which is magnetized to saturation along the major axis (shape anisotropy). The elongated particles are aligned by a field during manufacturing so that their long axes are parallel with the length of the tape or the track. The most popular magnetic material has been ferrimagnetic γ-Fe_2O_3. Its coercivity is 250–350 Oe (20–28 kA/m). More recently, ferromagnetic chromium dioxide has been used having a coercivity between 500 and 1000 Oe (40–80 kA/m) and a particle size of 0.05 μm by 0.4 μm. High coercivity and high remanence prevent self-demagnetization and accidental erasure, they provide strong signals and permit thinner coatings. A high H_c also allows tape duplication by "contact printing." However, CrO_2 has a relatively low Curie temperature (128° C compared to 600° C for γ-Fe_2O_3). Thus, chromium dioxide tapes which are exposed to excessive heat (glove compartment!) may lose their stored information. Lately, most video tapes use cobalt-doped-γ-Fe_2O_3

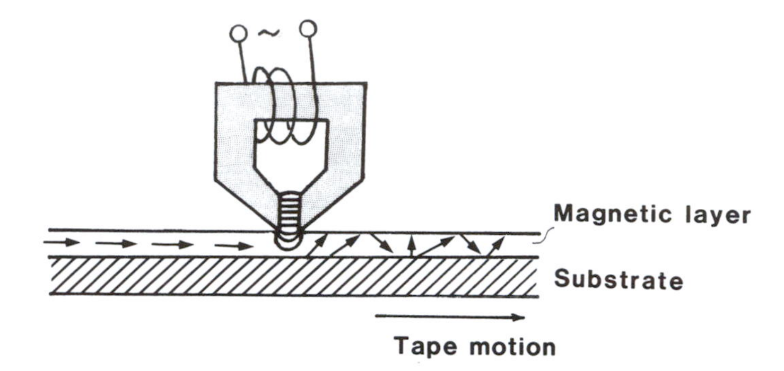

Figure 17.5. Schematic arrangement of a recording (playback) head and a magnetic tape. (Recording mode.) The gap width is exaggerated. The plastic substrate is about 25 μm thick.

which has a somewhat higher Curie temperature than chromium dioxide and a coercivity of 600 Oe (48 kA/m). Most recently, iron particles have been utilized ($H_c = 1500$ Oe, i.e., 120 kA/m). This technology requires, however, a surface coating of tin to prevent coalescence of the individual particles and corrosion.

The recording head of a tape machine consists of a laminated electromagnet made of permalloy or soft ferrite (Table 17.1) which has an air gap about 0.3 μm wide (Fig. 17.5). The tape is passed along this electromagnet whose fringing field redirects the spin moments of the particles in a certain pattern proportional to the current which is applied to the recording head. This leaves a permanent record of the signal. In the playback mode, the moving tape induces an alternating emf in the coil of the same head. The emf is amplified, filtered, and fed to a loudspeaker.

Some modern recording heads utilize conventional ferrites whose gap surfaces are coated with a micrometer-thick metal layer composed of aluminum, iron, and silicon (*Sendust*). This *metal-in-gap* (M-I-G) technology combines the superior high-frequency behavior and good wear properties of ferrites with the higher coercivity of ferromagnetic metals. Thus, fields two or three times as intense as for pure ferrites can be supported. Such high fields are necessary to record efficiently on high density (i.e., on high coercivity) media in which tiny regions of alternating magnetization are closely spaced and should not mutually demagnetize each other.

For ultrahigh recording densities (extremely small bit sizes) the signal strength produced in the heads diminishes considerably. Thus, the latest head technology utilizes a thin *magnetoresistive element*, made out of permalloy, which senses the slight change in resistance (about 2%) that occurs as the angle of magnetization is changed when the magnetized data bits pass beneath the head. 1.8 M bits/mm^2 have been achieved this way. In contrast to an inductive head (see above), whose output voltage is directly proportional to the tape speed, magnetoresistivity is governed by the flux density. This is an advantage for low speed applications (credit cards).

17.5. Magnetic Memories

Ferrite-core memories used to be the dominant devices for random access storage in computers. The principle is simple: a donut-shaped piece of ferrimagnetic material, having a nearly square-shaped hysteresis loop and a low coercivity, is threaded with a wire (Fig. 17.6(a)). If a sufficiently high current pulse is sent through this wire, then the core becomes magnetically saturated. Now, suppose the flux lines point clockwise. An opposite directed current pulse of sufficient strength magnetizes the ferrite core counterclockwise. These two magnetization directions constitute the two possible values (0 and 1) in a binary system (see Section 8.7.12). A toroid-shaped memory core is used because a close-flux structure reacts efficiently to currents from a center wire but is not disturbed by external stray fields.

The actual configuration of a complete memory system consists of a stack of identical memory planes, each of which contains a set of wires in the x- as well as in the y-directions. The toroids are placed at the intersections (Fig. 17.6(c)). In order to switch the, say, X_3/Y_2 core from *zero* to *one* a current proportional to half the saturation field ($H_s/2$) is sent through each of the X_3 and the Y_2 wires (Fig. 17.6(b)). This provides only the X_3/Y_2 core with the necessary field for switching—the other cores stay at their present state. The

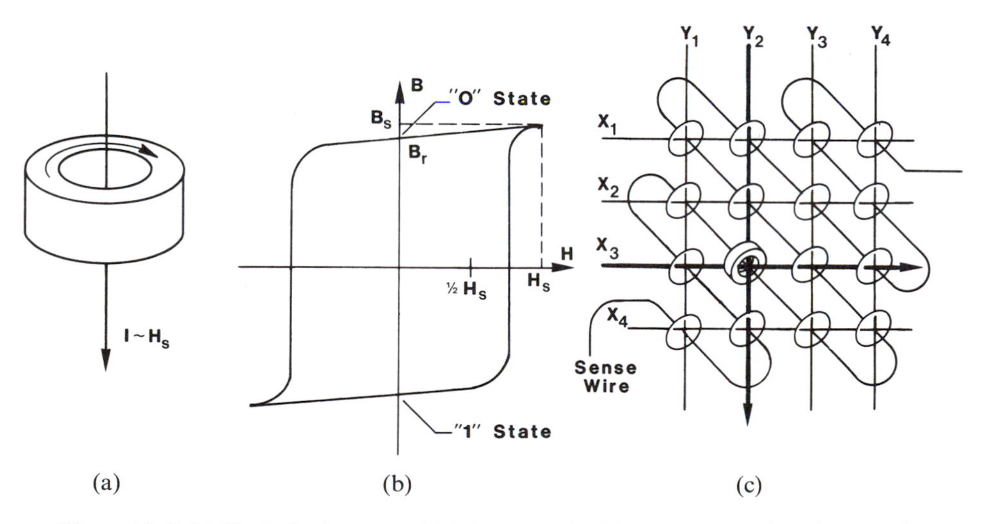

Figure 17.6. (a) Single ferrite core which is magnetized by a current-induced magnetic field; (b) square-shaped hysteresis loop of a soft ferrite memory core; and (c) one plane of a "coincident-current core memory device."

information thus permanently stored can be read by again sending a current pulse proportional to $H_s/2$ through the X_3/Y_2 wires. A third wire, the sensing wire, which passes through all the cores of a given plane, senses whether or not the core was switched during the reading process. Since the reading process destroys the stored information, a special circuit is needed to rewrite the information back into the core. Ferrite-core memories (like other magnetic storage devices) do not need an electrical current to maintain their stored information. The weight/bit-ratio for ferrite-core memories is, however, considerably larger than for electrical or optical storage devices. Thus, their usage is now limited to a few specialized applications.

Another magnetic storage device that has been heavily researched in the past, but is presently not much in use, is the *bubble domain memory*. Here, tiny cylindrical regions (as small as 1 μm in diameter) having a reversed magnetization compared to the matrix, are formed in thin crystals of "canted" antiferromagnetic oxides[3] ($BaFe_{12}O_{19}$, $YFeO_3$), or in amorphous alloyed films (Gd–Co, Gd–Fe), or in ferrimagnetic materials such as yttrium–iron–garnet ($Y_3Fe_5O_{12}$). These bubbles, whose easy axis is perpendicular to the plane of the film, can be generated, moved, replicated, or erased by electric currents. The crystals are transparent to red light. Thus, the domains can be visibly observed and optically read by the way in which they rotate the plane of polarization of polarized light (Faraday effect in transmission, or Kerr effect in reflection). Each such domain constitutes one bit of stored information.

Thin magnetic films consisting of Co–Ni–P or Co–Cr–Ta or Co75–Cr13–Pt12 are frequently used in hard-disk devices. They are laid down on an aluminum substrate and are covered by a 40 nm thick carbon layer for lubrication and corrosion resistance. The coercivities range between 750 and 1500 Oe (60–120 kA/m). Thin film magnetic memories can be easily fabricated (vapor deposition, sputtering, or electroplating), they can be switched rapidly, and they have a small unit size. Thin-film recording media are not used for tapes, however, because of their rapid wear. They have a density of 1.8 Mbits/mm^2 with a track separation of 3 μm and a bit length of 150 nm.

Magneto-optic memories possess the advantage of having no mechanical contact between medium and beam. Thus, no wear is encountered. When certain materials such as MnBi, EuO, amorphous Gd–Co, or GdFe–garnets are heated for a microsecond above the Curie temperature by a narrow and intense laser or electron beam, and then cooled in an external or self-field, a selected region becomes reversely magnetized. The information is then read by making use of the Faraday or Kerr effect. (See, in this context, Section 13.10.) Magneto-optic disks have a one thousand times larger storage density than common floppy disks (600 MBytes compared to 350 kBytes for a 5.25 in. disk) and a ten times faster access time.

Despite this relatively large number of possible magnetic storage devices, semiconductor technology (Section 8.7.12) is presently preferred for short-

[3] See Section 15.1.4.

term information storage, mainly because of price, easy handling capability, fast access time, and size. On the other hand, magnetic disks (for random access) or tapes (mainly for music recordings, etc.) are the choices for long-term, large-scale information storage, particularly since no electric energy is needed to retain the information. It should be noted in closing that tapes and floppy disks make direct contact with the recording (and playback) head, and are therefore subject to wear, whereas hard drive systems utilize a "flying head" which hovers a few micrometers or less above the recording medium on an air cushion, caused by the high speed of the disk. On the other hand, the signal to noise ratio for contact recording is 90 dB whereas for noncontact devices the signal to noise ratio is only 40 dB or lower. Magnetic recording is a $80 billion annual business worldwide with a 12–14% growth rate and is said to be the biggest consumer of high purity materials.

Problems

1. Calculate the energy expended during one full hysteresis cycle of a magnetic material having a rectangular hysteresis loop. Assume $H_c = 0.5$ Oe, $B_s = 20$ kG and $V = 0.25$ cm^3. What units can be used?
2. Which core material should be utilized to supply a large scale and constant magnetic field in a synchrotron? Justify your choice.
3. Pick an actual motor of your choice and find out (by analysis, by means of a data sheet, or by writing to the manufacturer) which type of electrical steel was used as core material. Also, find out the core loss (often given in watts/lb).
4. Find from a manufacturer's data sheet the price of several qualities of electrical steel.
5. Inspect the magnets contained in a gasket of a modern refrigerator. How does the magnet work? Where are the south and north poles?
6. It was said in the text that transformers suffer eddy current as well as hysteresis losses. What other types of losses can be expected in a transformer? How can those losses be reduced?

Suggestions for Further Reading (Part IV)

R.M. Bozorth, *Ferromagnetism*, Van Nostrand, New York (1951).
S. Chikazumi, *Physics of Magnetism*, Wiley, New York (1964).
B.D. Cullity, *Introduction to Magnetic Materials*, Addison-Wesley, Reading, MA (1972).
J.D. Jackson, *Classical Electrodynamics*, Wiley, New York (1962).
D. Jiles, *Magnetism and Magnetic Materials*, Chapman and Hall, London, (1991).
E. Kneller, *Ferromagnetismus*, Springer-Verlag, Berlin (1962).

J.C. Mallionson, *The Foundation of Magnetic Recording*, Academic Press, San Diego (1987).

F.W. Sears, *Electricity and Magnetism*, Addison-Wesley, Reading, MA (1953).

T. Smit and H.P.T. Wijn, *Ferrites*, Wiley, New York (1959).

H.H. Stadelmaier and E.Th. Henig, Permanent magnetic materials—Developments during the last 18 months, *Journal of Metals*, **43** (1991), 32.

R.S. Tebble and D.J. Craik, *Magnetic Materials*, Wiley–Interscience, London (1969).

J.K. Watson, *Applications of Magnetism*, Wiley–Interscience, New York (1980).

H.P.J. Wijn (Editor), Magnetic properties of metals, *d*-elements, alloys, and compounds, in *Data in Science and Technology*, Springer-Verlag, Berlin (1991).

PART V

THERMAL PROPERTIES OF MATERIALS

CHAPTER 18

Introduction

Heat was considered to be an invisible fluid, called *caloric*, until late into the eighteenth century. It was believed that a hot piece of material contained more caloric than a cold one and that an object would become warmer by transferring caloric into it. In the mid-1800s, Mayer, Helmholtz, and Joule discovered independently that heat is simply a form of energy. They realized that when two bodies have different temperatures, thermal energy is transferred from the hotter to the colder one when brought into contact. Count Rumford discovered, by observing the boring of cannons, that mechanical work expended in the boring process was responsible for the increase in temperature. He concluded that mechanical energy could be transformed into thermal energy. This observation lead eventually to the concept of a *mechanical heat equivalent*. Today, these results are treated in a different, more rigorous, scientific language (see next chapter).

The thermal properties of materials are important whenever heating and cooling devices are designed. Thermally induced expansion of materials has to be taken into account in the construction industry as well as in the design of precision instruments. Heat conduction plays a large role in thermal insulation, e.g., in homes, industry, and spacecraft. Some materials such as copper or silver conduct heat very well; other materials, like wood or rubber, are poor heat conductors. Good electrical conductors are generally also good heat conductors. This was discovered in 1853 by Wiedemann and Franz who found that the ratio between heat conductivity and electrical conductivity (divided by the temperature) is essentially constant for all metals.

The thermal conductivity of materials only varies within four orders of magnitude (Fig. 18.1). This is in sharp contrast to the variation in electrical conductivity, which spans about twenty-five orders of magnitude (Fig. 7.1).

The thermal conductivity of metals and alloys can be readily interpreted

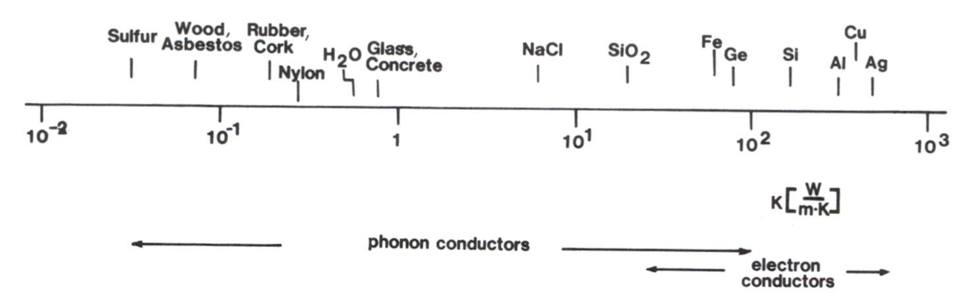

Figure 18.1. Room temperature thermal conductivities for some materials.

by making use of the electron theory which was developed in Part I of this book. The electron theory postulates that free electrons in the hot part of a metal bar pick up energy by interactions with the vibrating lattice atoms. This thermal energy is eventually transmitted to the cold end of the bar by a mechanism which we will treat in Chapter 21.

In electrical insulators, in which no *free* electrons exist, the conduction of thermal energy must occur by a different mechanism. This new mechanism was found by Einstein at the beginning of the century. He postulated the existence of *phonons* or lattice vibration quanta, which are thought to be created in large numbers in the hot part of a solid and partially eliminated in the cold part. Transferral of heat in dielectric solids is thus linked to a flow of phonons from hot to cold.

Figure 18.1 indicates that in a transition region both electrons as well as phonons may contribute to thermal conduction. Actually, phonon-induced thermal conduction occurs even in metals, but its contribution is negligible to that of the electrons.

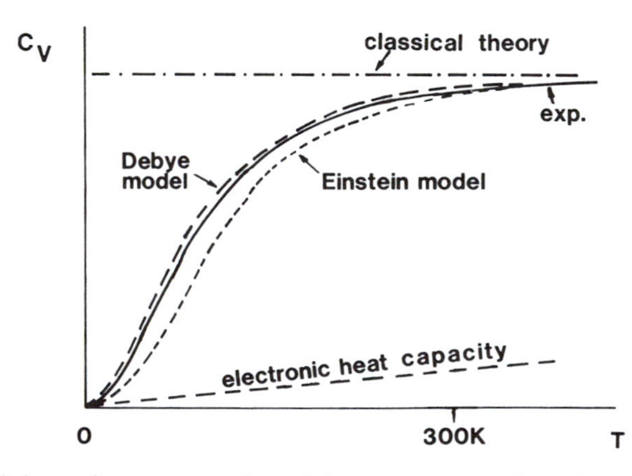

Figure 18.2. Schematic representation of the temperature dependence of the molar heat capacity, experimental and according to four models.

Another thermal property which will receive considerable attention in the following chapters is the specific heat capacity, as well as a related property, the molar heat capacity. Their importance can best be appreciated by the following experimental observations: Two substances with the same mass but different values for the specific heat capacity require different amounts of thermal energy to reach the same temperature. Aluminum, for example, which has a relatively high specific heat capacity needs more thermal energy to reach a given temperature than, say, copper or lead of the same mass.

The molar heat capacity is the product of the specific heat capacity and the molar mass. Its experimentally observed temperature dependence, as shown in Fig. 18.2, has stimulated various theories, among them the phonon model. Figure 18.2 shows schematically how the various theories for the interpretation of the heat capacity compare with the experimental findings. We will discuss these models in the chapters to come.

CHAPTER 19

Fundamentals of Thermal Properties

Before we discuss the atomistic and quantum mechanical theories of the thermal properties of materials, we need to remind the reader of some relevant fundamental concepts and definitions which he might have been exposed to before in courses of physics and thermodynamics.

19.1. Heat, Work, and Energy

When two bodies of different temperatures are brought in contact with each other, heat Q flows from the hotter to the colder substance. Actually, an increase in temperature can be achieved in a number of ways, such as by mechanical work (friction), electrical work (resistive heating), radiation, or by the just-mentioned direct contact with a hotter medium. The change in energy, ΔE, of a "system" can be expressed by the *first law of thermodynamics*

$$\Delta E = W + Q, \tag{19.1}$$

where W is the work done to the system and Q is the heat received by the system from the environment. The focus of this and subsequent chapters is on the thermal properties of materials. Thus, we limit our considerations to processes for which W can be considered to be zero so that

$$\Delta E = Q. \tag{19.1a}$$

Energy, work, and heat have the same unit. The SI unit is the joule (J), which is related to the now obsolete thermomechanical calorie (cal) by

$$1 \text{ cal} = 4.184 \text{ J}, \tag{19.2}$$

i.e.,

$$1 \text{ J} = 0.239 \text{ cal}. \tag{19.2a}$$

A technique which links thermal energy and mechanical energy was proposed by Joule in 1850. The experiment involves rotating paddles which raise the temperature of a given amount of water by means of friction. The paddles are driven by the mechanical work provided by descending weights.

19.2. Heat Capacity, C'

Different substances need different amounts of heat to raise their temperature by a given temperature interval. For example, it takes 4.18 J to raise 1 g of water by 1 K. But the same heat raises the temperature of 1 g of copper by about 11 K. In other words, water has a large heat capacity compared to copper. (The large heat capacity of water is incidentally the reason for the balanced climate in coastal regions and the heating of north European countries by the Gulf Stream.)

The heat capacity C' is the amount of heat dQ which needs to be transferred to a substance in order to raise its temperature by a certain temperature interval. Units for the heat capacity are J/K (or cal/K).

The heat capacity is not defined uniquely, i.e., one needs to specify the conditions under which the heat is added to the system. Even though several choices for the heat capacities are possible, one is generally interested in only two: the *heat capacity at constant volume* C'_v and the *heat capacity at constant pressure* C'_p. The former is the most useful quantity because C'_v is obtained immediately from the energy of the system. The heat capacity at constant volume is defined as

$$C'_v = \left(\frac{\partial E}{\partial T}\right)_v. \tag{19.3}$$

On the other hand, it is much easier to measure the heat capacity of a solid at constant pressure than at constant volume. Fortunately, the difference between C'_p and C'_v *for solids* vanishes at low temperatures and is only about 5% at room temperature. C'_v can be calculated from C'_p[1] if the volume expansion coefficient α and the compressibility κ of a material are known, by applying

$$C'_v = C'_p - \frac{\alpha^2 TV}{\kappa}, \tag{19.4}$$

where V is the volume of the solid. Equation (19.4) is derived in textbooks on thermodynamics.

[1] The heat capacity at constant pressure is defined as $C'_p = (\partial H/\partial T)_p$ where $H = U + P \cdot V$ is the enthalpy.

19.3. Specific Heat Capacity, c

The specific heat capacity is the heat capacity *per unit mass*

$$c = \frac{C'}{m}. \tag{19.5}$$

It is a material constant and it is temperature-dependent. Characteristic values for the specific heat capacity (c_v and c_p) are given in Table 19.1. The unit of the specific heat capacity is J/g · K (or cal/g · K). We note from Table 19.1 that values for the specific heat capacity of solids are considerably smaller than the specific heat capacity of water.

Combining (19.1a), (19.3), and (19.5) yields

$$\Delta E = Q = m\Delta T c_v, \tag{19.6}$$

which expresses that the thermal energy (or heat) which is transferred to a system equals the product of mass, increase in temperature, and specific heat capacity.

19.4. Molar Heat Capacity, C_v

A further useful material constant is the heat capacity *per mole* (i.e., per amount of substance of a phase, n). It compares materials that contain the same number of molecules or atoms. The molar heat capacity is obtained by

Table 19.1. Experimental Thermal Parameters of Various Substances at Room Temperature and Ambient Pressure.

Substance	Specific heat capacity, (c_p) $\left(\frac{J}{g \cdot K}\right)$	Molar (atomic) mass $\left(\frac{g}{mol}\right)$	Molar heat capacity (C_p) $\left(\frac{J}{mol \cdot K}\right)$	Molar heat capacity (C_v) $\frac{J}{mol \cdot K}$
Al	0.897	27.0	24.25	23.01
Fe	0.449	55.8	25.15	24.68
Ni	0.456	58.7	26.8	24.68
Cu	0.385	63.5	24.48	23.43
Pb	0.129	207.2	26.85	24.68
Ag	0.235	107.9	25.36	24.27
C (graphite)	0.904	12.0	10.9	9.20
Water	4.184	18.0	75.3	

multiplying the specific heat capacity c_v (or c_p) by the molar mass, M (see Table 19.1)

$$C_v = \frac{C'_v}{n} = c_v \cdot \mathrm{M}. \tag{19.7}$$

The units are J/mol · K (or cal/mol · K). The amount of substance is

$$n = N/N_0, \tag{19.7a}$$

where N is the number of particles (atoms, molecules, etc.), and N_0 is the Avogadro constant ($N_0 = 6.022 \times 10^{23}$ mol^{-1}).

We see from Table 19.1 that the room temperature molar heat capacity at constant volume is approximately 25 J/mol · K (6 cal/mol · K) for most solids. This was experimentally discovered in 1819 by Dulong and Petit. We shall attempt to interpret this interesting result in a later section.

The experimental molar heat capacities for some materials are depicted in Fig. 19.1 as a function of temperature. We notice that some materials, such as carbon, reach the Dulong–Petit value of 6 cal/mol · K only at high temperatures. Some other materials such as lead reach 6 cal/mol · K at relatively low temperatures.

All heat capacities are zero at $T = 0$ K. The C_v values near $T = 0$ K climb in proportion to T^3 and reach 96% of their final value at a temperature θ_D which is defined to be the Debye temperature. We shall see later that θ_D is an approximate dividing point between a high temperature region, where classical models can be used for the interpretation of C_v, and a low temperature region, where quantum theory needs to be applied. Selected Debye temperatures are listed in Table 19.2.

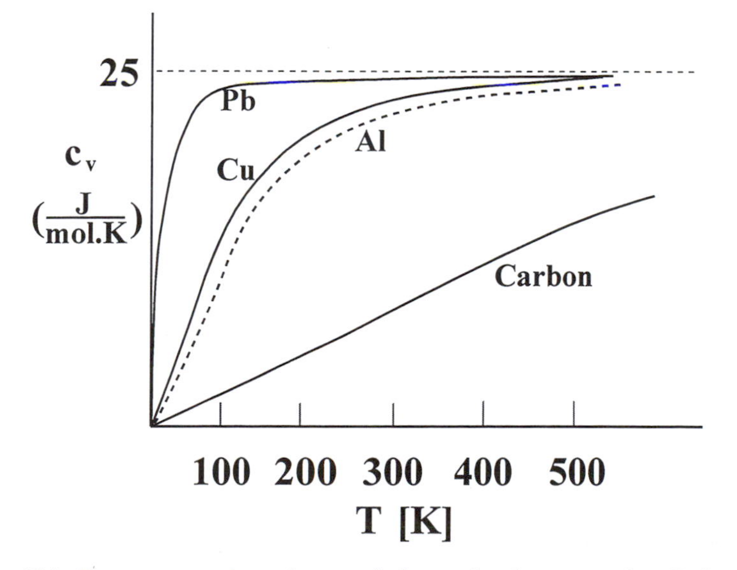

Figure 19.1. Temperature dependence of the molar heat capacity C_v for some materials.

Table 19.2. Debye Temperatures of Some Materials.

Substance	θ_D (K)
Pb	95
Au	170
Ag	230
W	270
Cu	340
Fe	360
Al	375
Si	650
C	1850
GaAs	204
InP	162

19.5. Thermal Conductivity, *K*

Heat conduction (or thermal conduction) is the transfer of thermal energy from a hot body to a cold body when both bodies are brought into contact. For best visualization we consider a bar of a material of length x whose ends are held at different temperatures. The heat that flows through a cross section of the bar divided by time and area, (i.e., the heat flux, J_Q) is proportional to the temperature gradient dT/dx. The proportionality constant is called the thermal conductivity K (or λ). We thus write

$$J_Q = -K \frac{dT}{dx}. \tag{19.8}$$

The negative sign indicates that the heat flows from the hot to the cold end (Fourier Law, 1822). Possible units for the heat conductivity are (J/m · s · K) or (W/m · K) or (cal/m · s · K). The heat flux J_Q is measured in (J/m^2 · s) or (cal/m^2 · s). Table 19.3 gives some characteristic values for K. The thermal conductivity decreases slightly with increasing temperature. For example, K for copper decreases by 20% within a temperature span of 1000° C. In the same temperature region K for iron decreases by 10%.

19.6. The Ideal Gas Equation

Free electrons in metals and alloys can often be considered to behave like an ideal gas. An ideal gas is an abstraction which is frequently used in thermodynamics. It is usually defined to be a gas whose density is low enough in order

Table 19.3. Thermal Conductivities at Room Temperature.[a]

Substance	$K\left(\frac{W}{m \cdot K}\right) \equiv \left(\frac{J}{s \cdot m \cdot K}\right)$
Diamond type IIa	2.3×10^2
SiC	4.9×10^2
Silver	4.29×10^2
Copper	4.01×10^2
Aluminum	2.37×10^2
Silicon	1.48×10^2
Brass (leaded)	1.2×10^2
Iron	8.02×10^1
GaAs	5×10^1
Ni-Silver[b]	2.3×10^1
Al_2O_3 (sintered)	3.5×10^1
SiO_2 (fused silica)	1.4
Concrete	9.3×10^{-1}
Soda-lime glass	9.5×10^{-1}
Water	6.3×10^{-1}
Polyethylene	3.8×10^{-1}
Teflon	2.25×10^{-1}
Snow (0°C)	1.6×10^{-1}
Wood (oak)	1.6×10^{-1}
Sulfur	2.0×10^{-2}
Cork	3×10^{-2}
Glass wool	5×10^{-3}
Air	2.3×10^{-4}

[a] See also Figure 18.1. Source: Handbook of Chemistry and Physics, CRC Press. Boca Raton, FL (1994).
[b] 62% Cu, 15% Ni, 22% Za.

for it to obey the equation

$$PV = nRT, \tag{19.9}$$

where P is the pressure of the gas, V is its volume, n is the amount of substance, T is the thermodynamic (absolute) temperature, R is the universal gas constant, and k_B is the Boltzmann constant. The gas constant is

$$\begin{aligned} R = k_B N_0 &= 8.314\ (\text{J/mol} \cdot \text{K}) \\ &= 1.986\ (\text{cal/mol} \cdot \text{K}). \end{aligned} \tag{19.10}$$

Equation (19.9) is a combination of two experimentally obtained thermodynamic laws: One, discovered by Boyle and Mariotte (PV = const. at constant T) and the other, discovered by Gay-Lussac ($V \sim T$, at constant P). The reader who has taken classes in physics or thermodynamics is undoubtedly familiar with these equations.

19.7. Kinetic Energy of Gases

In the chapters to come, we need to know the kinetic energy of atoms, molecules, or electrons at a given temperature from a classical point of view. The calculation which is summarized below is usually contained in textbooks on thermodynamics.

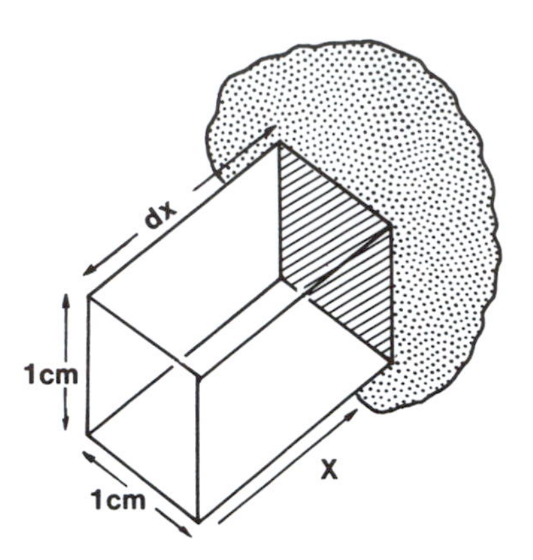

Figure 19.2. Diagram for the derivation of the kinetic energy of gases.

We commence by quoting the number of molecules in a gas which interact in the unit time, t, with the end face of unit area of a bar which has the length dx. We assume that, because of thermal agitation, one-third of the particles move in the x-directions, i.e., one-sixth in the positive x-direction. The volume element, shown in Fig. 19.2, is

$$dV = A\,dx = Av\,dt, \tag{19.11}$$

where A is the unit area and v is the velocity of the particles which fly in the x-direction. The number of particles reaching the end face is naturally proportional to the number of particles n_v in the given volume, i.e.,

$$z' = \tfrac{1}{6}n_v \cdot dV = \tfrac{1}{6}n_v Av\,dt.$$

The number of particles per unit time and unit area which hit the end face is, consequently,

$$z = \tfrac{1}{6}n_v v, \tag{19.12}$$

where

$$n_v = \frac{N}{V} \tag{19.13}$$

is the *number of particles per unit volume*. Each particle transfers the momentum $2mv$ during its collision with the wall and subsequent reflection. The *momentum per unit time and unit area* is then

$$p^* = z2mv = \tfrac{1}{6}n_v v2mv = \frac{1}{3}\frac{N}{V}mv^2. \tag{19.14}$$

This yields, for the pressure,

$$P = \frac{F}{A} = \frac{ma}{A} = \frac{d(mv)/dt}{A} = \frac{dp/dt}{A} = p^* = \frac{1}{3}\frac{N}{V}mv^2. \tag{19.15}$$

With

$$PV = nRT = nk_BN_0T = Nk_BT \tag{19.16}$$

(see (19.9), (19.7a), and (19.10)), we obtain from (19.15)

$$PV = \tfrac{1}{3}Nmv^2 = k_BNT. \tag{19.17}$$

Inserting

$$E_{kin} = \tfrac{1}{2}mv^2 \tag{19.18}$$

into (19.17) yields

$$k_BNT = \tfrac{1}{3}N2\tfrac{1}{2}mv^2 = \tfrac{2}{3}NE_{kin}, \tag{19.19}$$

which finally yields for the kinetic energy of a particle

$$E_{kin} = \tfrac{3}{2}k_BT. \tag{19.20}$$

A more precise calculation which considers the mutual collisions of the particles, and thus a velocity distribution, replaces the kinetic energy in (19.20) by an average kinetic energy.

Problems

Note: The problems in this chapter contain engineering applications in order to make the student aware of the importance of thermal properties in daily life.

1. Calculate the number of gas molecules which are left in an ultrahigh vacuum of 10^{-9} Pa ($\sim 7.5 \times 10^{-12}$ Torr) at room temperature.
2. Calculate the rate of heat loss in a 5 mm thick window glass when the exterior temperature is 0° C and the room temperature is 20° C. Compare your result with the heat loss in an aluminum and a wood frame of 10 mm thickness. How can you decrease the heat loss through the window?
3. A block of copper, whose mass is 100 g, is quenched directly from an annealing furnace into a 200 g glass container which holds 500 g of water. What is the temperature of the furnace when the water temperature rises from 0° to 15° C? ($c_{glass} = 0.5$ J/g · K.)
4. Explain in simple terms why wood has a smaller heat conductivity than copper.
5. What are the implications for the semiconductor industry that silicon has a relatively good heat conductivity?
6. Why is the fiberglass insulation used for buildings, etc., loose rather than compact? (*Hint*: Compare K for glass and air. Discuss also heat convection.)
7. Find in a handbook the relationship between J/g · K and BTU/lb °F.

CHAPTER 20

Heat Capacity

20.1. Classical (Atomistic) Theory of Heat Capacity

This section attempts to interpret the thermal properties of materials using atomistic concepts. In particular, an interpretation of the experimentally observed molar heat capacity at high temperatures, $C_v = 25$ (J/mol · K) = 6 (cal/mol · K), is of interest.

We postulate that each atom in a crystal is bound to its site by a harmonic force. A given atom is thought to be capable of absorbing thermal energy, and in doing so it starts to vibrate about its point of rest. The amplitude of the oscillation is restricted by electrostatic repulsion forces of the nearest neighbors. The extent of this thermal vibration is therefore not more than 5 or 10% of the interatomic spacing, depending on the temperature. In short, we compare an atom with a sphere which is held at its site by two springs (Fig. 20.1(a)). The thermal energy which a harmonic oscillator of this kind can absorb is proportional to the absolute temperature of the environment. The proportionality factor has been found to be the Boltzmann constant k_B (see below). The average energy of the oscillator is then

$$E = k_B T. \tag{20.1}$$

Now, solids are three dimensional. Thus, a given atom in a cubic crystal also responds to the harmonic forces of lattice atoms in the other two directions. In other words, it is postulated that each atom in a cubic crystal represents three oscillators (Fig. 20.1(b)), each of which absorbs the thermal energy $k_B T$. Therefore, the *average energy per atom* is

$$E = 3k_B T. \tag{20.2}$$

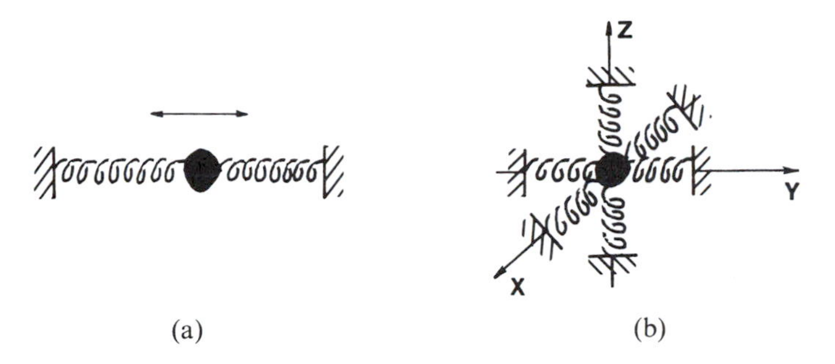

Figure 20.1. (a) A one-dimensional harmonic oscillator and (b) a three-dimensional harmonic oscillator.

We note in passing that the same result is obtained by using the kinetic theory of gases. It was shown in (19.20) that the average kinetic energy of a particle (or in the present case, an atom) is

$$E_{\text{kin}} = \tfrac{3}{2} k_B T. \tag{20.3}$$

Now, each elastic vibration in a solid involves not only kinetic energy but also potential energy, which has the same average energy as the kinetic energy. The total energy of a vibrating lattice atom is thus

$$E = 2 \cdot \tfrac{3}{2} k_B T, \tag{20.4}$$

which is the result of (20.2).

We consider now all N_0 atoms per mole. Then the *total internal energy per mole* is

$$E = 3N_0 k_B T. \tag{20.5}$$

Finally, the molar heat capacity is given by combining (19.3), (19.7), and (20.5) which yields

$$\boxed{C_v = \left(\frac{\partial E}{\partial T}\right)_v = 3N_0 k_B.} \tag{20.6}$$

Inserting the numerical values for N_0 and k_B into (20.6) yields

$$C_v = 25 \text{ J/mol} \cdot \text{K} \qquad \text{or} \qquad 5.98 \text{ cal/mol} \cdot \text{K},$$

quite in agreement with the experimental findings at high temperatures (Figs. 18.2 and 19.1).

It is satisfying to see that a simple model involving three harmonic oscillators per atom can readily explain the experimentally observed heat capacity. However, one shortcoming is immediately evident: the calculated molar heat capacity turned out to be temperature-independent, according to (20.6), and also independent of the material. This discrepancy with the observed behavior (see Fig. 18.2) was puzzling to scientists at the turn of this century and had to await quantum theory to be properly explained.

20.2. Quantum Mechanical Considerations—the Phonon

20.2.1. Einstein Model

Einstein postulated, in 1907, that the energies of the above-mentioned classical oscillators should be quantized, i.e., he postulated that only certain vibrational modes should be allowed, quite in analogy to the allowed energy states of electrons. These lattice vibration quanta were called *phonons*.

The term *phonon* stresses an analogy with *electrons* or *photons*. As we know from Chapter 2, photons are quanta of electromagnetic radiation, i.e., photons describe (in the appropriate frequency range) classical light. Phonons, on the other hand, are quanta of the ionic displacement field which (in the appropriate frequency range) describe classical sound.

The phonon describes the *particle* nature of an oscillator. A phonon has, in analogy to the de Broglie relation (2.3), the momentum $p = \hbar/\lambda$.

Furthermore, Einstein postulated a particle–wave duality. This suggests *phonon waves* which propagate through the crystal with the speed of sound. Phonon waves are *not* electromagnetic waves: they are *elastic waves*, vibrating in a longitudinal and/or in a transversal mode.

In analogy to the electron case shown in Part I of this book, one can describe the properties of phonons in terms of band diagrams, Brillouin zones, or density of states curves. Small differences exist however. For example, the energy in the band diagram of an electron is replaced in a phonon band diagram by the vibrational frequency ω of the phonon. The branches in the phonon band diagram are sinusoidal in nature (compared to parabolic in the free electron case). The individual phonon bands are no longer called valence or conduction bands, but more appropriately *acoustic bands* and *optical bands*, mainly because the frequencies in which the branches are situated are in the acoustical and optical ranges, respectively. The density of states or, better, the *density of vibrational modes* $D(\omega)$ for the phonon case is defined so that $D(\omega) \cdot d\omega$ is the number of modes whose frequencies lie in the interval ω and $\omega + d\omega$. For a continuous medium the density of modes is

$$D(\omega) = \frac{3V}{2\pi^2} \frac{\omega^2}{v_s^3}, \tag{20.7}$$

where v_s is the sound velocity. This equation can be derived quite similarly as demonstrated in Section 6.3.

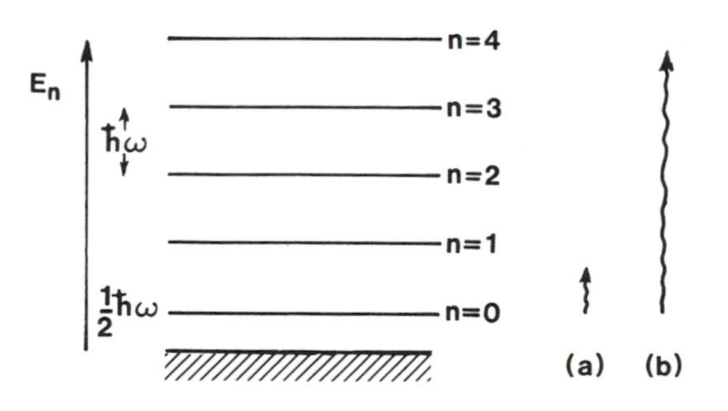

Figure 20.2. Allowed energy levels of a phonon: (a) average thermal energy at low temperatures and (b) average thermal energy at high temperatures.

The allowed energies of a single oscillator are

$$E_n = n\hbar\omega, \tag{20.8}$$

similarly as in Section 4.2, where n is an integer.[2] A schematic energy level diagram for the allowed phonon energies is shown in Fig. 20.2.

One important difference between phonons and electrons needs to be emphasized. Phonons are created by *raising* the temperature, and eliminated by lowering it, i.e., the number of phonons is *not* conserved as we shall show momentarily. (In contrast to this, the number of electrons is constant.) Einstein postulated that with increasing temperature more and more phonons are created, each of which has the same energy $\hbar\omega$ or the same frequency of vibration ω. The average number of phonons $\overline{N}_{ph}$ at a given temperature was found by Bose and Einstein to obey a special type of statistics

$$\overline{N}_{ph} = \frac{1}{\exp\left(\dfrac{\hbar\omega}{k_B T}\right) - 1}. \tag{20.10}$$

This equation is similar in form to the Fermi distribution function (6.1).

We note in passing that for high phonon energies $\hbar\omega$, the exponential term in (20.10) becomes large when compared to unity so that the number of phonons can be approximated by Boltzmann statistics, i.e., by the laws of classical thermodynamics,

$$\overline{N}_{ph} \approx e^{-\hbar\omega/k_B T}. \tag{20.11}$$

We already made a similar statement in Section 6.2. We also see from (20.10)

[2] Since the ground state ($n = 0$) still has a zero point energy of $\frac{1}{2}\hbar\omega$ we should, more appropriately, write

$$E_n = n\hbar\omega + \tfrac{1}{2}\hbar\omega = (n + \tfrac{1}{2})\hbar\omega. \tag{20.9}$$

The zero point energy is, however, of no importance for the present considerations.

that the number of phonons decreases rapidly when the temperature approaches 0 K.

The average energy of an isolated oscillator is then the average number of phonons times the energy of a phonon

$$E_{osc} = \hbar\omega \overline{N}_{ph} = \frac{\hbar\omega}{\exp\left(\frac{\hbar\omega}{k_B T}\right) - 1}. \tag{20.12}$$

The *thermal energy of a solid* can now be calculated by taking into account (as in Section 20.1) that a mole of a substance contains $3N_0$ oscillators. This yields for the thermal energy per mole

$$E = 3N_0 \frac{\hbar\omega}{\exp\left(\frac{\hbar\omega}{k_B T}\right) - 1}. \tag{20.13}$$

The molar heat capacity is, finally,

$$\boxed{C_v = \left(\frac{\partial E}{\partial T}\right)_v = 3N_0 k_B \left(\frac{\hbar\omega}{k_B T}\right)^2 \frac{\exp\left(\frac{\hbar\omega}{k_B T}\right)}{\left(\exp\left(\frac{\hbar\omega}{k_B T}\right) - 1\right)^2}.} \tag{20.14}$$

We discuss C_v for two special temperature regions. For large temperatures the approximation $e^x \simeq 1 + x$ can be applied, which yields $C_v \simeq 3N_0 k_B$ (see Problem 5) in agreement with (20.6), i.e., we obtain the classical Dulong–Petit value. For $T \to 0$, C_v approaches zero, again in agreement with experimental observations. Thus, the temperature dependence of C_v is now in qualitative accord with the experimental findings. One minor discrepancy has to be noted however: At very small temperatures the experimental C_v decreases by T^3, as stated in Section 19.4. The Einstein theory predicts, instead, an exponential reduction. The Debye theory which we shall discuss below alleviates this discrepancy by postulations that the individual oscillators interact with each other.

By inspecting (20.14), we observe that this relation contains only one adjustable parameter, namely, the angular frequency, which we shall redesignate for this particular case by ω_E. By fitting (20.14) to experimental curves, the frequency of the phonon waves can be obtained. For copper, the angular frequency ω_E has been found in this way to be $2.5 \times 10^{13}\ s^{-1}$, which yields $\nu_E = 4 \times 10^{12}\ s^{-1}$. It is customary to call frequencies up to $10^5\ s^{-1}$ sound waves, frequencies between 10^5 and $10^9\ s^{-1}$ ultrasonics, and frequencies above $10^9\ s^{-1}$ thermal waves. The Einstein frequency is thus situated very appropriately in the thermal wave region.

Occasionally, the *Einstein temperature* θ_E is quoted, which is defined by equating the phonon energy

$$\hbar\omega_E = k_B\theta_E, \tag{20.15}$$

which yields

$$\theta_E = \frac{\hbar\omega_E}{k_B}. \tag{20.16}$$

Characteristic values for θ_E are between 200 K and 300 K. From the above-quoted ω_E value for copper, θ_E^{Cu} can be calculated to be 240 K.

20.2.2. Debye Model

We now refine the Einstein model by taking into account that the atoms in a crystal interact with each other. Consequently, the oscillators are thought to vibrate interdependently. We recall that the Einstein model considered only *one* frequency of vibration, ω_E. When interactions between the atoms occur, many more frequencies are thought to exist which range from about the Einstein frequency down to frequencies of the acoustical modes of oscillation. We postulate that these vibrational modes are quantized (Fig. 20.2). The total displacement of a given atom in a crystal during the oscillation is found by summing up all vibrational modes. This has been done by Debye who modified the Einstein equation (20.13) by replacing the $3N_0$ oscillators of a single frequency by the number of modes in a frequency interval $d\omega$ and by summing up over all allowed frequencies. The total energy of vibration for the solid is then

$$E = \int E_{osc} D(\omega)\, d\omega, \tag{20.17}$$

where E_{osc} is the energy of one oscillator given in (20.12), and $D(\omega)$ is the density of modes given in (20.7). Inserting (20.7) and (20.12) into (20.17) yields

$$E = \frac{3V}{2\pi^2 v_s^3} \int_0^{\omega_D} \frac{\hbar\omega^3}{\exp\left(\frac{\hbar\omega}{k_B T}\right) - 1}\, d\omega. \tag{20.18}$$

The integration is performed between $\omega = 0$ and the Debye frequency ω_D, because above the Debye frequency the oscillators are defined to behave in a classical manner, see Section 19.5.

The molar heat capacity C_v is obtained, as usual, by performing the derivative of (20.18) with respect to temperature. This yields

$$C_v = \frac{3V\hbar^2}{2\pi^2 v_s^3 k_B T^2} \int_0^{\omega_D} \frac{\omega^4 \exp\left(\frac{\hbar\omega}{k_B T}\right)}{\left(\exp\left(\frac{\hbar\omega}{k_B T}\right) - 1\right)^2}\, d\omega \tag{20.19}$$

or

$$C_v = 9k_B N_0 \left(\frac{T}{\theta_D}\right)^3 \int_0^{\theta_D/T} \frac{x^4 e^x}{(e^x - 1)^2} dx, \tag{20.20}$$

where

$$x = \frac{\hbar\omega}{k_B T} \tag{20.21}$$

varies with the angular frequency ω, and

$$\theta_D = \frac{\hbar\omega_D}{k_B} \tag{20.22}$$

is called the Debye temperature. Values for θ_D can be obtained again by curve-fitting, particularly at low temperatures. They have been listed in Table 19.2. For low temperatures, i.e., for $T < \theta_D$, the upper limit of the integral in (20.20) can be approximated by infinity. Then (20.20) can be evaluated and it becomes

$$C_v = \frac{12\pi^4}{5} N_0 k_B \left(\frac{T}{\theta_D}\right)^3. \tag{20.23}$$

From both equations, (20.20) as well as (20.23), it can be seen that C_v decreases proportionally to T^3 at low temperatures, which is quite in agreement with the experimental observations.

In summary, the main difference between the two theories is that the Debye model takes the low frequency modes into account whereas the Einstein model does not. We have to realize, however, that the excitation of oscillators at low temperatures occurs only with a small probability, because at low temperatures only a few oscillators can be raised to the next higher level. This is a consequence of the fact that the energy difference between levels is comparatively large for the available small thermal energies, as schematically illustrated in Fig. 20.2.

It should be noted that even (20.20) is only an approximation, because the underlying model does not take into consideration the periodicity of the atoms in a crystal lattice. Thus, a refinement of the Debye model needs to utilize the actual density of modes function $D(\omega)$ for a given material. This has been done by scientists with good success. Equation (20.20) is, however, a fairly good approximation (see Fig. 18.2).

20.3. Electronic Contribution to the Heat Capacity

In the previous sections we have digressed considerably from the principal theme of this book, namely, the description of the *electronic* properties of materials. We now return to our main topic by discussing the contributions

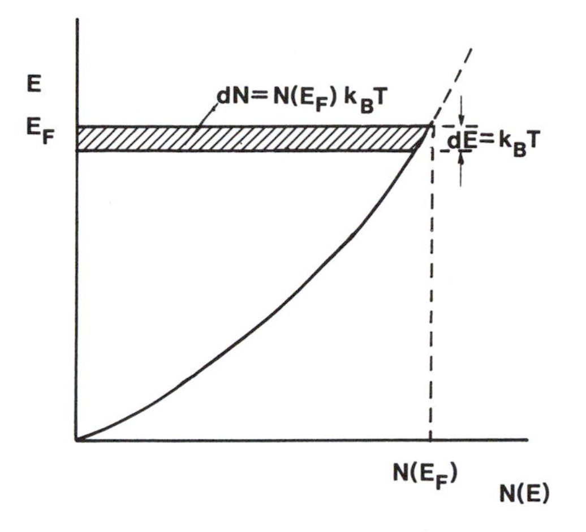

Figure 20.3. Population density as a function of energy for a metal. The electrons within the shaded area below E_F can be excited by a thermal energy $k_B T$.

which the electrons provide to the specific heat. We will quickly see that this contribution is relatively small compared to that of the phonons.

First, we need to remember that only the kinetic energy of the *free* electrons can be raised with increasing temperature. Consequently, our present discussion is restricted to metals and alloys which have, as we know, partially filled bands and thus free electrons. Second, we need to remember that only those electrons which lie within an energy interval $k_B T$ of the Fermi energy can be excited into higher states because only these electrons find empty energy states after their excitation. We know that the number dN of these excitable electrons depends on the population density of the metal under consideration (Section 6.4). In other words, dN is the product of the population density at the Fermi level, $N(E_F)$, and the energy interval $k_B T$, as indicated by the shaded area in Fig. 20.3.

We postulate that the electrons which are excited by thermal energy behave like a monatomic gas. We have already shown in (19.20) that the mean kinetic energy of gas molecules, or in the present case, the mean kinetic energy of the electrons above the Fermi energy is $\frac{3}{2}k_B T$. Thus, the thermal energy at a given temperature is

$$E_{kin} = \tfrac{3}{2}k_B T\, dN = \tfrac{3}{2}k_B T N(E_F) k_B T. \tag{20.24}$$

The heat capacity of the electrons is then, as usual,

$$C_v^{el} = \left(\frac{\partial E}{\partial T}\right)_v = 3k_B^2 T N(E_F). \tag{20.25}$$

We need now an expression for $N(E_F)$. We obtain this by combining (6.8) and (6.11) for $E < E_F$ (see Fig. 20.3 and Problem 9), which yields

$$N(E_F) = \frac{3N^*}{2E_F}\left(\frac{\text{electrons}}{\text{J}}\right), \tag{20.26}$$

where N^* is the number of electrons which have an energy equal to or smaller than E_F (Section 6.4). Inserting (20.26) into (20.25) yields

$$C_v^{el} = \frac{9}{2}\frac{N^* k_B^2 T}{E_F}\left(\frac{J}{K}\right). \tag{20.27}$$

So far we assumed that the thermally excited electrons behave like a classical gas. In reality, the excited electrons must obey the Pauli principle. If this is taken into consideration properly, (20.27) changes slightly and reads

$$\boxed{C_v^{el} = \frac{\pi^2}{2}\frac{N^* k_B^2 T}{E_F} = \frac{\pi^2}{2} N^* k_B \frac{T}{T_F}.} \tag{20.28}$$

Let us assume now a monovalent metal in which we can reasonably assume one free electron per atom (see Part I). Then, N^* can be equated to the number of atoms per mole N_0, and (20.28) becomes the heat capacity per mole

$$\boxed{C_v^{el} = \frac{\pi^2}{2}\frac{N_0 k_B^2 T}{E_F} = \frac{\pi^2}{2} N_0 k_B \frac{T}{T_F} \quad \left(\frac{J}{K \cdot mol}\right).} \tag{20.29}$$

We see from (20.29) that C_v^{el} is a linear function of the temperature and is zero at $T = 0$ K, quite in agreement with the experimental observations, see Fig. 18.2. The room temperature contribution of the electronic specific heat to the total specific heat is less than 1% (see Problem 3). There are, however, two temperature regions where the electronic specific heat plays an appreciable role. This is at very low temperatures, i.e., at $T < 5$ K (see Fig. 18.2). Second, we have learned in the previous sections that the lattice heat capacity levels off above the Debye temperature. Thus, the electron heat capacity can give at high temperatures a small contribution to the Dulong–Petit value.

An interesting aspect is added: Equation (20.25) may be rewritten in the following form:

$$C_v^{el} = \gamma T, \tag{20.30}$$

where

$$\gamma = 3 k_B^2 N(E_F). \tag{20.31}$$

Furthermore, (20.20) can be rewritten as

$$C_v^{ph} = \beta T^3. \tag{20.32}$$

Below the Debye temperature, the heat capacity of metals is the sum of electron and phonon contributions, i.e.,

$$C_v^{tot} = C_v^{el} + C_v^{ph} = \gamma T + \beta T^3, \tag{20.33}$$

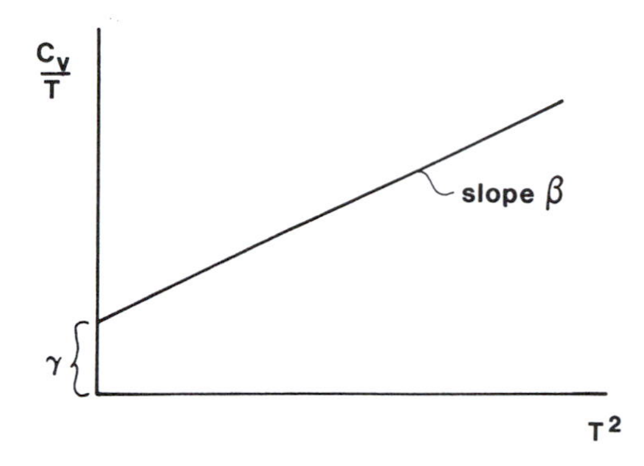

Figure 20.4. Schematic representation of an experimental plot of C_v/T versus T^2.

which yields

$$\frac{C_v^{tot}}{T} = \gamma + \beta T^2. \tag{20.34}$$

A plot of experimental values for C_v^{tot}/T versus T^2 provides the material constants γ (intercept) and β (slope), see Fig. 20.4. Heat capacity measurements thus serve as a means to obtain the electron population density at the Fermi surface by using (20.31).

Some calculated and observed values for γ are given in Table 20.1. From the slight discrepancy between observed and free-electron γ-values, a *thermal effective mass* can be calculated which is defined as

$$\frac{m_{th}^*}{m_0} = \frac{\gamma \text{ (obs.)}}{\gamma \text{ (calc.)}}. \tag{20.35}$$

The deviation between the two γ-values is interpreted to stem from neglecting electron–phonon and electron–electron interactions.

Table 20.1. Calculated and Observed Values for the Constant γ, see (20.31).

Substance	γ, observed		γ, calculated		
	$\left(\frac{\text{cal}}{\text{mol}\cdot\text{K}^2}\right)$	$\left(\frac{\text{J}}{\text{mol}\cdot\text{K}^2}\right)$	$\left(\frac{\text{cal}}{\text{mol}\cdot\text{K}^2}\right)$	$\left(\frac{\text{J}}{\text{mol}\cdot\text{K}^2}\right)$	$\frac{m_{th}^*}{m_0}$
Ag	1.54×10^{-4}	0.646×10^{-3}	1.54×10^{-4}	0.645×10^{-3}	1.0
Al	3.22×10^{-4}	1.35×10^{-3}	2.18×10^{-4}	0.912×10^{-3}	1.48
Au	1.74×10^{-4}	0.729×10^{-3}	1.53×10^{-4}	0.642×10^{-3}	1.14
Na	3.11×10^{-4}	1.3×10^{-3}	2.37×10^{-4}	0.992×10^{-3}	1.31
Fe	1.19×10^{-3}	4.98×10^{-3}	—	—	—
Ni	1.68×10^{-3}	7.02×10^{-3}	—	—	—

Problems

1. How many electrons (in percent of the total number of electrons per mole) lie $k_B T$ (eV) below the Fermi energy? Take $E_F = 5$ eV and $T = 300$ K.

2. Calculate C_v at high temperatures (500 K) by using the quantum mechanical equation derived by Einstein. Assume an Einstein temperature of 250 K, and convince yourself that C_v approaches the classical value at high temperatures.

3. Calculate the electronic specific heat for $E_F = 5$ eV and $T = 300$ K. How does your result compare with the experimental value of 25 (J/mol K), i.e., 6 (cal/mol K)?

4. Calculate the population density at the Fermi level for a metal whose electronic specific heat at 4 K was measured to be 8.37×10^{-3} (J/mol K).

5. Confirm that (20.14) reduces for large temperatures to the Dulong–Petit value.

6. At what temperature would the electronic contribution to C_v of silver eventually become identical to the Dulong–Petit value? (*Hint*: Use proper units for the heat capacity! Take $E_F = 5$ eV, $N_f = 10^{28}$ el/m^3.)

7. Discuss thermal expansion in materials from an atomistic point of view.

8. Show that for small temperatures, (20.20) reduces to (20.23) and that for large temperatures (20.20) reduces to the Dulong–Petit value.

9. Derive (20.26) for $E < E_F$ as shown in Fig. 20.3 by combining (6.8) and (6.11) and eliminating the Planck constant.

10. *Computer problem.* Plot the Einstein equation (20.14) and the Debye equation (20.20) as a function of temperature by utilizing different values for ω_E, ω_D, and θ_D. Why can (20.23) not be used in the entire temperature range?

CHAPTER 21

Thermal Conduction

We stated in Chapter 19 that heat conduction can be described as the transfer of thermal energy from the hot to the cold part of a piece of material. We shall discuss now the mechanisms which are involved in this transfer of thermal energy.

We postulate that the heat transfer in solids may be provided by *free electrons* as well as by *phonons*. We understand immediately that in insulators, which do not contain any free electrons, the heat must be conducted exclusively by phonons. In metals and alloys, on the other hand, the heat conduction is dominated by electrons because of the large number of free electrons in metals. Thus, the phonon contribution is usually neglected in this case.

One particular point should be clarified right at the beginning. Electrons in metals travel in equal numbers from hot to cold and from cold to hot in order that the charge neutrality be maintained. Now, the electrons in the hot part of a metal possess and transfer a high energy. In contrast to this, the electrons in the cold end possess and transfer a lower energy. The heat transferred from hot to cold is thus proportional to the difference in the energies of the electrons.

The situation is quite different in phonon conductors. We know from Section 20.2.1 that the number of phonons is larger at the hot end than at the cold end. Thermal equilibrium thus involves in this case a net transfer of phonons from the hot into the cold part of a material.

21.1. Thermal Conduction in Metals and Alloys—Classical Approach

We now attempt to calculate the heat conductivity K (see (19.8)). The train of thought is borrowed from the kinetic theory of gases because the same arguments hold true for electrons as for gas molecules.

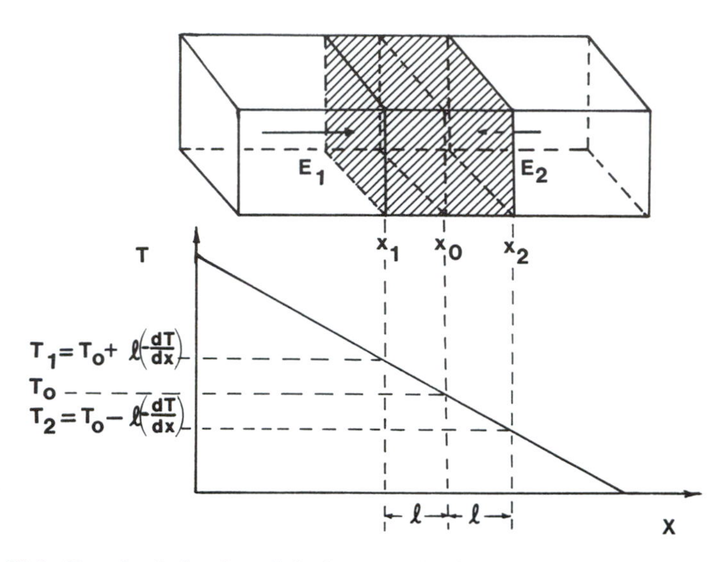

Figure 21.1. For the derivation of the heat conductivity in metals. Note that (dT/dx) is negative for the case shown in the graph.

Consider a bar of metal whose left side is hot and whose right side is cold (Fig. 21.1). Thus, a temperature gradient dT/dx exists in the x-direction of the bar. Consider also a volume at the center of the bar whose faces have the size of a unit area and whose length is $2l$, where l is the mean free path between two consecutive collisions between an electron and lattice atoms. We assume that at the distance l from the center x_0 the average electron has had its last collision and has picked up the energy of this place. We calculate first the energy E_1 per unit time and unit area of the electrons which drift from the left into the above-mentioned sample volume. This energy E_1 equals the number of electrons z, times the energy of one of the electrons. The latter is, according to (19.20), $\frac{3}{2}k_B T_1$, where T_1 can be taken from Fig. 21.1, and z is given in (19.12)

$$E_1 = z \cdot \frac{3}{2} k_B \left(T_0 + l\left(-\frac{dT}{dx} \right) \right) = \frac{n_v v}{6} \frac{3}{2} k_B \left(T_0 - l\frac{dT}{dx} \right). \tag{21.1}$$

The same amount of electrons is drifting from right to left through the volume under consideration. These electrons, however, carry a lower energy E_2 because of the lower temperature of the particles at the site of interaction. Thus,

$$E_2 = \frac{n_v v}{6} \cdot \frac{3}{2} k_B \left(T_0 + l\frac{dT}{dx} \right). \tag{21.2}$$

The excess thermal energy transferred per unit time into the unit volume is therefore

$$J_Q = E_1 - E_2 = -\frac{n_v v}{6} \frac{3}{2} k_B \left(2l\frac{dT}{dx} \right) = -\frac{n_v v}{2} k_B l \frac{dT}{dx}. \tag{21.3}$$

We compare (21.3) with (19.8)

$$J_Q = -K\frac{dT}{dx}. \tag{21.4}$$

Then we obtain for the heat conductivity of the electrons

$$\boxed{K = \frac{n_v v k_B l}{2}.} \tag{21.5}$$

The heat conductivity is thus larger the more electrons are involved, the larger their velocity, and the larger the mean free path between two consecutive electron–atom collisions. This result intuitively makes sense.

We now seek a connection between the heat conductivity and C_v^{el}. We know from (19.20) the kinetic energy of all n_v electrons per unit volume:

$$E = n_v \tfrac{3}{2} k_B T. \tag{21.6}$$

From this we obtain the heat capacity *per volume*

$$C_v^{el} = \left(\frac{dE}{dT}\right)_v = n_v \frac{3}{2} k_B. \tag{21.7}$$

Combining (21.5) with (21.7) yields

$$K = \tfrac{1}{3} C_v^{el} v l. \tag{21.8}$$

All three variables contained in (21.8) are temperature-dependent, but while C_v^{el} increases with temperature, l and, to a small degree, also v are decreasing. Thus, K should change very little with temperature, which is indeed experimentally observed. As mentioned in Section 19.5, the thermal conductivity decreases about 10^{-5} W/m·K per degree. K also changes at the melting point and when a change in atomic packing occurs.

21.2. Thermal Conduction in Metals and Alloys—Quantum Mechanical Considerations

The question arises as to what velocity the electrons (that participate in the heat conduction process) have? Further, do all the electrons participate in the heat conduction? We have raised a similar question in Section 20.3, see Fig. 20.3. We know from there that only those electrons which have an energy close to the Fermi energy, E_F, are able to participate in the conduction process. Thus, the velocity in (21.5) and (21.8) is essentially the Fermi velocity, v_F, which can be calculated with

$$E_F = \tfrac{1}{2} m v_F^2 \tag{21.9}$$

if the Fermi energy is known (see Appendix).

Second, the number of participating electrons contained in (21.5) is proportional to the population density at the Fermi energy, $N(E_F)$, i.e., in first approximation, by the number of free electrons N_f per volume. Inserting the quantum mechanical expression for C_v^{el} (20.28)

$$C_v^{el} = \frac{\pi^2}{2} \frac{N_f k_B^2 T}{E_F} \left(\frac{J}{K \cdot m^3} \right) \tag{21.10}$$

into (21.8) yields

$$K = \frac{\pi^2 N_f k_B^2 T v_F l_F}{6 E_F}, \tag{21.11}$$

which reduces with (21.9) and $l_F = \tau v_F$ (7.15a) (τ = relaxation time) to

$$\boxed{K = \frac{\pi^2 N_f k_B^2 T \tau}{3 m^*}.} \tag{21.12}$$

This is the result we were seeking. Again, the heat conductivity is larger the more *free* electrons are involved and the smaller the (effective) mass of the electrons.

Next we return to a statement which we made in Chapter 18. We pointed out there that Wiedemann and Franz observed that good electrical conductors are also good thermal conductors. We are now in a position to compare the thermal conductivity (21.12) with the electrical conductivity (7.15)

$$\sigma = \frac{N_f e^2 \tau}{m^*}. \tag{21.13}$$

The ratio of K and σ is proportional to a constant called the Lorentz number L which is a function of two universal constants k_B and e

$$\frac{K}{\sigma T} = L = \frac{\pi^2 k_B^2}{3 e^2}. \tag{21.14}$$

The Lorentz number is calculated to be 5.838×10^{-9} ($\Omega \cdot$ cal/K$^2 \cdot$ s) or 2.443×10^{-8} (J $\cdot \Omega$/K$^2 \cdot$ s) (see Problem 2). Experiments for most metals confirm this number quite well.

21.3. Thermal Conduction in Dielectric Materials

Heat conduction in dielectric materials occurs by a flow of phonons. The hot end possesses more phonons than the cold end causing a drift of phonons down a concentration gradient.

The thermal conductivity can be calculated similarly as in the previous section, which leads to the same equation as (21.8),

$$K = \tfrac{1}{3} C_v^{ph} v l. \tag{21.15}$$

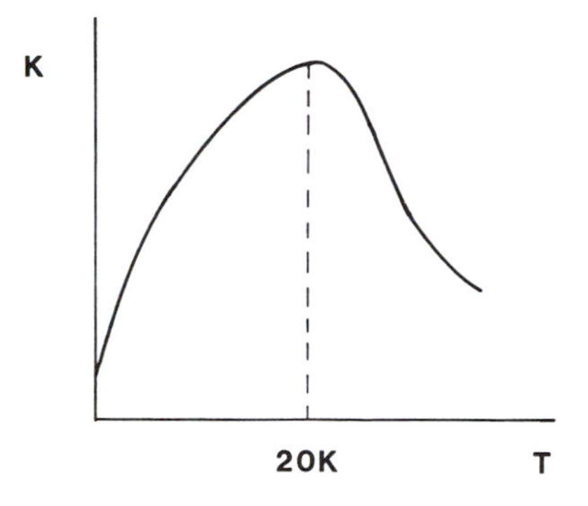

Figure 21.2. Schematic representation of the thermal conductivity in dielectrical materials as a function of temperature.

In the present case C_v^{ph} is the (lattice) heat capacity per volume of the phonons, v is the phonon velocity, and l the phonon mean free path. A typical value for v is about 5×10^5 cm/s (sound velocity) with v being relatively temperature-independent. In contrast, the mean free path varies over several orders of magnitude, i.e., from about 10 nm at room temperature to 10^4 nm near 20 K. The drifting phonons interact on their path with lattice imperfections, with external boundaries, and with other phonons. These interactions constitute a thermal resistivity which is quite analogous to the electrical resistivity. Thus, we may treat the thermal resistance just as we did in Part II; i.e., in terms of interactions between particles (here phonons) and matter, or in terms of the scattering of phonon waves on lattice imperfections.

At low temperatures, where only a few phonons exist, the thermal conductivity depends mainly on the heat capacity, C_v^{ph}, which increases with the third power of increasing temperature according to (20.20) (see Fig. 21.2). At low temperatures, the phonons possess small energies, i.e., long wavelengths which are too long to be scattered by lattice imperfections. The mean free path l becomes thus a constant and is virtually identical to the dimensions of the material.

More effective are the phonon–phonon interactions which are dominant at higher temperatures since, as we know, the phonon density increases with increasing T. Thus, the mean free path and consequently, the thermal conductivity decreases for temperatures above about 20 K (Fig. 21.2).

Another mechanism which impedes the flow of phonons at higher temperatures has been discovered. We explain this mechanism in quantum mechanical terms. When two phonons collide, a third phonon results in a proper manner to conserve momentum. Now, phonons (just like electrons) can be represented to travel in k-space. The same arguments, as discussed in Chapter 5, may then apply here. We need to consider Brillouin zones which represent the areas in which the phonon interactions occur. In the example of Fig. 21.3, the resultant vector $\mathbf{a}_1 + \mathbf{a}_2 = \mathbf{a}_3$ is shown to be outside the first Brillouin zone. We project this vector back to a corresponding place *inside* the first

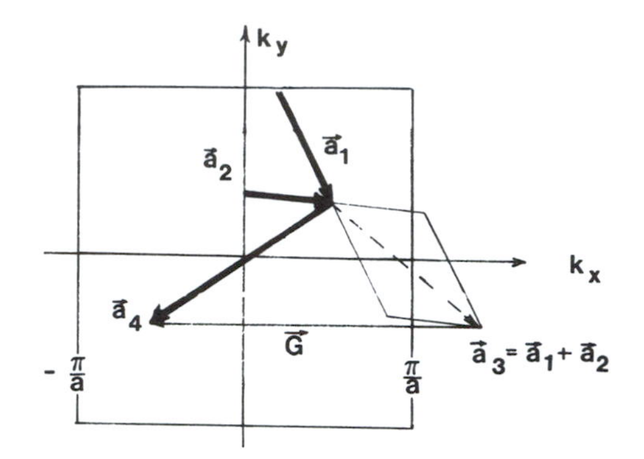

Figure 21.3. First Brillouin zone in a reciprocal square lattice. Two phonons $\mathbf{a}_1$ and $\mathbf{a}_2$ are shown to interact. In the example, the resultant vector $\mathbf{a}_3$ lies outside the first Brillouin zone.

Brillouin zone by applying a similar vector relationship as in (5.34)

$$\mathbf{a}_1 + \mathbf{a}_2 = \mathbf{a}_3 + \mathbf{G}, \tag{21.16}$$

where **G** is again a translational vector which has in the present case the modulus $-2\pi/a$ (see Fig. 21.3). As a consequence, the resultant phonon of vector $\mathbf{a}_4$ proceeds after the collision in a direction which is almost opposite to $\mathbf{a}_2$, which constitutes, of course, a resistance against the flow of phonons. This mechanism is called *umklapp* process (German for "flipping over" process).

Phonon collisions in which $\mathbf{a}_1$ and $\mathbf{a}_2$ are small, so that the resultant vector $\mathbf{a}_3$ stays inside the first Brillouin zone (i.e., $\mathbf{G} = 0$), are called *normal processes*. A normal process has no effect on the thermal resistance since the resultant phonon proceeds essentially in the same direction.

Problems

1. Calculate the thermal conductivity for a metal assuming $\tau = 3 \times 10^{-14}$ s, $T =$ 300 K, and $N_f = 2.5 \times 10^{22}$ el/cm^3.

2. Calculate the Lorentz number from values of e and k_B. Show how you arrived at the correct units!

3. Calculate the mean free path of electrons in a metal, such as silver, at room temperature from heat capacity and heat conduction measurements. Take $E_F =$ 5 eV, $K = 4.1 \times 10^2$ J/s·m·K, and $C_v^{el} = 1\%$ of the lattice heat capacity. (*Hint*: Remember that the heat capacity in (21.8) is given per volume!)

4. Why is the thermal conduction in dielectric materials two or three orders of magnitude smaller than in metals?

5. Why does the thermal conductivity span only 4 orders of magnitude whereas the electrical conduction spans nearly 25 orders of magnitude?
6. Is there a theoretical possibility for a thermal superconductor?
7. Discuss why the thermal conductivity of alloys is lower than that of the pure constituents.

CHAPTER 22

Thermal Expansion

The length L of a rod increases with increasing temperature. Experiments have shown that in a relatively wide temperature range the linear expansion ΔL is proportional to the increase in temperature ΔT. The proportionality constant is called the coefficient of linear expansion α_L. The observations can be summarized in

$$\frac{\Delta L}{L} = \alpha_L \Delta T. \tag{22.1}$$

Experimentally observed values for α_L are given in Table 22.1.

The expansion coefficient has been found to be proportional to the molar heat capacity C_v, i.e., the temperature dependence of α_L is similar to the temperature dependence of C_v. As a consequence, the temperature dependence of α for dielectric materials follows closely the $C_v = f(T)$ relationship predicted by Debye and shown in Fig. 18.2. Specifically, α approaches a constant value for $T > \theta_D$ and vanishes at T^3 for $T \to 0$. The thermal expansion coefficient for metals, on the other hand, decreases at very small temperatures in proportion to T, and depends, in other temperature regions, on the sum of the heat capacities of phonons and electrons.

We turn now to a discussion of possible mechanisms which may explain thermal expansion from an atomistic point of view. We postulate, as in the previous chapters, that the lattice atoms absorb thermal energy by vibrating about their equilibrium position. In doing so, a given atom responds with increasing temperature and vibrational amplitude to the repulsive forces of the neighboring atoms. Let us consider for a moment two adjacent atoms only, and let us inspect their potential energy as a function of internuclear separation (Fig. 22.1). We understand that as two atoms move closer to each other, strong repulsive forces are experienced between them. As a conse-

Table 22.1. Linear Expansion Coefficients α_L for Some Solids Measured at Room Temperature.

Substance	α_L in 10^{-5} [K^{-1}]
Hard rubber	8.00
Lead	2.73
Aluminum	2.39
Brass	1.80
Copper	1.67
Iron	1.23
Glass (ordinary)	0.90
Glass (pyrex)	0.32
NaCl	0.16
Invar (Fe–36% Ni)	0.07
Quartz	0.05

quence, the potential energy curve rises steeply with decreasing r. On the other hand, we know that two atoms also attract each other somewhat. This results in a slight decrease in $U(r)$ with decreasing r.

Now, for small temperatures, a given atom may rest in its equilibrium position r_0, i.e., at the minimum of potential energy. If, however, the temperature is raised, the amplitude of the vibrating atom increases too. Since the amplitudes of the vibrating atom are symmetric about a median position and since the potential curve is not symmetric, a given atom moves further apart from its neighbor, i.e., the average position of an atom moves to a larger r, say, r_T, as shown in Fig. 22.1. In other words, the thermal expansion is a direct

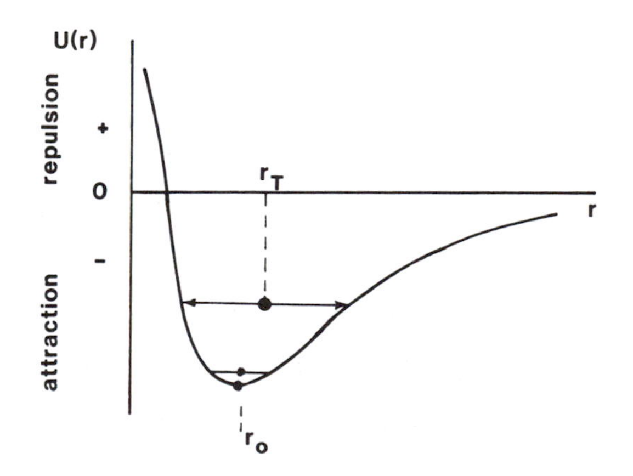

Figure 22.1. Schematic representation of the potential energy, $U(r)$, for two adjacent atoms as a function of internuclear separation, r.

consequence of the asymmetry of the potential energy curve. The same arguments hold true if all atoms in a solid are considered.

A few substances are known to behave differently from that described above. They contract during a temperature increase. This happens, however, only within a narrow temperature region. For its explanation, we need to realize that longitudinal as well as transverse vibrational modes may be excited by thermal energy (see Section 20.2). The lattice is expected to contract if transverse modes predominate. Interestingly enough, only one known liquid substance, namely water, behaves in a limited temperature range in this manner. Specifically, water has its largest density at 4° C. (Furthermore, the density of ice is smaller than the density of water at the freezing point.) As a consequence, water of 4° C sinks to the bottom of a lake during winter, while ice stays on top. This prevents the freezing of a lake at the bottom and thus enables aquatic life to survive during the winter. This exceptional behavior of water suggests that the laws of physics do not just "happen," but rather they were created by a superior being. I want to conclude my book with this thought.

Problems

1. Estimate the force which is exerted by the end of a 1 m long iron rod of 1 cm^2 cross section which was heated to 100° C.
2. Calculate the gap which has to be left between two 10 m long railroad tracks when they are installed at 0° C and if no compression is allowed at 40° C.
3. Explain some engineering applications of thermal expansion, such as the bimetal thermal switch, metal thermometer, etc.
4. What happens if a red hot piece of glass is immersed into cold water? What happens if the same experiment is done with quartz?

Suggestions for Further Reading (Part V)

A.J. Dekker, *Solid State Physics*, Prentice-Hall, Englewood Cliffs, NJ (1957).

C. Kittel and H. Kroemer, *Thermal Physics*, 2nd ed., W.H. Freeman, San Francisco, CA (1980).

P.G. Klemens and T.K. Chu, eds., *Thermal Conductivity*, Vols. 1–17, Plenum Press, New York.

T.F. Lee, F.W. Sears, and D.L. Turcotte, *Statistical Thermodynamics*, Addison-Wesley, Reading, MA (1963).

J.M. Ziman, *Electrons and Phonons*, Oxford University Press, Oxford (1960).

APPENDICES

APPENDIX 1

Periodic Disturbances

A *vibration* is a time-dependent *or* space-dependent periodic disturbance. We restrict our discussion to harmonic vibrations. In this case, the space-dependence of time is represented by a simple sine or cosine function or, equivalently, because of the Euler equations (see Appendix 2), by an exponential function.

A.1.1. Undamped Vibration

(a) Differential equation for *time-dependent periodicity*:

$$m\frac{d^2u}{dt^2} + \kappa u = 0 \tag{A.1}$$

(m = mass, κ = retracting force parameter). A solution is

$$u = Ae^{i\omega t}, \tag{A.2}$$

where

$$\omega = \sqrt{\frac{\kappa}{m}} = 2\pi\nu \tag{A.3}$$

is the angular frequency and A is a constant called maximum amplitude.

(b) Differential equation for *space-dependent periodicity* (in one dimension):

$$a\frac{d^2u}{dx^2} + bu = 0. \tag{A.4}$$

Solution:

$$u = Ae^{i\alpha x} + Be^{-i\alpha x}, \tag{A.5}$$

where A and B are constants, and

$$\alpha = \sqrt{\frac{b}{a}}. \tag{A.6}$$

A.1.2. Damped Vibration

(a) Differential equation for *time-dependent periodicity*:

$$m\frac{d^2u}{dt^2} + \gamma\frac{du}{dt} + \kappa u = 0 \tag{A.7}$$

(γ is the damping constant). The solution is

$$u = Ae^{-\beta t} \cdot e^{i(\omega_0 t - \phi)}, \tag{A.8}$$

where

$$\omega_0 = \sqrt{\frac{\kappa}{m} - \beta^2} \tag{A.9}$$

is the resonance frequency,

$$\beta = \frac{\gamma}{2m} \tag{A.10}$$

is the damping factor, and ϕ is the phase (angle) difference. In a damped vibration, the amplitude $Ae^{-\beta t}$ decreases exponentially.

(b) Differential equation for *space-dependent periodicity*:

$$\frac{d^2u}{dx^2} + D\frac{du}{dx} + Cu = 0. \tag{A.11}$$

Solution:

$$u = e^{-Dx/2}(Ae^{i\rho x} + Be^{-i\rho x}), \tag{A.12}$$

where

$$\rho = \sqrt{C - \frac{D^2}{4}} \tag{A.13}$$

and A, B, C, and D are constants.

A.1.3. Forced Vibration (Damped)

Differential equation for *time-dependent periodicity*:

$$m\frac{d^2u}{dt^2} + \gamma\frac{du}{dt} + \kappa u = K_0 e^{i\omega t}. \tag{A.14}$$

The right-hand side is the time-periodic excitation force. The solution consists of the start-up vibration and a steady-state part. The steady-state solution is

$$u = \frac{K_0}{\sqrt{m^2(\omega_0^2 - \omega^2)^2 + \gamma^2\omega^2}} e^{i(\omega t - \phi)}, \tag{A.15}$$

where

$$\omega_0 = \sqrt{\frac{\kappa}{m}} \tag{A.16}$$

is the resonance frequency of the undamped, free oscillation.

The tangent of the phase difference ϕ between the excitation force and the forced vibration is

$$\tan\phi = \frac{\gamma\omega}{m(\omega_0^2 - \omega^2)}. \tag{A.17}$$

A.1.4. Wave

A wave is a space- *and* time-dependent periodic disturbance.

(a) The differential equation for the undamped wave is

$$v^2\nabla^2 u = \frac{\partial^2 u}{\partial t^2}, \tag{A.18}$$

where

$$\nabla^2 = \frac{\partial^2}{\partial x^2} + \frac{\partial^2}{\partial y^2} + \frac{\partial^2}{\partial z^2}. \tag{A.19}$$

The differential equation for the plane wave is

$$v^2 \frac{\partial^2 u}{\partial x^2} = \frac{\partial^2 u}{\partial t^2}, \tag{A.20}$$

whose solution is

$$u(t, x) = e^{i\omega t}(Ae^{i\alpha x} + Be^{-i\alpha x}), \tag{A.21}$$

or

$$u(t, x) = Ae^{i(\omega t + \alpha x)} + Be^{i(\omega t - \alpha x)}, \tag{A.22}$$

or

$$u(t, x) = Ae^{i\omega(t + (x/v))} + Be^{i\omega(t - (x/v))}. \tag{A.23}$$

(b) Damped wave:

$$v^2 \frac{\partial^2 u}{\partial x^2} = a\frac{\partial u}{\partial t} + b\frac{\partial^2 u}{\partial t^2}. \tag{A.24}$$

The solution is possible by using (A.23). The general wave equation

$$v^2 \nabla^2 u = a \frac{\partial u}{\partial t} + b \frac{\partial^2 u}{\partial t^2} \tag{A.25}$$

can be solved with

$$u(t, x, y, z) = A e^{i(\omega t - \mathbf{k} \cdot v)}, \tag{A.26}$$

where

$$|\mathbf{k}| = \frac{2\pi}{\lambda} \tag{A.27}$$

is the wave number vector. It has the unit of a reciprocal length.

APPENDIX 2

Euler Equations

$$\cos\phi = \tfrac{1}{2}(e^{i\phi} + e^{-i\phi}), \tag{A.28}$$

$$\sin\phi = \frac{1}{2i}(e^{i\phi} - e^{-i\phi}), \tag{A.29}$$

$$\sinh\phi = \tfrac{1}{2}(e^{\phi} - e^{-\phi}) = \frac{1}{i}\cdot\sin i\phi, \tag{A.30}$$

$$\cosh\phi = \tfrac{1}{2}(e^{\phi} + e^{-\phi}) = \cos i\phi, \tag{A.31}$$

$$e^{i\phi} = \cos\phi + i\sin\phi, \tag{A.32}$$

$$e^{-i\phi} = \cos\phi - i\sin\phi. \tag{A.33}$$

APPENDIX 3

Summary of Quantum Number Characteristics

The energy states of electrons are characterized by four quantum numbers. The main quantum number, n, determines the overall energy of the electrons, i.e., essentially the radius of the electron distribution. It can have any integral value. For example, the electron of a hydrogen atom in its ground state has $n = 1$.

The quantum number, l, is a measure of the angular momentum L of the electrons and is determined by $|\mathbf{L}| = \sqrt{l(l+1)}\hbar$, where l can assume any integral value between 0 and $n - 1$.

It is common to specify a given energy state by a symbol which utilizes the n- and l-values. States with $l = 0$ are called s-states; with $l = 1$, p-states; and with $l = 2$, d-states, etc. A $4d$-state, for example, is one with $n = 4$ and $l = 2$.

The possible orientations of the angular momentum vector with respect to an external magnetic field are again quantized and are given by the magnetic quantum number m. Only m values between $+l$ and $-l$ are permitted.

The electrons of an atom fill the available states starting with the lowest state and obeying the Pauli principle which requires that each state can be filled only with two electrons having opposite spin ($|\mathbf{s}| = \pm\frac{1}{2}$). Because of the just-mentioned multiplicity, the maximal number of electrons in the s-states is 2, in the p-states 6, in the d-states 10, and in the f-states 14.

The electron bands in solids are named by using the same nomenclature as above, i.e., a $3d$-level in the atomic state widens to a $3d$-band in a solid. The electron configurations of some isolated atoms are listed on the next page.

		K	*L*		*M*			*N*				*O*			
Z	Element	1*s*	2*s*	2*p*	3*s*	3*p*	3*d*	4*s*	4*p*	4*d*	4*f*	5*s*	5*p*	5*d*	5*f*
1	H	1													
2	He	2													
3	Li	2	1												
4	Be	2	2												
5	B	2	2	1											
6	C	2	2	2											
7	N	2	2	3											
8	O	2	2	4											
9	F	2	2	5											
10	Ne	2	2	6											
11	Na	2	2	6	1										
12	Mg	2	2	6	2										
13	Al	2	2	6	2	1									
14	Si	2	2	6	2	2									
15	P	2	2	6	2	3									
16	S	2	2	6	2	4									
17	Cl	2	2	6	2	5									
18	Ar	2	2	6	2	6									
19	K	2	2	6	2	6		1							
20	Ca	2	2	6	2	6		2							
21	Sc	2	2	6	2	6	1	2							
22	Ti	2	2	6	2	6	2	2							
23	V	2	2	6	2	6	3	2							
24	Cr	2	2	6	2	6	5	1							
25	Mn	2	2	6	2	6	5	2							
26	Fe	2	2	6	2	6	6	2							
27	Co	2	2	6	2	6	7	2							
28	Ni	2	2	6	2	6	8	2							
29	Cu	2	2	6	2	6	10	1							
30	Zn	2	2	6	2	6	10	2							
31	Ga	2	2	6	2	6	10	2	1						
32	Ge	2	2	6	2	6	10	2	2						
33	As	2	2	6	2	6	10	2	3						
34	Se	2	2	6	2	6	10	2	4						
35	Br	2	2	6	2	6	10	2	5						
36	Kr	2	2	6	2	6	10	2	6						
37	Rb	2	2	6	2	6	10	2	6			1			
38	Sr	2	2	6	2	6	10	2	6			2			
39	Y	2	2	6	2	6	10	2	6	1		2			
40	Zr	2	2	6	2	6	10	2	6	2		2			
41	Nb	2	2	6	2	6	10	2	6	4		1			
42	Mo	2	2	6	2	6	10	2	6	5		1			
43	Tc	2	2	6	2	6	10	2	6	5		2			

APPENDIX 4

Tables

The International System of Units (SI or mksA System)

In the SI unit system, essentially four base units, the meter, the kilogram (for the mass), the second, and the ampere are *defined*. Further base units are the Kelvin, the mole (for the amount of substance), and the Candela (for the luminous intensity). All other units are *derived units* as shown in the table below. Even though the use of the SI unit system is highly recommended, other unit systems are still widely used.

Quantity	Name	Symbol	Expression in terms of	
			Other SI units	SI base units
Force	Newton	N	—	$kg \cdot m/s^2$
Energy, work	Joule	J	$N \cdot m = V \cdot A \cdot s$	$kg \cdot m^2/s^2$
Pressure	Pascal	Pa	N/m^2	$kg/m \cdot s^2$
El. charge	Coulomb	C	J/V	$A \cdot s$
Power	Watt	W	J/s	$kg \cdot m^2/s^3$
El. potential	Volt	V	W/A	$kg \cdot m^2/A \cdot s^3$
El. resistance	Ohm	Ω	V/A	$kg \cdot m^2/A^2 \cdot s^3$
El. conductance	Siemens	S	A/V	$A^2 \cdot s^3/kg \cdot m^2$
Magn. flux	Weber	Wb	$V \cdot s$	$kg \cdot m^2/A \cdot s^2$
Magn. induction	Tesla	T	Wb/m^2	$kg/A \cdot s^2$
Inductance	Henry	H	Wb/A	$kg \cdot m^2/A^2 \cdot s^2$
Capacitance	Farad	F	C/V	$A^2 \cdot s^4/kg \cdot m^2$

Physical Constants (cgs and SI units)

Mass of electron (free electron mass; rest mass)	$m_0 = 9.11 \times 10^{-28}$ (g) $= 9.11 \times 10^{-31}$ (kg)
Charge of electron	$e = 1.602 \times 10^{-19}$ (C)
	$= 4.803 \times 10^{-10}$ (statcoul) $\equiv$ ($cm^{3/2} \cdot g^{1/2}/s$) $\equiv$ (erg/stat V)
	$= 1.602 \times 10^{-20}$ (abcoul) $\equiv$ ($g^{1/2} \cdot cm^{1/2}$)
Velocity of light in vac.	$c = 2.998 \times 10^{10}$ (cm/s) $= 2.998 \times 10^{8}$ (m/s)
Planck constant	$h = 6.626 \times 10^{-27}$ ($g \cdot cm^2/s$) $= 4.136 \times 10^{-15}$ (eV·s)
	$= 6.626 \times 10^{-34}$ (J·s)
	$\hbar = 1.054 \times 10^{-27}$ ($g \cdot cm^2/s$) $= 6.582 \times 10^{-16}$ (eV·s)
	$= 1.054 \times 10^{-34}$ (J·s)
Avogadro constant	$N_0 = 6.022 \times 10^{23}$ (atoms/mol)
Boltzmann constant	$k_B = 1.381 \times 10^{-16}$ (erg/K) $= 8.616 \times 10^{-5}$ (eV/K)
	$= 1.381 \times 10^{-23}$ (J/K)
Bohr magneton	$\mu_B = 9.274 \times 10^{-21} \left(\frac{\text{erg}}{\text{Oe}}\right) \equiv (\text{Oe} \cdot \text{cm}^3) \equiv (g^{1/2}\ cm^{5/2}/s)$
	$= 9.274 \times 10^{-24}$ (J/T)
Gas constant	$R = 8.314$ (J/mol·K) $= 1.986$ (cal/mol·K)
Permittivity of empty space	$\varepsilon_0 = 1/\mu_0 c^2 = 8.854 \times 10^{-12}$ (F/m) $\equiv$ A·s/V·m
Permeability of empty space	$\mu_0 = 4\pi \times 10^{-7} = 1.257 \times 10^{-6}$ (H/m) $\equiv$ V·s/A·m

Useful Conversions

1 (eV) $= 1.602 \times 10^{-12}$ ($g \cdot cm^2/s^2$) $= 1.602 \times 10^{-19}$ ($kg \cdot m^2/s^2$) $= 1.602 \times 10^{-19}$ (J)
$= 3.829 \times 10^{-20}$ (cal)

$$1\ (\text{J}) = 1\left(\frac{\text{kg} \cdot \text{m}^2}{\text{s}^2}\right) = 10^7\ (\text{erg}) = 10^7\left(\frac{\text{g} \cdot \text{cm}^2}{\text{s}^2}\right) = 2.39 \times 10^{-1}\ (\text{cal})$$

1 (Rydberg) = 13.6 (eV)
1 (1/Ωcm) $= 9 \times 10^{11}$ (1/s)
1 (1/Ωm) $= 9 \times 10^{9}$ (1/s)
1 (C) = 1 (A·s) = 1 (J/V)
1 (Å) $= 10^{-10}$ (m)
1 (torr) $\equiv$ 1 (mm Hg) = 133.3 (N/m^2) $\equiv$ 133.3 (Pa)
1 (bar) $= 10^5$ (N/m^2) $\equiv 10^5$ (Pa)
1 (Pa) = 10 (dyn/cm^2)
1 cal $= 2.6118 \times 10^{19}$ (eV)

1 (mm) (milli) $= 10^{-3}$ (m)	1 km (Kilo) $= 10^3$ m
1 (μm) (micro) $= 10^{-6}$ (m)	1 Mm (Mega) $= 10^6$ m
1 (nm) (nano) $= 10^{-9}$ (m)	1 Gm (Giga) $= 10^9$ m
1 (pm) (pico) $= 10^{-12}$ (m)	1 Tm (Tera) $= 10^{12}$ m
1 (fm) (femto) $= 10^{-15}$ (m)	1 Pm (Peta) $= 10^{15}$ m
1 (am) (atto) $= 10^{-18}$ (m)	1 Em (Exa) $= 10^{18}$ m

Electronic Properties of Some Metals

Material	Effective mass $\left(\frac{m^*}{m_0}\right)_{el}$	$\left(\frac{m^*}{m_0}\right)_{opt}$	Fermi energy, E_F [eV]	Number of free electrons, N_{eff} $\left[\frac{electrons}{m^3}\right]$	Work function (photoelectric) ϕ [eV]	Resistivity ρ [$\mu\Omega$ cm] at 20° C
Ag		0.95	5.5	6.1×10^{28}	4.7	1.59
Al	0.97	1.08	11.8	16.7×10^{28}	4.1	2.65
Au		1.04	5.5	5.65×10^{28}	4.8	2.35
Be	1.6		12.0		3.9	4.0
Ca	1.4		3.0		2.7	3.91
Cs			1.6		1.9	20.0
Cu	1.0	1.42	7.0	6.3×10^{28}	4.5	1.67
Fe	1.2				4.7	9.71
K	1.1		1.9		2.2	6.15
Li	1.2		4.7		2.3	8.55
Na	1.0		3.2		2.3	4.20
Ni	2.8				5.0	6.84
Zn	0.85		11.0	3×10^{28}	4.3	5.91

Electronic Properties of Some Semiconductors

	Material	Gap energy E_g [eV]		Conductivity $\sigma\left[\frac{1}{\Omega \cdot m}\right]$	Mobility of electrons $\mu_e\left[\frac{m^2}{V \cdot s}\right]$	Mobility of holes $\mu_h\left[\frac{m^2}{V \cdot s}\right]$	Work function (photoelectric) ϕ [eV]	Effective mass at 4 K	
		0 K	300 K					$\frac{m_e^*}{m_0}$	$\frac{m_h^*}{m_0}$
Element	C (diamond)	5.48	5.47	10^{-12}	0.18	0.12	4.8	0.2	0.25
	Ge	0.74	0.66	2.2	0.39	0.19	4.6	1.64[a] 0.08[b]	0.04[c] 0.28[d]
	Si	1.17	1.12	9×10^{-4}	0.15	0.045	3.6	0.98[a] 0.19[b]	0.16[c] 0.49[d]
	Sn (gray)	0.08		10^6	0.14	0.12	4.4		
III–V	GaAs	1.52	1.42	10^{-6}	0.85	0.04		0.067	0.082
	InAs	0.42	0.36	10^4	3.30	0.046		0.023	0.40
	InSb	0.23	0.17		8.00	0.125		0.014	0.40
	GaP	2.34	2.26		0.01	0.007		0.82	0.60
IV–IV	α-SiC	3.03	2.99		0.04	0.005		0.60	1.00
II–VI	ZnO	3.42	3.35		0.02	0.018		0.27	
	CdSe	1.85	1.70		0.08			0.13	0.45

[a] Longitudinal effective mass.
[b] Transverse effective mass.
[c] Light-hole effective mass.
[d] Heavy-hole effective mass.

Ionization Energies for Various Dopants in Semiconductors (Experimental)

Donor ionization energies are given from the donor levels to the bottom of the conduction band. Acceptor ionization energies are given from the top of the valence band to the acceptor levels.

Semiconductor	Dopant		Ionization energy (eV)
	Type	Element	
Ge	Donors	Sb	0.0096
		P	0.012
		As	0.013
	Acceptors	B	0.01
		Al	0.01
		Ga	0.011
		In	0.011
Si	Donors	Sb	0.039
		P	0.045
		As	0.054
	Acceptors	B	0.045
		Al	0.067
		Ga	0.072
		In	0.16
GaAs	Donors	Si	0.0058
		Ge	0.006
		Sn	0.006
	Acceptors	Be	0.028
		Mg	0.028
		Zn	0.031

Physical Properties of Si and GaAs

	Si	GaAs
Lattice constant (Å)	5.431	5.654
Atoms (cm^{-3})	5.00×10^{22}	4.43×10^{22}
Band gap (eV) at 25 °C	1.11	1.43
Temperature dependence of band gap (eV $°C^{-1}$)	-2.4×10^{-4}	-4.3×10^{-4}
Specific gravity (g cm^{-3})	2.33	5.32
Dielectric constant	11.8	10.9
Electron lattice mobility (cm^2 V^{-1} s^{-1})	1.5×10^3	8.5×10^3
Hole lattice mobility (cm^2 V^{-1} s^{-1})	4.8×10^2	4×10^2
Number of intrinsic electrons (cm^{-3}) at 25 °C	1.5×10^{10}	1.1×10^6
Coefficient of linear thermal expansion ($°C^{-1}$) at 25 °C	2.33×10^{-6}	6.86×10^{-6}
Thermal conductivity (W $°C^{-1}$ m^{-1})	147	46

Optical Constants of Si and GaAs (from *Handbook of Optical Constants of Solids*, *Academic Press*, 1985)

		Si		GaAs	
E (eV)	λ (nm)	n	k	n	k
4.96	250	1.580	3.632	2.654	4.106
3.54	350	5.442	2.989	3.513	1.992
3.10	400	5.570	0.387	4.373	2.146
2.48	500	4.298	0.073	4.305	0.426
2.07	600	3.943	0.025	3.914	0.228
1.55	800	3.688	0.006	3.679	0.085
0.91	1370	3.5007	$\rightarrow 0$	3.3965	$\rightarrow 0$

Magnetic Units

Name	Symbol	em-cgs units	mks (SI) units	Conversions
Magnetic field strength	**H**	$\text{Oe} \equiv \frac{\text{g}^{1/2}}{\text{cm}^{1/2} \cdot \text{s}}$	$\frac{\text{A}}{\text{m}}$	$1\frac{\text{A}}{\text{m}} = \frac{4\pi}{10^3}\text{Oe}$
Magnetic induction	**B**	$\text{G} \equiv \frac{\text{g}^{1/2}}{\text{cm}^{1/2} \cdot \text{s}}$	$\frac{\text{Wb}}{\text{m}^2} = \frac{\text{kg}}{\text{s} \cdot \text{C}} \equiv \text{T}$	$1\ \text{T} = 10^4\ \text{G}$
Magnetization	**M**	$\frac{\text{Maxwell}}{\text{cm}^2} \equiv \frac{\text{g}^{1/2}}{\text{cm}^{1/2} \cdot \text{s}}$	$\frac{\text{A}}{\text{m}}$	$1\frac{\text{A}}{\text{m}} = \frac{4\pi}{10^3}\frac{\text{Maxwells}}{\text{cm}^2}$
Magnetic flux	$\boldsymbol{\phi}$	$\text{Maxwell} \equiv \frac{\text{cm}^{3/2} \cdot \text{g}^{1/2}}{\text{s}}$	$\text{Wb} = \frac{\text{kg} \cdot \text{m}^2}{\text{s} \cdot \text{C}} = \text{V} \cdot \text{s}$	$1\ \text{Wb} = 10^8\ \text{Maxwells}$
Susceptibility	χ	Unitless	Unitless	$\chi_{\text{mks}} = 4\pi\chi_{\text{cgs}}$
(Relative) permeability	μ	Unitless	Unitless	Same value
Energy product	BH	MGOe	$\frac{\text{kJ}}{\text{m}^3}$	$1\frac{\text{kJ}}{\text{m}^3} = \frac{4\pi}{10^2}\text{MGOe}$

Conversions Between Various Unit Systems

SI	Electrostatic cgs (esu) units	Electromagnetic cgs (emu) units	emu-esu conversion
1 (C)	3×10^9 (statcoul) $\equiv \left(\frac{cm^{3/2} \cdot g^{1/2}}{s}\right)$	$\frac{1}{10}$(abcoul) $\equiv (g^{1/2} \cdot cm^{1/2})$	1 (abcoul) = f (statcoul)
1 (V)	$\frac{1}{300}$(statvolts) $\equiv \left(\frac{cm^{1/2} \cdot g^{1/2}}{s}\right)$	10^8 (abvolts) $\equiv \left(\frac{cm^{3/2} \cdot g^{1/2}}{s^2}\right)$	1 (abvolt) = $\frac{1}{f}$ (statvolts)
1 (A)	3×10^9 (statamps) $\equiv \left(\frac{cm^{3/2} \cdot g^{1/2}}{s^2}\right)$	$\frac{1}{10}$(abamps) $\equiv \left(\frac{cm^{1/2} \cdot g^{1/2}}{s}\right)$	1 (abamp) = f (statamps)
1 (Ω)	$\frac{1}{9 \times 10^{11}}$ (statohms) $\equiv \left(\frac{s}{cm}\right)$	10^9 (abohms) $\equiv \left(\frac{cm}{s}\right)$	1 (abohm) = $\frac{1}{f^2}$ (statohms)

Note: The factor "f" in column four of this table has the value 3×10^{10} (s/cm).

Conventions from the Gaussian Unit System into the SI Unit System

The equations given in this book can be converted from the cgs (Gaussian) unit system into the SI (mks) system and vice versa by replacing the symbols in the respective equations with the symbols listed in the following table. Symbols which are not listed here remain unchanged. It is imperative that consistent sets of units are utilized.

Quantity	mks (SI)	cgs (Gaussian)
Magnetic induction	B	B/c
Magnetic flux	Φ_B	Φ_B/c
Magnetic field strength	H	$cH/4\pi$
Magnetization	M	cM
Magnetic dipole moment	μ_m	$c\mu_m$
Permittivity constant	ε_0	$1/4\pi$
Permeability constant	μ_0	$4\pi/c^2$
Electric displacement	D	$D/4\pi$

$\mu_0 = 4\pi \times 10^{-7} = 1.257 \times 10^{-6}$ (V·s/A·m) ≡ (Kg·m/C²) ≡ (H/m).
$\varepsilon_0 = 8.854 \times 10^{-12}$ (A·s/V·m) ≡ (F/m).

PERIODIC TABLE OF THE ELEMENTS

Group	Atomic number	Symbol	Name	Atomic mass (g/mol)	Oxidation states	Boiling point, °C	Melting point, °C	Density (g/ml)	Electron structure
IA	1	H	Hydrogen	1.00797	1	−252.7	−259.2	0.071	$1s^1$
IA	3	Li	Lithium	6.939	1	1330	180.5	0.53	$1s^2 2s^1$
IA	11	Na	Sodium	22.9898	1	892	97.8	0.97	$[Ne]3s^1$
IA	19	K	Potassium	39.102	1	760	63.7	0.86	$[Ar]4s^1$
IA	37	Rb	Rubidium	85.47	1	688	38.9	1.53	$[Kr]5s^1$
IA	55	Cs	Cesium	132.905	1	690	28.7	1.90	$[Xe]6s^1$
IA	87	Fr	Francium	(223)	1	—	(27)	—	$[Rn]7s^1$
IIA	4	Be	Beryllium	9.0122	2	2770	1277	1.85	$1s^2 2s^2$
IIA	12	Mg	Magnesium	24.312	2	1107	650	1.74	$[Ne]3s^2$
IIA	20	Ca	Calcium	40.08	2	1440	838	1.55	$[Ar]4s^2$
IIA	38	Sr	Strontium	87.62	2	1380	768	2.6	$[Kr]5s^2$
IIA	56	Ba	Barium	137.34	2	1640	714	3.5	$[Xe]6s^2$
IIA	88	Ra	Radium	(226)	2	—	700	5.0	$[Rn]7s^2$
IIIB	21	Sc	Scandium	44.956	3	2730	1539	3.0	$[Ar]3d^1 4s^2$
IIIB	39	Y	Yttrium	88.905	3	2927	1509	4.47	$[Kr]4d^1 5s^2$
IIIB	57	La ★	Lanthanum	138.91	3	3470	920	6.17	$[Xe]5d^1 6s^2$
IIIB	89	Ac ★★	Actinium	(227)	3	—	1050	—	$[Rn]6d^1 7s^2$
IVB	22	Ti	Titanium	47.90	4,3	3260	1668	4.51	$[Ar]3d^2 4s^2$
IVB	40	Zr	Zirconium	91.22	4	3580	1852	6.49	$[Kr]4d^2 5s^2$
IVB	72	Hf	Hafnium	178.49	4	5400	2222	13.1	$[Xe]4f^{14} 5d^2 6s^2$
IVB	104								
VB	23	V	Vanadium	50.942	5,4,3,2	3450	1900	6.1	$[Ar]3d^3 4s^2$
VB	41	Nb	Niobium	92.906	5,3	3300	2468	8.4	$[Kr]4d^4 5s^1$
VB	73	Ta	Tantalum	180.948	5	5425	2996	16.6	$[Xe]4f^{14} 5d^3 6s^2$
VIB	24	Cr	Chromium	51.996	6,3,2	2665	1875	7.19	$[Ar]3d^5 4s^1$
VIB	42	Mo	Molybdenum	95.94	6,5,4,3,2	5560	2610	10.2	$[Kr]4d^5 5s^1$
VIB	74	W	Wolfram	183.85	6,5,4,3,2	5930	3410	19.3	$[Xe]4f^{14} 5d^4 6s^2$
VIIB	25	Mn	Manganese	54.938	7,6,4,2,3	2150	1245	7.43	$[Ar]3d^5 4s^2$
VIIB	43	Tc	Technetium	(98)	7	—	2140	11.5	$[Kr]4d^5 5s^2$
VIIB	75	Re	Rhenium	186.2	7,6,4,2,−1	5900	3180	21.0	$[Xe]4f^{14} 5d^5 6s^2$
VIII	26	Fe	Iron	55.847	2,3	3000	1536	7.86	$[Ar]3d^6 4s^2$
VIII	44	Ru	Ruthenium	101.07	2,3,4,6,8	4900	2500	12.2	$[Kr]4d^7 5s^1$
VIII	76	Os	Osmium	190.2	2,3,4,6,8	5500	3000	22.6	$[Xe]4f^{14} 5d^6 6s^2$
VIII	27	Co	Cobalt	58.933	2,3	2900	1495	8.9	$[Ar]3d^7 4s^2$
VIII	45	Rh	Rhodium	102.905	2,3,4	4500	1966	12.4	$[Kr]4d^8 5s^1$
VIII	77	Ir	Iridium	192.2	2,3,4,6	5300	2454	22.5	$[Xe]4f^{14} 5d^7 6s^2$
VIII	28	Ni	Nickel	58.71	2,3	2730	1453	8.9	$[Ar]3d^8 4s^2$
VIII	46	Pd	Palladium	106.4	2,4	3980	1552	12.0	$[Kr]4d^{10} 5s^0$
VIII	78	Pt	Platinum	195.09	2,4	4530	1769	21.4	$[Xe]4f^{14} 5d^9 6s^1$
IB	29	Cu	Copper	63.54	2,1	2595	1083	8.96	$[Ar]3d^{10} 4s^1$
IB	47	Ag	Silver	107.870	1	2210	960.8	10.5	$[Kr]4d^{10} 5s^1$
IB	79	Au	Gold	196.967	3,1	2970	1063	19.3	$[Xe]4f^{14} 5d^{10} 6s^1$
IIB	30	Zn	Zinc	65.37	2	906	419.5	7.14	$[Ar]3d^{10} 4s^2$
IIB	48	Cd	Cadmium	112.40	2	765	320.9	8.65	$[Kr]4d^{10} 5s^2$
IIB	80	Hg	Mercury	200.59	2,1	357	−38.4	13.6	$[Xe]4f^{14} 5d^{10} 6s^2$
IIIA	5	B	Boron	10.811	3	—	(2030)	2.34	$1s^2 2s^2 2p^1$
IIIA	13	Al	Aluminum	26.9815	3	2450	660	2.70	$[Ne]3s^2 3p^1$
IIIA	31	Ga	Gallium	69.72	3	2237	29.8	5.91	$[Ar]3d^{10} 4s^2 4p^1$
IIIA	49	In	Indium	114.82	3	2000	156.2	7.31	$[Kr]4d^{10} 5s^2 5p^1$
IIIA	81	Tl	Thallium	204.37	3,1	1457	303	11.85	$[Xe]4f^{14} 5d^{10} 6s^2 6p^1$
IVA	6	C	Carbon	12.01115	±4,2	4830	3727	2.26	$1s^2 2s^2 2p^2$
IVA	14	Si	Silicon	28.086	4	2680	1410	2.33	$[Ne]3s^2 3p^2$
IVA	32	Ge	Germanium	72.59	4	2830	937.4	5.32	$[Ar]3d^{10} 4s^2 4p^2$
IVA	50	Sn	Tin	118.69	4,2	2270	231.9	7.30	$[Kr]4d^{10} 5s^2 5p^2$
IVA	82	Pb	Lead	207.19	4,2	1725	327.4	11.4	$[Xe]4f^{14} 5d^{10} 6s^2 6p^2$
VA	7	N	Nitrogen	14.0067	±3,5,4,2	−195.8	−210	0.81	$1s^2 2s^2 2p^3$
VA	15	P	Phosphorus	30.9738	±3,5,4	280w	44.2w	1.82w	$[Ne]3s^2 3p^3$
VA	33	As	Arsenic	74.922	±3,5	613*	817	5.72	$[Ar]3d^{10} 4s^2 4p^3$
VA	51	Sb	Antimony	121.75	±3,5	1380	630.5	6.62	$[Kr]4d^{10} 5s^2 5p^3$
VA	83	Bi	Bismuth	208.980	3,5	1560	271.3	9.8	$[Xe]4f^{14} 5d^{10} 6s^2 6p^3$
VIA	8	O	Oxygen	15.9994	−2	−183	−218.8	1.14	$1s^2 2s^2 2p^4$
VIA	16	S	Sulfur	32.064	±2,4,6	444.6	119.0	2.07	$[Ne]3s^2 3p^4$
VIA	34	Se	Selenium	78.96	−2,4,6	685	217	4.79	$[Ar]3d^{10} 4s^2 4p^4$
VIA	52	Te	Tellurium	127.60	−2,4,6	989.8	449.5	6.24	$[Kr]4d^{10} 5s^2 5p^4$
VIA	84	Po	Polonium	(210)	2,4	—	254	(9.2)	$[Xe]4f^{14} 5d^{10} 6s^2 6p^4$
VIIA	9	F	Fluorine	18.9984	−1	−188.2	−219.6	1.505	$1s^2 2s^2 2p^5$
VIIA	17	Cl	Chlorine	35.453	±1,3,5,7	−34.7	−101.0	1.56	$[Ne]3s^2 3p^5$
VIIA	35	Br	Bromine	79.909	±1,5	58	−7.2	3.12	$[Ar]3d^{10} 4s^2 4p^5$
VIIA	53	I	Iodine	126.904	±1,5,7	183	113.7	4.94	$[Kr]4d^{10} 5s^2 5p^5$
VIIA	85	At	Astatine	(210)	±1,3,5,7	—	(302)	—	$[Xe]4f^{14} 5d^{10} 6s^2 6p^5$
VIIIA	2	He	Helium	4.0026		−268.9	−269.7	0.126	$1s^2$
VIIIA	10	Ne	Neon	20.183		−246	−248.6	1.20	$1s^2 2s^2 2p^6$
VIIIA	18	Ar	Argon	39.948		−185.8	−189.4	1.40	$[Ne]3s^2 3p^6$
VIIIA	36	Kr	Krypton	83.80		−152	−157.3	2.6	$[Ar]3d^{10} 4s^2 4p^6$
VIIIA	54	Xe	Xenon	131.30		−108.0	−111.9	3.06	$[Kr]4d^{10} 5s^2 5p^6$
VIIIA	86	Rn	Radon	(222)		(−61.8)	(−71)	—	$[Xe]4f^{14} 5d^{10} 6s^2 6p^6$

★ Lanthanides

Atomic number	Symbol	Name	Atomic mass (g/mol)	Oxidation states	Boiling point, °C	Melting point, °C	Density (g/ml)	Electron structure
58	Ce	Cerium	140.12	3,4	3468	795	6.67	$[Xe]4f^2 5d^0 6s^2$
59	Pr	Praseodymium	140.907	3,4	3127	935	6.77	$[Xe]4f^3 5d^0 6s^2$
60	Nd	Neodymium	144.24	3	3027	1024	7.00	$[Xe]4f^4 5d^0 6s^2$
61	Pm	Promethium	(147)	3	—	(1027)	—	$[Xe]4f^5 5d^0 6s^2$
62	Sm	Samarium	150.35	3,2	1900	1072	7.54	$[Xe]4f^6 5d^0 6s^2$
63	Eu	Europium	151.96	3,2	1439	826	5.26	$[Xe]4f^7 5d^0 6s^2$
64	Gd	Gadolinium	157.25	3	3000	1312	7.89	$[Xe]4f^7 5d^1 6s^2$
65	Tb	Terbium	158.924	3,4	2800	1356	8.27	$[Xe]4f^9 5d^0 6s^2$
66	Dy	Dysprosium	162.50	3	2600	1407	8.54	$[Xe]4f^{10} 5d^0 6s^2$
67	Ho	Holmium	164.930	3	2600	1461	8.80	$[Xe]4f^{11} 5d^0 6s^2$
68	Er	Erbium	167.26	3	2900	1497	9.05	$[Xe]4f^{12} 5d^0 6s^2$
69	Tm	Thulium	168.934	3,2	1727	1545	9.33	$[Xe]4f^{13} 5d^0 6s^2$
70	Yb	Ytterbium	173.04	3,2	1427	824	6.98	$[Xe]4f^{14} 5d^0 6s^2$
71	Lu	Lutetium	174.97	3	3327	1652	9.84	$[Xe]4f^{14} 5d^1 6s^2$

★★ Actinides

Atomic number	Symbol	Name	Atomic mass (g/mol)	Oxidation states	Boiling point, °C	Melting point, °C	Density (g/ml)	Electron structure
90	Th	Thorium	232.038	4	3850	1750	11.7	$[Rn]5f^0 6d^2 7s^2$
91	Pa	Protactinium	(231)	5,4	—	(1230)	15.4	$[Rn]5f^2 6d^1 7s^2$
92	U	Uranium	238.03	6,5,4,3	3818	1132	19.07	$[Rn]5f^3 6d^1 7s^2$
93	Np	Neptunium	(237)	6,5,4,3	—	637	19.5	$[Rn]5f^4 6d^1 7s^2$
94	Pu	Plutonium	(242)	6,5,4,3	3235	640	—	$[Rn]5f^6 6d^0 7s^2$
95	Am	Americium	(243)	6,5,4,3	—	—	11.7	$[Rn]5f^7 6d^0 7s^2$
96	Cm	Curium	(247)	3	—	—	—	$[Rn]5f^7 6d^1 7s^2$
97	Bk	Berkelium	(247)	4,3	—	—	—	$[Rn]5f^8 6d^1 7s^2$
98	Cf	Californium	(249)	3	—	—	—	$[Rn]5f^{10} 6d^0 7s^2$
99	Es	Einsteinium	(254)	—	—	—	—	$[Rn]5f^{11} 6d^0 7s^2$
100	Fm	Fermium	(253)	—	—	—	—	$[Rn]5f^{12} 6d^0 7s^2$
101	Md	Mendelevium	(256)	—	—	—	—	$[Rn]5f^{13} 6d^0 7s^2$
102	No	Nobelium	(254)	—	—	—	—	$[Rn]5f^{14} 6d^0 7s^2$
103	Lw	Lawrencium	(257)	—	—	—	—	$[Rn]5f^{14} 6d^1 7s^2$

KEY

30 — ATOMIC NUMBER; 65.37 — ATOMIC MASS (g/mol); 2 — OXIDATION STATES (Bold most stable); 906 — BOILING POINT, °C; 419.5 — MELTING POINT, °C; 7.14 — DENSITY (g/ml); Zn — SYMBOL; $[Ar]3d^{10}4s^2$ — ELECTRON STRUCTURE; Zinc — NAME

APPENDIX 5

About Solving Problems

There are two types of exercises contained at the end of each chapter of this book; both of them are provided for the students to deepen their understanding of the material covered in the text. About 25% of the problems are concerned with conceptual reviews. These usually do not seem to be any major stumbling block to the reader. In contrast to this, however, the numerical problems are the ones which seem to provide some challenges. The goal of this section is to sketch a systematic approach for the solution of numerical problems and to give an actual example.

The first task is, of course, to find one or several equations which can be applied to the problem at hand. As a rule, however, the equations to be used are not yet provided in a form which lists the unknown variable on the left side of the equation and all the known variables plus a handful of constants on the right side. Thus, algebraic manipulations need to be applied until this goal has been achieved. (Under no circumstances should one insert numerical values immediately into the starting equations, in particular, if these variables are given in different unit systems.)

Once a final equation (containing the unknown quantity on the left side) has eventually been obtained, a unit check should be attempted by listing all known quantities in *one* unit system and inserting these units into the final equation. This provides a simple check on whether the algebraic manipulation was done correctly and in what unit the numerical result will turn out. Only then is a numerical calculation in place. At the end of each calculation the student should ask, "Does the result make sense?". A comparison with tabulated values in one of the appendices or with information given in the text can, most of the time, quickly answer this question. If the result seems to be off by several orders of magnitude, a recalculation should definitely be performed.

Example (Problem 2/1)

$$\left.\begin{aligned} \lambda &= \frac{h}{p}, \\ E &= \frac{p^2}{2m}, \end{aligned}\right\} \qquad \lambda = \frac{h}{\sqrt{2Em}}\left(\frac{\text{kg m}^2\text{ s s}}{\text{s}^2\text{ kg}^{1/2}\text{ m kg}^{1/2}}\right) \equiv (m),$$

$$E = 4\,(\text{eV}) = 4 \times 1.602 \times 10^{-19}\,(\text{J}) \equiv \left(\frac{\text{kg m}^2}{\text{s}^2}\right),$$

$$h = 6.626 \times 10^{-34}\,(\text{J s}),$$

$$m = 9.11 \times 10^{-31}\,(\text{kg}),$$

$$\lambda = 0.613 \times 10^{-9}\,(\text{m}) = 6.13\,(\text{Å}).$$

Solutions to Numerical Problems

Chapter 2

1. $\lambda = 6.13\,(\text{Å})$.
2. $E = 4.18 \times 10^{-6}\,(\text{eV})$.
4. $E = 2.07\,(\text{eV})$.
5. $\lambda = 2.38 \times 10^{-24}\,(\text{Å})$.

Chapter 4

6. $E = 13.6\,(\text{eV})$.

Chapter 5

1. $L_1 \approx 14\,(\text{eV})$; $L_2' \approx 8\,(\text{eV})$; $L_2' - L_1 \approx 6\,(\text{eV})$.
2. $E = 1.1\,(\text{eV})$.
3. $\Delta E = \hbar^2\pi^2/ma^2$; or $E_{111}/E_{100} = 3$.
5. (a) $X = 0 \rightsquigarrow E = 4C$; $X = \pi/a \rightsquigarrow E = 9C$ $(C = \pi^2\hbar^2/2ma^2)$;
   (b) $X = 0 \rightsquigarrow E = 16C$; $X = \pi/a \rightsquigarrow E = 9C$.
6. (a) $X = 0 \rightsquigarrow E = 4C$; $X = 1 \rightsquigarrow E = 1C$ $(C = 2\hbar^2\pi^2/ma^2)$;
   (b) $X = 0 \rightsquigarrow E = 2C$; $X = 1 \rightsquigarrow E = 5C$;
   (c) $X = 0 \rightsquigarrow E = 2C$; $X = 1 \rightsquigarrow E = 5C$.
7. $b_1 = (1/a)(\bar{1}11)$; $b_2 = (1/a)(1\bar{1}1)$; $b_3 = (1/a)(11\bar{1})$.
8. (a) $X = 0 \rightsquigarrow E = 0$; $X = 1 \rightsquigarrow E = \frac{1}{2}C$ $(C = 2\hbar^2\pi^2/ma^2)$;
   (b) $X = 0 \rightsquigarrow E = 2C$; $X = 1 \rightsquigarrow E = 1\frac{1}{2}C$;
   (c) $X = 0 \rightsquigarrow E = 4C$; $X = 1 \rightsquigarrow E = 2\frac{1}{2}C$.

Chapter 6

1. $v_F = 1.38 \times 10^6\,(\text{m/s})$.
3. $T = 290.5\,(\text{K})$.
4. $E_F = 5.64\,(\text{eV})$.

5. $Z(E) = 5.63 \times 10^{46}$ (electron states/J).
   $= 9.03 \times 10^{27}$ (electron states/eV).
6. For entire band: $N^*/V = 8.42 \times 10^{22}$ (1/cm^3).
7. $\eta = 2.5 \times 10^{23}$ (energy states).
8. (a) $N^* = 8.42 \times 10^{22}$ (electrons/cm^3);
   (b) $N_a = 8.49 \times 10^{22}$ (atoms/cm^3);
   (c) Not exactly one free electron per atom.
9. 0.88%.
10. $F(E) = \frac{1}{2}$.

12. (a)

n	1	2	3	4	5	6
error (%)	27	12	5	2	0.7	0.2

   (b) $E = 5.103$ (eV).

Chapter 7

1. $N_f = 5.9 \times 10^{22}$ (electrons/cm^3).
2. See Fig. 7.9.
3. $\tau = 2.5 \times 10^{-14}$ (s); $l = 393$ (Å).
5. $N_f = 2.73 \times 10^{22}$ (electrons/cm^3) or 1.07 (electrons/atom).
7. $N(E) = 1.95 \times 10^{47}$ (electrons/J cm^3) $\equiv 3.12 \times 10^{22}$ (electrons/eV cm^3). The joule is a relatively large energy unit for the present purpose.

Chapter 8

1. $N' = 9.77 \times 10^9$ (electrons/cm^3).

2.

T (K)	300	400	500	600	700
N_e (electrons/cm^3)	6.2×10^{-15}	2.4×10^{-6}	3.7×10^{-1}	1.1×10^3	3.5×10^5

3. $E_F = -E_g/2$ (using the bottom of the conduction band as the origin of the energy scale).
4. $T = 19{,}781$ (K) (!).
6. $E_g = 0.396$ (eV).
7. $E_F = -0.16$ eV; $\sigma = 31.2$ (1/Ω cm).
8. $(N_e)_{300^\circ C} = 7.88 \times 10^{14}$ (electrons/cm^3);
   $(N_e)_{350^\circ C} = 2.22 \times 10^{15}$ (electrons/cm^3).
   See also Fig. 8.9 (watch scale!).
9. (a) extrinsic $N_e = 1 \times 10^{13}$ (electrons/cm^3);
   (b) intrinsic $N_e = 9.95 \times 10^{10}$ (electrons/cm^3).
10. $E = 0.043$ (eV).

12.

Metal	Ag	Al	Au	Cu
ϕ_M (eV)	4.7	4.1	4.8	4.5

$\phi_M > \phi_S$,

$\phi_{Si} = 3.6$ (eV).

15. (a) $E_{light} > E_{gap}$;
   (b) $N = 1.6 \times 10^{14}$ (pairs/s).

16. $I_S = 2.97 \times 10^{-3}$ (A);
    $I_{net} = 3.2 \times 10^2$ (A).
20. $E = 2.58 \times 10^{-2}$ (eV).
22. 24 (nm).
23. $\eta = 0.97$.

Chapter 9

1. $\mu_{ion} = 3.32 \times 10^{-16}$ (m²/Vs);
   $\mu_{S.C.} = 0.1$ (m²/Vs).
2. $N_{ion} = 6.2 \times 10^{14}$ (sites/cm³).
3. $Q = 0.83$ (eV).
4. $\sigma_{ion} = 1.35 \times 10^{-15}$ (1/Ω cm).

Chapter 10

2.	el. (DC)	opt. (AC)
ρ	1.67×10^{-6} (Ω cm)	3.85×10^{-3} (Ω cm)
σ	5.99×10^{5} (Ω^{-1} cm^{-1})	2.6×10^{2} (Ω^{-1} cm^{-1})
σ	5.39×10^{17} (s^{-1})	2.34×10^{14} (s^{-1})

4. $Z = 27.8$ (nm).
5. $R_{Ag} = 98.88\%$; $R_{glass} = 5.19\%$.
6. $Z = 7.81$ (nm).
8. $T_2 = 94.3\%$.

Chapter 11

1. ν (s^{-1})	n	R (%)
1.43×10^{15}	0	100
1.44×10^{15}	0.1176	62
1.53×10^{15}	0.3556	23
2.0×10^{15}	0.6991	3.1
3.0×10^{15}	0.8791	0.4

2. $(\nu_1)_K = 1.03 \times 10^{15}$ (s^{-1});
   $(\nu_1)_{Li} = 1.92 \times 10^{15}$ (s^{-1}).
3. $(N_{eff})_{Na} = 1$; $(N_{eff})_K = 0.85$.
5. $R = 99.03\%$.
6. $\nu_1 = 2 \times 10^{15}$ (s^{-1}); $\nu_2 = 3.56 \times 10^{12}$ (s^{-1}).

7. λ (μm)	0.3	0.4	0.5	0.7	1	2	5
n_{calc}	1.599	1.460	1.416	1.384	1.369	1.359	1.356
ν (in 10^{14} s^{-1})	9.99	7.49	5.99	4.28	2.99	1.5	0.51

12. $N_{\text{eff}} = 5.49 \times 10^{22}$ (electrons/cm^3);
$N_a = 5.86 \times 10^{22}$ (atoms/cm^3);
$N_{\text{eff}}/N_a = 0.94$ (electrons/atom).

Chapter 12

3.

	(a)	(b)	(c)	(d)	(e)
A	Metal (Ni) High R in IR (intraband tr.)	None	1.5 eV (weak) 3 eV (strong)	yes	partially filled bands
B	Semiconductor (GaAs) Low R in IR (no intraband transitions)	IR	1.5 eV (Band gap)	no	filled bands

4. 1.5 ($\Sigma_3 \rightarrow \Sigma_1$ and $W_2' \rightarrow W_1$).

Chapter 13

1. $n_2 - n_3 \approx 10^{-2}$.
2. $t = 2$ (μm).
3. (a) $\beta_T = 16.1°$;
(b) $\beta_T = 41.8°$.
4. 1.15×10^{18} (cm^{-3}).
5. $\alpha = 1.6$ (cm^{-1}) $= 6.9$ (dB/cm).
7. $\lambda_{\text{disc}} = 503.2$ (nm); $\lambda/4 = 126$ nm.
8. $E = 1.4$ (eV); $\lambda = 886$ (nm)

Chapter 14

2. $H = 2.51 \times 10^2$ (Oe);
$H = 2 \times 10^4$ (A/m).

Chapter 15

1. $\chi = -0.78 \times 10^{-6}$ (for $Z = 2$).
6. $\zeta = 6.71 \times 10^{-3}$.
7. $\chi = 6.91 \times 10^{-5}$ (about one magnetic moment per atom).
8. $FeO \cdot Fe_2O_3$, $\mu_m = 4\ \mu_B$;
$CoO \cdot Fe_2O_3$, $\mu_m = 3\ \mu_B$.
10. $H_M = 2.1 \times 10^7$ (Oe).
11. Sure! (No tricks please.)

Chapter 16

1. $\chi_{\text{para}} = 4.3 \times 10^{-6}$ (Al?).
(Diamagnetism *not* taken into consideration).
3. Fe: 7.8 out of 10; Co: 8.2 out of 10.

4. Ferro Fe: $\mu_m = 2.2\ \mu_B$; Co: $\mu_m = 1.8\ \mu_B$;
   Ferri Fe: $\mu_m = 4\ \mu_B$; Co: $\mu_m = 3\ \mu_B$;
   The number of Bohr magnetons for a single iron atom is zero. Ferromagnetism needs interaction with other atoms.

Chapter 17

1. $E = 10^4$ (erg) $= 10^{-3}$ (J).
2. For a synchrotron a steady magnetic field is used. No eddy current! High flux multiplication needed. Consult Table 17.1.
6. Joule heating in wires.

Chapter 19

1. $N_a = 2.4 \times 10^{11}$ (atoms/m^3).
2. $-J_Q$: Glass 3.2×10^3 (J/s m^2);
   Al 4.0×10^5 (J/s m^2);
   Wood 1.6×10^2 (J/s m^2).
3. $T = 1{,}142$ (K) $= 869$ (°C).
5. Proper heat dissipation is essential in semiconductor devices.

Chapter 20

1. $\Delta N/N_{tot} = 0.566\%$.
2. $C_v = 24.4$ (J/K mol) $= 5.84$ (cal/K mol).
3. $C_v^{el} = 0.212$ (J/K mol) $= 0.051$ (cal/K mol).
4. $N(E_F) = 3.66 \times 10^{42}$ (energy states/mol J) $= 5.86 \times 10^{23}$ (energy states/mol eV); $N(E_F)$ per cubic centimeter is about one order of magnitude smaller.
6. $T = 2.1 \times 10^5$ (K).

Chapter 21

1. $K = 1.55 \times 10^2$ (J/s m K).
2. $L = 2.44 \times 10^{-8}$ (J Ω/K^2 s).
3. $l = 374$ (Å).

Chapter 22

1. $F = 2{,}600$ (N) (!).
2. $\Delta L = 4.9$ (mm).
4. Compare expansion coefficients (Table 22.1).

Index

Absorbance 182, 260–261
Absorption 180
  frequency dependence 194, 199
  of photon energy 206
  spectroscopic ellipsometry 218
Absorption bands 187, 188, 191, 199
  interband transitions 206
  silver spectrum 222
  small damping 200
  visible and ultraviolet spectrum 204
Absorption constant 180
Absorption product 180, 199
Acceptor impurities 106
Acceptor levels 106, 107–108, 109
Acoustic bands 342
Activation energy 163, 164, 165, 171
Adsorbed layers 214
Alkali halides 162, 164, 165, 237
Alkali metals 190–191
  band models 66, 67
Alloys
  optical spectra 226–229
  resistivity 86–87
Alnico alloys 319, 320
Aluminum
  band diagram 54, 59
  energy loss function 225–226
  Fermi energy 54, 60, 372
  as ohmic contact material 118–119
  reflection spectrum 224
  spectral dependence of polarization and absorption 224
Aluminum–antimonide 113
Aluminum–gallium–arsenide 113
Aluminum–gallium–arsenide–antimonide 113
Amber 75
Amorphous materials (metallic glasses)
  electrical conduction 167–173
  selenium 171–172
  silicon 124, 169, 171–172, 173
  zirconium–copper 168–169
Amorphous metals 317–318
Amorphous semiconductors 167–168, 170, 172, 173
Amorphous solid 167
Ampère, André 276
Ampere (unit) 370, 377
Amplification
  bipolar transistor 128, 129
  junction field-effect transistors 130–131
Amplitude
  electric field strength 179
  maximum 363
Amplitude modulation 250, 251, 308
Analog storage devices 261
Analog switches 131–132
Analyzer 216–217, 218
AND gate (device) 146, 147, 148
AND logic circuit 265–266
Angle of incidence 178
Angular frequency 4, 8, 179, 344–346, 363

Angular momentum quantum number 245, 368
Antiferromagnetism 272, 274, 286–288
 molecular field theory 299
 quantum theory 307–311
Arrhenius equation 164
Atomic mass 78, 90
Atomistic theory of the optical properties 186–202
Attenuation 182, 260
 index 180
Avalanche photodiode 125
Avalanching 122
 GaAs MESFET 135
Avogadro constant 78
Azimuth 215–216

Band(s) 31–32, 50–52
 forbidden 31, 38, 101, 160, 166
Band diagrams (structures) 4, 39
 for aluminum 54, 59
 for copper 54, 55, 59, 206
 for gallium–arsenide 56, 134
 for germanium 234
 for silicon 55, 56, 234
 insulators 156
 polymers 158, 160
 quantum dot 253
 quantum semiconductor devices 138
 semiconductors 156
Band gap 39, 54, 56
Band-gap energy 247
 Franz–Keldysh effect 257
 temperature dependence 248
Band model, consequences 66–67
Band overlapping 42, 58, 59, 66, 67, 304, 309
Barium ferrite 319, 320
Barkhausen effect 286
Base 127–128, 129
BCS theory of superconductivity (Bardeen, Cooper, and Schrieffer) 93–94
Beats 9, 308
Beer's equation 183
Bernal model 168
Bernal–Polk model 168
Bethe–Slater curve 309
Biasing 116
Bias voltage 117
BIFETs 132
Binary digits 145
Binding energy 22, 235–236
Binding strength 237
Biosensors 162
Bipolar junction transistor 127–130
Birefringent material 216
Bit 145
Bitter lines 286
Bivalent metals
 band models 66, 67
 highest electron energies 84
Black body radiation 178
Bloch function 28
Bloch theorem 168
Bloch wall 286
Bohr, Niels 23
Bohr magneton 280, 304, 306
 ferrites 289
 ferromagnetic metals 309–310
Boltzmann constant 61, 293, 337, 340
Boltzmann distribution function 61
Boltzmann factor 61–62
Boltzmann statistics 293, 343
Boltzmann tail 62, 102
Born's postulate 11
Borosilicate/phosphosilicate glass 238, 239
Bose–Einstein statistics 243
Bound electron(s) 222
 classical electron theory of dielectric materials 196–199
 quasi-elastic 197
Bound electron theory, small damping 200
Boundary conditions 15, 19, 20
 finite potential barrier 24, 25
Boundary problems 15
Boyle–Mariotte law 337
Bragg planes 40
Bragg reflection of an electron wave 43, 44
Bragg relation 42
Bragg rings 167
Branched polymers 156
Bravais lattices 44
Breakdown 122–123
 voltage 121
Brewster angle 245
Bridgman technique 141, 142
Brillouin function 296
Brillouin zones (BZ) 4, 38, 65, 66, 219
 boundary 42, 58
 first
  for bcc crystal structure 51
  for copper 58
  for fcc crystal structure 50, 53

for ordered structure 230, 231
interband transitions 207
one- and two-dimensional 40–43
phonon flow in dielectric materials 355–356
quasi-continuous bands 208
three-dimensional 44, 49–50
two-dimensional 40–43, 49, 57
vs. momentum of a photon wave vector 204
Bubble domain memory 324
Byte 145

Cadmium sulfide 113
Cadmium telluride 113
Calcia-stabilized zirconia 165
Caloric 329
Calorie 332–333
Canted antiferromagnetics 287–288
Carbon dioxide laser 244, 245
Carrier recombination 118
Ceramic superconductors 90, 91–92, 95–96
Channel (in field-effect transistors) 130, 131, 133
Characteristic penetration depth 181
Charge
of an electron 76, 78, 81–82, 188–189, 371
waveguide structure 256
Charge transfer 159–160, 162
complexes 162
salts 162
Chip 139, 144–145, 150
Chromium dioxide tapes 321
cis-$(CH)_x$ 158
Cladding 239, 248
Classical dispersion equation 212
Classical electron theory 4, 83
of dielectric materials 196–199
Classical free electron theory of metals 191–194
Classical infrared absorption 207
Classical oscillators 222
Classical theory of heat capacity 340–349
Closure domain structures 285
Cluster model approach 168, 169
CMOS technology 146
CMOSFET 132
Cobalt–samarium 319
Coercive field 282
Coercive force 312–313
Coercivity 282
ferromagnetic metals 322, 323, 324
permanent magnets 318, 319, 321
soft magnetic materials 314
Coherent scattering 77, 87, 155
Collector 127–128, 129, 263
Collimation 240
Color 177
Compact disk 262
Compass 271–272
Compensator 216–217, 218
Complementary MOSFET (CMOSFET) 132
Complex dielectric constant 180, 193, 221
Complex index of refraction 179–180
Compositional disorder 167
Compound semiconductors 112–113, 247, 253
fabrication 248
lattice constant 248
Compressibility 333
Conducting polymer 158
Conduction band 39, 98–102, 104
alloys 229
extrinsic semiconductors 105–109
ionic conductors 163
number of electrons 101, 103
optical computer 266
photodiodes 123
p–n rectifiers 120
polymers 156, 158, 160, 162
quantum semiconductor devices 136, 138
quantum well lasers 252
rectifying contacts 115, 116
semiconductors 155–156
extrinsic 106–109
silicon 234–236
tunnel diode 126, 127
Zener diodes 122
Conduction electrons, number of 111
Conductivity 83–84, 179, 192–193, 195, 317
alloys 86
classical electron theory 78, 80
extrinsic semiconductors 108–109
intrinsic semiconductors 104
quantum mechanical considerations 80–84
relation to reflectivity 184–185
semiconductors 76
Conjugated organic polymer 158
Conjugated radical ion 160

Constantan, thermocouple material 88
Constructive interference 264–265
Contact potential 88, 116
Contact printing of magnetic tapes 321
Continuous mode 243
Continuous random network 168
Continuum theory 3, 186
Conventional unit cell 44
of bcc lattice 46, 47
of fcc structure 45
Conversions between various unit systems 377
Cooper pair (electrons) 94, 95, 96
Copper
band diagram (structure) 54, 55, 59, 206
Fermi energy 55, 60
Fermi surface 58
first Brillouin zone 58
optical spectra 222–223
thermocouple material 88
Copper-based alloys
schematic band structure 229
threshold energies 228–229
Copper–gold alloys
optical spectrum of ordered vs. disordered 230
ordering effect 87–88
resistivity 87
Copper oxide 166
differential reflectograms 232
rectifiers 166
Copper–zinc alloys 226–227
optical spectra 226–227
Core losses 317
Corona wire 172
Corrosion
alloys 232
permanent magnets 321
Coulombic potential 22
Coupling of optical devices 258–260
by butting 258
by monolithic integration of optical components 259
by phase coherent energy transfer 258–259
by prism coupler 258
Covalent bonds 99, 100, 105, 156–157
Covalent elements 99
Critical magnetic field strength 90–91, 92
Critical point transitions 233
Critical points in the band structure 219
Critical temperature in superconductors 89, 90, 95
Crosstalk 254
Crucible pulling process 140
Cryotrons 90
Crystals
electron behavior 26–33
energy bands 35–59
imperfections, and electrical resistance 78
Cubic primitive lattice 46, 47
Curie constant 279, 296
Curie law 279, 295–296
Curie temperature 282–284, 288, 299, 309
chromium dioxide tapes 321–322
ferrimagnetism 290
ferromagnetic metals 310
permanent magnets 321
Curie–Weiss law 279, 282–283, 287, 293, 296–297
molecular field theory 299
susceptibility of paramagnetic metals 305–306
Current density 76, 80–82, 192
Current-voltage characteristic 117, 121, 126, 129, 132, 133, 139
Curves of equal energy 57
Czochralski method 140, 141, 145

Damped amplitude 181
Damped vibration 28–29, 364
Damping 237
constant 178–182, 197, 200, 364
waveguide structure 256
factor 191–192
force 79
frequency 193–195, 222
of electrons 188–191
Dangling bonds 168, 169, 171
Data-processing 145
Davisson, C. 7
d-bands 54, 56, 227–228, 280, 283
ferromagnetic metals 307
ordering 230
de Broglie relation 7, 342
Debye frequency 345
Debye temperature 248, 335, 346, 348–349
Debye theory 344–346, 358
Decibels 182, 260
Deep-lying localized electron states 166
Defect electron 69–70

Degenerate energy state 23
Demagnetization curve 318
Demagnetization field 318
Dense random packing 168
Density 78
  of modes 345
  of states 4, 62–64, 70, 81, 102
    amorphous materials 169, 170
    complete function within a band 65–66
    ferromagnetic metals 307
    optical spectra of materials 208
    paramagnetism and diamagnetism 303, 304–305
    schematic 83, 84
    superconductors 95
  of states tails 170
  of vibrational modes 342
Depletion layer 113–114, 116–117, 120, 134, 135
  electro-optical waveguides 256
  homojunction lasers 249
  junction field-effect transistor 130
  photovoltaic cell 125
  quantum efficiency of transverse photodiode 260
  semiconductor laser 245
  Zener diode 122
Depletion-type electro-optical waveguide 257
Depletion-type field-effect transistor 130, 134–135
Depletion-type metal–oxide semiconductor field-effect transistor (MOSFET) 131, 132
Diamagnetism 272–274, 276–278, 280–281
  quantum theory 302–306
Diatomic solids 237–238
Dielectric constant 179, 189, 235–236
  complex 180, 193, 221
Dielectric materials 237–239
Differential reflectograms 218, 219, 228
  copper–gold alloy 230, 231
  copper oxide 232
  copper–zinc alloys 226–227
  corrosion 232
Differential reflectometer 218–220
Diffraction 6
Diffusion coefficient 163–164
Diffusion current 114, 116
Diffusion theory 163–164
Digital circuits 145–152
Digital optical memory 262
Diodes 119–121
Dipole moment 197
Dipole momentum 189
Direct band-gap materials 112, 246–247
Direct interband transitions 204–205, 206
Dislocations 123
Dispersion 6, 10, 178, 208–213
Divergence of an emerging beam 250
Domain size 285
Domain wall 286
Donor electrons 105–106
  mobility of 108
  number of 108
Donor levels 106–109, 112
Doping 105, 107, 108, 121
  amorphous materials 171
  bipolar transistor 129
  field-effect transistor 131, 133, 134
  photovoltaic cell 125
  polymers 155, 159–161
  quantum semiconductor devices 136, 137, 138
  semiconductor device fabrication 140, 142, 143, 144, 145
  tunnel diodes 126
  waveguide structure 257
  Zener diodes 122
Drain 130, 131
  current 132, 133
Drift current 116, 120
Drift velocity 192
Drude, Paul 4, 185
Drude equations 4, 195, 199, 222
  for the optical constants 193
Drude model (atomistic model) 186–187
Drude theory of conduction 78, 81
Dry etching 144
Dulong–Petit value 335, 344, 348
Dynamic random-access memory (DRAM) 150

Easy direction 315, 316
Eddy current 288, 317
  loss 313, 315, 317
Effective mass (of electrons) 4, 67–70, 102–103, 222
  average 110
  extrinsic semiconductors 109–110
  gallium–arsenide 112, 134
  polymers 158

Effective mass (of electrons) (*cont.*)
  thermal conduction 354
  waveguide structure 256
Effective number of free electrons 190–191
Efficiency of a photovoltaic device 124
Eigenfunctions 15, 37
  normalized 15
  unperturbed *i*th 210
Eigenvalue problems 15
Einstein, Albert 6, 330
Einstein equation 345
Einstein frequency 344, 345
Einstein model 344, 345
  low frequency modes 346
  particle-wave duality model 342
Einstein relation 120, 163
Einstein temperature 344–345
Einstein's mass–energy equivalence 4
Elastic waves 342
Electric dipole 196, 197
Electric field force 79
Electric field strength 76, 179, 188–189, 192, 255
  dispersion, 208–209
  reflected, perpendicular components 216
Electric field vector 82–83
Electric motors 312, 313
Electric power storage 89
Electrical conduction 3, 75–96
  amorphous materials (metallic glasses) 167–173
  history 75
  metal oxides 165–167
  organic metals 155
  polymers 155–161
Electrical current 76
Electrical steels (soft magnetic materials) 312–318
  amorphous ferromagnetics 317–318
  core losses 313–315
  core material composition 317
  grain orientation 315–316
Electrically erasable-programmable read-only memory device (EEPROM) 151
Electromagnetic cgs (emu) units 274, 275
Electromagnetic wave 11, 179
  equation 179
Electromagnets 312
Electromet reduction 140
Electromigration 119
Electromodulation 220
Electromotive force 313
Electron(s)
  bound 19–23
  crystal behavior 26–33
  detection of 11
  donor 105–106, 108
  free 17, 78, 188–195
  free mobility 7
  "hot" 246
  in a potential well 19–23
  lifetime of 120–121
  mass of 14, 188–189
  number per unit volume 65, 76
  wave-particle duality 7
Electron affinity 115, 116
Electron–atom collisions 353
Electron bands 54–56, 234
  free 39, 52, 53
  nomenclature 368–369
Electron charge 76, 78, 81–82, 188–189
Electron collisions 76, 82
Electron density 120
Electron diffraction 7
Electron gas 62, 78
Electron hole 69–70
Electron–hole pairs 70, 96, 116
  amorphous materials 172
  avalanche photodiode 125
  photodiodes 123
  polymers 160
  semiconductors 235
Electron in a box 23
Electron–orbit paramagnetism 278, 279–280
Electron plasma 225
Electron theory of solids 3ff
Electron velocity 81–82
Electron wave 11
Electronic conduction 67
Electronic structure of solids 229
Electronic switch 129
Electro-optical waveguides (EOW) 256–257
Electrophotography 171–173
Electroreflectance 219, 220, 233, 234
Electrostatic force 78, 79
Electrotransport 119, 191
Elementary magnets 271
Ellipsometer 217
Elliptically polarized light 215, 216, 217
Emitter 127–128, 129, 263

Enamels 238
Energy, thermal 332–333
  total, of system 14
Energy bands. *See* Band diagrams
Energy barrier 163
Energy continuum 18
Energy gap 59, 95, 98–99, 103, 121
  in semiconductors 56
Energy levels 20, 22, 23
  distribution over a band 62–64
  quantization of 4
Energy loss
  by absorption 260
  by interband transitions 260
  by intraband transitions 260
  by radiation 261
  by scattering 260
Energy loss function 225–226
Energy quantization 20
Energy states 61–62, 94
  number of 63
Energy zone 31
  forbidden 67
Enhancement-type electro-optical waveguide 257
Enhancement-type MOSFET 131, 132, 133, 146
Enthalpy 333
Epitaxial layer 248
Erasable magneto-optic discs 262
Erasable-programmable read-only memory (EPROM) 151
Erbium-doped fiber amplifier 252
Ethylene 156
Euler's equations 367
Exchange energy 285, 307
Exchange force 297–298
Exchange integral 309, 311
Excitation energy 20
Excitation force 188–189, 365
Exciton 70, 96, 235–238
Exciton levels 235–237
Exciton wave 235
Expansion 329
  coefficient 333, 358–360
Extended zone scheme 37–38, 40
Extinction coefficient 180
Extrinsic region 164
Extrinsic semiconductors 105–109, 119–120, 236
  conductivity 108–109
  paramagnetism 305
  temperature dependence of the number of carriers 106–108
Fabry–Perot interferometer 263–264
Faraday effect 324
Faraday, Michael 272
Fermi–Dirac statistics 61
Fermi distribution function 4, 60, 61–62, 64
  semiconductors 100, 101, 102
Fermi energy 4, 62, 81, 95, 120, 227–228
  alloys 228–229
  amorphous materials 169
  definition 60
  extrinsic semiconductor 109
  ferromagnetism 307, 311
  metals 54, 55, 60, 372
  paramagnetism and diamagnetism (quantum theory) 303, 304, 305
  p–n rectifiers 119
  polymers 158
  rectifying contacts 114
  rigid band model 226
  semiconductors 100–101
  thermal conduction 353–354
  tunnel diode 126, 127
Fermi level. *See* Fermi energy
Fermi surface (or sphere) 58, 60, 81, 83–84, 227–228
  copper 58
  electron population density 349
  superconductivity 94
Fermi velocity 81, 82, 83, 353
Ferrimagnetism 272, 274, 288–291
  molecular field theory 299
Ferrite-core memories 323
Ferrites 313
  soft 322
Ferromagnetism 272, 273, 274, 281–286
  magnetic moments 310
  molecular field theory 297, 299
  quantum theory 307–311
Fiber-optics 239
Field-effect transistor 130–131, 162
Field ion microscope 25
Filamentary casting 167
First-derivative techniques 219
First law of thermodynamics 332
Flash memory
  cards (device) 151–152
Flip-flop 148, 149, 150
Floating gate 151
Float-zone technique 140–141
Floppy disk 324, 325
Fluorescence 239

Flux meter 281, 282
Flux multiplication 313
Flying head 325
Folded-chain model 157
Forbidden energy zones 42
Forward bias 116–117, 118, 120, 121
  bipolar junction transistor 128
  light-emitting diode 253
  semiconductor laser 245, 246
  tunnel diode 126, 127
Forward current 117
Fourier analysis 10
Fourier Law 336
Fourier transformation 8
Four-level laser 242
Franz–Keldysh effect 257, 258
Free carrier density 256–257
Free electron bands 38–39, 50–53
Free electron mass 67–68, 69
Free electron model 17, 78, 187–195
Free electrons 4, 17–18, 62, 277
  classical infrared absorption 207
  collisions with atoms 191
  contributions to optical constants 201–202
  effective number of 190–191
  in reduced zone scheme 38
  number of 78, 80, 85
  thermal conduction 351, 354
  with damping (classical free electron theory of metals) 191–194
  without damping 188–191
Frenkel defect 165
Frequency 6–8
  modulation 251
  response 133
Friction 332
  force 78, 79, 187, 192
Friedel, J. 228
Fringing 320
Fundamental edge 206
Fundamental vectors 46–47
Fused quartz 238

Gallium–aluminum–arsenide laser 263
Gallium–arsenide 112, 116
  absorbance 235
  band diagram 56
  crystal structure 112
  electron mobility 125
  MESFET 134, 135
  physical properties (table) 374, 375
  quantum efficiency 246
  quantum semiconductor devices 136–138
  semiconductor laser 244, 245
Gallium–arsenide metal–semiconductor field-effect transistor (GaAs MESFET) 134–136
Gallium–phosphide 113, 247
γ-Fe_2O_3 321
Gap energy 95, 99–100
  semiconductors 233, 235, 236, 373
Gas sensors 162
Gate 130, 131
  voltage 131, 132, 133
Gating input 149–150
Gauss 275
Gaussian unit system, conversions into SI unit system 378
Gay–Lussac law 337
General wave equation 366
Generation current 117
Generators 312
Germanium
  band diagram 234
  electroreflectance spectrum 234
  junction transistor 139
Germer, L. 7
Gettering 140
g-factor 280
Gigascale integration (GSI) 139, 140
Glass
  reflectivity 183
  transition temperature 157
Glass fibers 237–239
Glassy amorphous solid 157
Glassy metals 167
Glazes 238
Goethe, von, Johann Wolfgang 177
Gold–manganese alloys, ordering 87
Grain boundaries 77, 85, 123, 165
Graphite 162
Group velocity 10, 68

Hagen–Rubens equation 184–187, 195
Hall constant 111, 191
Hall effect 110–111
Hamiltonian operators 14–15
Hard drive systems 325
Hard magnetic materials 282, 318–321
Harmonic oscillators 187, 188, 196–197, 340–341
  contributions to optical constants 201–202

Harmonic vibrations 363
Harmonic wave 8
Headers 144
Heat 329, 332–333
Heat capacity 331, 333, 335
  at constant pressure 333
  at constant volume 333
  classical theory of 340–349
  electronic contribution 346–349
  per mole 348
  per volume 353
  quantum theory 342
  specific 330–331, 334–335
  thermal conduction 355
Heat conduction 329
Heat conductivity 336, 351–354
Heat flux 336
Heat treatments 85, 88
  polymers 157, 159
Heavy holes 110
Heisenberg's uncertainty principle 10, 241
Helium–neon laser 243, 244, 245, 263
Helmholtz, von, Hermann 329
Hertz, H. 6
Heterojunction lasers 249–250, 263
Heusler ferromagnetic alloys 309
High-energy particle accelerators 89
High-purity silicon 140
Hole density 103
Homojunction lasers 249–250
Hopping 158–159, 163, 164, 235
  amorphous materials 171
"Hot" electron 246
Hund's rule 280, 289, 309–310
Hybrid bands 99
Hybrid states 98–99
Hydrogen atom, wave mechanical properties 22
Hydrogenated amorphous silicon 171
Hysteresis
  loop 282, 286, 312–313, 316, 318, 323
  losses 313, 315, 317

Ideal crystal 77
Ideal diode law 120
Ideal gas equation 336–337
Ideal resistivity 85
Impact ionization 121
Impedance 131
Imperfect antiferromagnetics 288
Impurities 123
Impurity atoms 77
Impurity levels 107
Impurity states 106
Incident intensity 215
Incoherent scattering 77, 84, 87
  amorphous semiconductors 171
  extrinsic semiconductors 108
Index of refraction 177–179, 181, 183, 189, 199
  coupling and device integration 258–259
  optical computer 266
  photoelastic effect 258
  transphasor 265
  waveguide structure 256–257
Indirect band-gap materials 112, 246–247
Indirect interband transitions 205–206, 235, 236, 246, 247
Indium–antimonide 113
Indium–arsenide 113
Indium–gallium–arsenide–phosphide laser 248, 252, 253
Indium–phosphide 113
Indium–tin–oxide (ITO) 166
Induction 274, 275
Inertia effects 7
Information processing 145
Information storage 145
Infrared region 185, 186
Inhomogeneous magnetic field 272, 273
Input impedance 131, 132
Input transistor 146
Insulators 3, 42, 70, 237–239, 280, 287
  band model, consequences 66–67
  core materials 313
  described by harmonic oscillators 188
  electrical 76, 155
  electron bands 84
  gap energies 100
  interband transitions 237
  intraband transitions 237
  metal oxides 166
  optical properties 196, 207, 237
  thermal 330, 336
  valence bands 98
Integrated circuit 139, 144
Intensity 182
Interatomic force constant 237
Interband transitions 100, 204–208, 213–215, 220, 222, 230
  aluminum–copper 223, 224
  copper–zinc alloys 226–227
  energy losses 260
  germanium 234

Interband transitions (*cont.*)
insulators 237
semiconductors 233
silver 226
International System of Units 370–379
Internuclear separation 359
Intraband transitions
definition 207
energy losses 260
in semiconductors 232–233, 236
Intrinsic region 165
Intrinsic semiconductors 100–105, 280
conductivity 104
diamagnetism (quantum theory) 305
hole density 103
Inverse spinel structure 289–290
Inversion layer 131
Inverter circuit 146
Ion etching 144
Ion implantation 144
Ionic bonds 166
Ionic conduction 162–165
Ionic crystals 162–163, 165, 237–238, 280
Ionization energy 22, 112
dopants in semiconductors 374
rectifying contacts 114
Iron ferrite 272
Iron, magnetic properties 281–286, 307–309, 312–325
Iron–neodymium–boron 319, 320
Iron–silicon alloys 317
Isomers 158
Isotope effect 90
ITO 166

Josephson effect 95, 96
Joule, James Prescott 329
Joule (unit) 332–333, 370
Junction field-effect transistor (JFET) 130–131, 132

Kerr effect 286, 324
Kilobit 150
Kinetic energy 4
of electrons 337–339
of gases 337–339
Kinetic theory of gases 341, 351
Kondo effect 87
Kramers–Kronig analysis 215, 221, 223
Kronecker-Delta symbol 46
Kronig–Penney model 27

k-space 50–51, 53–54, 57, 68–69, 82
phonons in dielectric materials 355
quasi-continuous allowed energies 59
k-value, critical 43
k-vector 8, 18
energy of electron related 39

Langevin function 296, 298, 299
Langevin theory
of diamagnetism 291–293
of paramagnetism 293–297
Langevin variable 298
Larmor precession 277
Laser(s) 112, 239–254
amplifier 251–252
carbon dioxide 244, 245
coating to provide feedback for stimulated emission 259
definition 240
erbium-doped fiber amplifier 252
four-level 242–243
gallium–aluminum–arsenide 263
helium–neon 243, 244, 263
heterojunction 263
indium–gallium–arsenide–phosphide 252, 253
material properties 244
modulation 250–251
quantum well 252–253
spectral line width 248
temperature 250
three-level 242–243
threshold current density 248–249, 250, 252
"traveling-wave" 252
Laser cavity 240, 243, 264, 265
Laser printer 173
Lattice constant 5
compound semiconductors 248
Lattice vibration quanta 330, 342
Lead halide 165
Lead selenide 113
Lead sulfide 113
Lead telluride 113
Leakage (magnetic) 320
Lenz's law 276, 277, 291
Light
absorption by interband and intraband transitions 204–208
intensity 7, 178
particle concept 6
speed of 4, 178

wave-particle duality 7
wave properties 6
Light-emitting diodes (LED) 112, 113, 239, 253–254
Light holes 110
Light quantum 7
Linde's rule 86
Line-shape analysis 220
Littrow prism 243, 245
Load transistor 146
Localized states 170
Lodestone (Fe_3O_4) 272
Logic circuits 131–132
London theory of superconductivity 93
Longitudinal effective mass 110
Longitudinal vibrational modes 360
Long-range ordering 230
Lorentz equations 199–201
Lorentz force 110
Lorentz model (optics) 187, 196
Lorentz number 354
Low carbon steel 317
magnetic constants 273
Luminescence 239

Macromolecules 156, 157, 159
Magnetic anisotropy 315
Magnetic constants 273, 274
Magnetic core materials 312
Magnetic disk 325
Magnetic domains 283–285, 316, 321
ferromagnetism (quantum theory) 307
molecular field theory 297
Magnetic field
of earth 271
strength 272, 274, 275, 278, 303
Magnetic flux 274
density 274
lines 278, 281, 315, 323
Magnetic hardness 286
Magnetic induction 274, 275
Magnetic materials 272
Magnetic memories 323–325
Magnetic moment 274, 276–280, 284, 303, 306
alloys 310–311
antiferromagnetism 286, 287–288
diamagnetism 291, 292
ferrimagnetism 288, 289
ferromagnetic metals 310
molecular field theory 297
paramagnetism 293, 294, 295
Magnetic phenomena, classical approach 276–299
Magnetic quantum number 368
Magnetic recording (tapes, disks, drums) 321–323
Magnetic resonance imaging 89
Magnetic sensors 318
Magnetic short-range order 284
Magnetic tapes 272
Magnetic units (table) 376
Magnetism
basic concepts 272–274
foundations of 271–275
units 274–275, 376, 378
Magnetite 272
Magnetization 274, 278, 281–284
alloys 311
antiferromagnetism 288
diamagnetism 292
magnetic properties and quantum theory 303
molecular field theory 297, 298, 299
paramagnetism 295
stress dependence of 282
temperature dependence of 282, 283
Magnetization curve 281
Magneto-optic memories 324
Magnetoresistive recording 322
Magnetostatic energy 285
Magnetostriction 283
Magnetostrictive transducers 318
Main quantum number 368
Majority carriers 106, 108, 165
Manganese (ferromagnetism and antiferromagnetism) 309, 311
Mass
effective (of electrons) 67–70
rest 67, 371
Mass susceptibility 273
Materials barrier 139
Matter wave 9, 10
Matthiessen's rule 85, 86, 87
Maximum energy product (permanent magnets) 318, 319
Maxwell equations 4, 179, 183
Maxwell relation 180
Mayer, Julius 329
M_1 critical point 233
Mean free path 80, 85, 87, 352–353, 355
metallic glasses 170
Mechanical hardness 286
Mechanical heat equivalent 329
Meissner effect 278

Melt spinning 167
Memory addressing system 149
Memory devices 145–152, 261, 321–325
Mercury sulfide 113
MESFET 132–133, 134–136
Metal-in-gap (MIG) technology 322
Metal–insulator–semiconductor field-effect transistor (MISFET) 132
Metal–oxide–semiconductor field-effect transistor (MOSFET) 131–134
Metal–oxide–semiconductor transistor (MOST) 132
Metal oxides, electrical conduction 165–167
Metal–semiconductor contacts 113–114
Metal–semiconductor field-effect transistor (MESFET) 134–136
Metallic glasses, electrical conduction 167–173
Metals
 electronic properties (table) 372
 resistivity 84–85
METGLAS 2826A 170
Microwave frequency detectors 118
Miller indices 5, 43, 316
Minority carrier 120, 121
MISFET 132
Mobility 104, 108, 109, 120
 amorphous semiconductors 170–171
 gallium arsenide 112, 125, 134, 135
 ionic conductors 163
 polymers 158–159
 semiconductors 105
Mode 255
MODFET 133
Modulated amplitude 8–10
Modulated light wave 181
Modulation-doped field-effect transistors (MODFET) 113, 133
Modulation spectroscopy 219
Modulator of light 257
Molar heat capacity 330–331, 334–336, 340–342, 344
 Debye model 345
Molar mass 334–335
Molecular current 276
Molecular field 297
 constant 297, 299
 theory 297–299
Molecular magnetic field strength 299
Momentum 4, 7, 18, 94
 of photons 204
Monochromatic light 240
Monochromatic wave 9–10
Monochromator 218–219
Monomer 156
MOSFET 146–147, 257
MOST 132
Mott, N.F. 226
Muffin tin potential 28
Multichip modules (MCM) 145
Multilayer dielectric mirrors 245
Multiplexing 254
Mumetal 317

NAND gate 147, 148
n-channel field-effect transistors 131
Néel, L. 288
Néel temperature 287
Newton, Isaac 6
Newton's law 4
Nickel ferrite 288
Nickel oxide 166–167
NMOSFET 132
Noncrossing rule, quantum mechanical 37
Nonlinear optical material 263–264, 265
Nonvolatile memories 150
NOR gate 147, 148
Nordheim's rule 86
Norm 15
Normal processes 356
Normal spinel structure 289–290
Normalized eigenfunction 15
Normally-off electro-optical waveguide 257
Normally-off GaAs MESFET 134–135
Normally-off MOSFET 131, 133, 146, 147
Normally-on electro-optical waveguide 257
Normally-on field-effect transistor 134
"Normally on" MOSFET 131, 132
"Normally-on" type field-effect transistor 130
North pole 272, 285, 318
NOT gate 146–148, 266
n-type semiconductors 105, 166
Number of atoms 78
Number of electrons 81–82
n-values 22, 23, 38, 63

Oersted, H.C. 271, 275

Ohmic contact 113, 118–119
  GaAs MESFET 134
Ohm's law 4, 76, 81, 113, 118, 192
1-2-3 compound 89
Opacifiers 239
Opacity 238
Opal glasses 238
Operational amplifiers 131–132
Optical AND gate 265
Optical bands 342
Optical compact disk 262
Optical computer 263–266
Optical constants 177–185, 197–198, 217
  metals (table) 183
Optical fibers 252
Optical integrated circuits (OICs) 254
Optical pumping 241
Optical spectra
  of alloys 226–229
  of materials 208
  of pure metals 220–226
Optical storage devices 261–263
Optical switches 257–258, 265
Optical telecommunication 250
Optical transistor 263–264
Optical tunneling 255, 258–259
Optical waveguide structure 250
Optoelectronics 166
  integrated 254–261
OR gate 148, 266
Orbital paramagnetism 280
Orbits 23
Order–disorder transition temperature 87–88
Ordering 87–88, 230–231
  conduction electrons 94
  long-range 230
  short-range 230, 231
Organic metals 162
Organic polymers 156
Organic superconductors 95–96
Oscillator 77, 342, 344–346
  allowed energies 343, 344
  equations 364, 365
  more than one 200–201
  strength 212–213
Oxidation 142, 143
Oxide etch 143–144

Packaging 144–145
Packing 168, 353
Paramagnetic Curie point 283
Paramagnetism 272–274, 278–282, 284
  of antiferromagnetic materials 287
  ferromagnetic metals 311
  quantum theory 302–306
Particle-wave duality 6–11, 342
Passive waveguides 254–255
Pauli principle 64, 66, 106, 207, 278
  paramagnetism 280
  quantum number characteristics 368
p-channel field-effect transistors 131
Peltier effect 88–89
Penetration depth 182
Periodic table of the elements 379
Periodic zone scheme 37
Permalloy 282–283, 317, 322
Permanent magnets 282, 318–321
Permeability 272, 275, 286, 317
  electrical steel 314, 315
  of free space 275
  soft magnetic materials 312, 313
Perturbation theory 210
Phase 77
  coherent 240
  difference 197, 215, 216, 217
    damped vibration 364
  incoherent 239
  jump 215
  velocity 9, 10
Phonon(s) 112, 205, 207, 235–237, 330–331
  Debye model 344, 345–346
  Einstein model 342–345
  energy levels 343
  heat conduction in dielectric materials 354–355
  superconductivity theory 94, 96
  thermal conduction 351
  thermalization 246
Phonon-assisted decay 242
Phonon energy 343, 344–345, 347
Phonon mean free path 355
Phonon–phonon interactions 355
Phonon velocity 355
Phonon waves 342
Phosphorescence 11, 239
Phosphors 11
Photodiodes 123–125, 259
Photoelastic effect 258
Photoelectric effect 6
Photolithography 142–143
Photon(s) 7, 204, 218, 236
  emission 239–241
    in laser amplification 251, 252

Photon energy 206–207, 221–222, 225
  semiconductors 235
Photonic computer 266
Photoreceptor 172
Photoresist 142–143, 144, 263
Photovoltaic devices 124, 171
Piezoreflectance 219
Pitch 9, 308
Planar epitaxial transistor 139
Planck constant 7, 371
Planck's hypothesis 7
Plane of incidence 215
Plane polarized wave 179
Plasma 78
  etching 144
  frequency 190–191, 225
  oscillation 190, 225–226
PMOSFET 132
p–n rectifier (diode) 119–121
Point contact transistor 139
Polarization 189, 192
  dispersion 209, 211–212
  frequency dependence 194, 195, 197, 199
  spectroscopic ellipsometry 218
Polarizers 216, 217
Polyacetylene 158–159
  doped 159–161
Polyaniline 162
Polychromatic materials 239
Polycrystalline silicon 140
Polyethylene 156
Polymer(s)
  electrical conduction 155–161
  semicrystalline 157
Polymerization 156
Polypyrrole 160, 162
Polystyrene 156
Poly(sulfur nitride) 161
Polyvinylchloride (PVC) 156
Population density 64–65, 81, 83–84, 102, 347
  at Fermi surface 349
  thermal conduction 354
Population inversion 239, 241–245, 248
  quantum well lasers 252
Porcelains 238
Positional disorder 167
Positive charge carriers 106
Potential barrier 17, 19, 107, 114–116
  bipolar junction transistor 127, 128–129
  extrinsic semiconductor 119, 120
  finite 23–26
  ionic conduction 163
  quantum semiconductor devices 137, 138
  tunnel diodes 126, 127
Potential barrier strength 30, 31–32
Potential difference 76
Potential energy 13, 17
  finite potential barrier 23
Precessions of the electron orbits 277
Precursor polymer 159
Primitive unit cell 44
Primitive vectors 46
Probability of
  locating a particle in space 15, 20–22
  an electron transition 213
Probe beam 265
Programmable read-only memory (PROM) 150
p-type semiconductor 106
Pulse modulation 250–251
Pulse wave 10
Pulsed mode 243
Pump band 242
Pumping 241, 245
  semiconductor laser 245
  efficiency 241–243

Quantum dot 136–137, 139
  band structure 253
Quantum efficiency 125, 246
  heterojunction lasers 250
  homojunction lasers 250
  transverse photodiode 260
Quantum number characteristics 368–369
Quantum number space 62
Quantum numbers 33
Quantum semiconductor devices 136–139
Quantum states, number of 63
Quantum structures 136–139
Quantum theory 4, 6–7, 178, 302–311
Quantum well 136, 139
Quantum well lasers 252–253
Quantum wire 136
Quartzite 140
Quasi-Fermi levels 120
Quenched orbital moments 279
Quenching 88, 167
Q-switching 242, 243

Radiation damage 167
Random access
memory 261
read only memory devices 262
storage 323–325
Rapid solidification techniques 167
Rare earth elements, ferromagnetism 311
Raw silicon 140
Reactive plasma etching 144
Read-only memory (ROM) 150, 261
Reciprocal lattice 39–40, 45–50
Reciprocal space 39, 40–42, 44
Recombination 120–121
Rectifiers 166
voltage–current characteristics 117
Rectifying contacts 113–119
Reduced zone scheme 37–40, 50
Reflected intensity 215
Reflection spectra 220–224
for aluminum 224
for copper 222–223
for silver 221, 222, 226
Reflectivity 177, 182–184, 189, 195–196
frequency dependence 195–196
measurement of 214, 215
spectrum
Refraction, index of. *See* Index of refraction
Refractive power 178
Relative permeability 275
Relaxation time 79–80, 82, 83, 85, 354
Remalloy 319
Remanence 282
permanent magnets 318, 319, 321
Remanent magnetization 282
Repeat units 156
Repeaters 252
Residual resistivity 84–85
Resistance
bipolar junction transistor 128
electrical 76–92
standards 170
Resistive heating 332
Resistivity (definition) 77
of alloys 86–87
due to vacancies 85
ideal 85
metals 84–85
ordered alloys 87–88
residual 84–85
soft magnetic materials 314
superconductors 89, 90, 94
temperature coefficient of 84
Resonance 138
frequency 187, 188, 197, 199–200, 237
forced vibration (damped) 365
multiple absorption bands 201
quantum mechanical treatment 212
Resonant energy transfer 245
Rest mass (of electron) 67, 371
Restoring force 237, 308
Retentivity 282
Reverse bias 116–118, 120–121
Rigid band model 226, 227, 228
R–S flip-flop with latch 149
Ruby, laser 244
Rumford, Benjamin Thompson, Count 329
Rutherford model 22

Saturation current 117, 120
Saturation induction, soft magnetic materials 314
Saturation magnetization 282–284
amorphous ferromagnetics 317
ferrimagnetism 290
ferromagnetic metals 310
s-bands 54, 56, 66, 228
Scalar product 46, 49
Scattering 77, 87, 93–94, 96, 238–239
of phonon waves 355
Schottky-barrier contacts 114–118, 256–257
Schottky defects 163
Schrödinger equation 7, 13–15
perturbed 209–211
solutions 17–33, 209
time-dependent 14–15
time-independent 13–14
Screening effects 228
Seebeck effect 88
Seed crystal 140, 141
Self-demagnetization 321
Semiconductor(s) 3, 42, 70, 98–152
compound 112–113, 247–248, 253
conductivity 76
devices 113–152
electron bands 55, 56, 234
electronic properties (table) 373
extrinsic 105–109, 119–120, 236, 305
Fermi energy 100–101
intrinsic 100–105, 280, 305
ionization energies for various dopants 374

Semiconductor(s) (*cont.*)
  n-type 105
  optical properties 232–236
  p-type 106
  switches for computer applications 136
Semiconductor device fabrication 139–145
Semiconductor lasers 245
  direct vs. indirect band-gap 246–247
  wavelength 247–248
Sendust 322
Sequential storage media 261
Shape anisotropy 320, 321
Shockley equation 120
Short-range ordering 88, 167–168, 230–231
Silicon
  band diagram (structure) 55, 56
  physical properties (table) 375
  fabrication 139–140
Silicon carbide 113, 254
Silicon dioxide 140, 142–143
Silicon nitride 142
Silicon rectifiers 139
Silicon transistor (historic) 139
Silver
  energy loss function 226
  reflection spectrum 221, 222, 226
  resistivity changes due to alloying 86
Single crystal growth 140
SI units system 274–275, 332, 370–379
Size quantization 136, 138
Skin effect 182, 313, 315
Slater and Bethe calculation, exchange integral 309
Small angle X-ray scattering 88
Snell's law 178
Sodium chloride 280
  spectral reflectivity 238
Soft magnetic materials 282, 312–318
Solar cells 123–125
Solenoid 275, 281
Sol–gel silica "glasses" 238
Soliton 160–161
Sound velocity 355
Sound waves 344
Source 130, 131
Space charge region 113, 120
Space quantization 296
Specific heat 347, 348
Specific heat capacity 330–331, 334–335
Specific resistance 77
Specific susceptibility 273
Spectroscopic ellipsometry 215–218
Spin 64, 94, 278, 280, 368
  magnetic moments 316
  paramagnetism 278–280, 293, 296–297
Spontaneous emission 239, 240, 248
Spring constant 197, 237
Sputtering 167
Square potential well 62
SRAM memory device 149
Stacked-gate avalanche injected MOS (SAMOS) 151
Standing waves 22, 42
Static random-access memory 149
Steady-state solution 365
Stimulated emission 239, 240, 242, 252
Strontium ferrite 320
Superconducting polymers 161
Superconductivity 89–96, 191
  critical temperatures (table) 89
  experimental results 90–93
  theory 93–96
Superconductors 89–96, 162, 281
  ceramic 90–92, 95–96
  crystal structure 93
  magnetic constants 273
  magnetic properties 92
  Meissner effect 278
  organic 95–96
  resistivity 90, 94
  type I 91, 92
  type II 91, 92
Superlattice 139, 161, 231
Supermalloy 317
Surface 214
  emitter 254
  of equal energy 58, 59
Susceptibility 272–273, 275–276, 279, 287
  due to electron–orbit paramagnetism 295
  due to spin paramagnetism 304, 305
  electrical steels 313
  ferrimagnetism 290
  Langevin theory
    of diamagnetism 292–293
    of paramagnetism 293, 295–296
  metals, temperature-dependence 305, 306
  molecular field theory 297
  quantum theory and magnetic properties 303
  specific 273

Switching devices 145
Switching speeds 134
Switching time 133

Tape duplication 321
Technical saturation 284
Telecommunication 125, 247–248, 250–251
Temperature dependence of conductivity 108
Tesla (unit) 275, 376
Tetracyanoquinodimethane (TCNQ) 162
Tetrathiafulvalene (TTF) 162
Texture 316
Thermal conduction 351–356
  in dielectric materials 354–356
  in metals and alloys
    classical approach 351–353
    quantum mechanical considerations 353–354
  temperature dependence 355
Thermal conductivity 329–330, 336–337, (Table)
Thermal effective mass 349
Thermal energy 100, 329, 340, 343, 346–347, 352
  at a given temperature 347
  of solids 344
Thermal excitation 236
Thermal expansion 358–360
  coefficient 359, (Table)
Thermal insulation 329
Thermal properties 332–339
Thermal resistivity 355
Thermal vibration, of atoms 77
Thermal waves 344
Thermocouples 88
Thermoelectric phenomena 88–89
Thermoelectric power 88
Thermoreflectance 219
Thin magnetic films 324
Thomson, G.P. 7
Three-level laser 242
Threshold energy for interband transitions 206
  copper 223, 227
  copper-based alloys 227, 228
  germanium 233
Tin oxide 166
Titanium oxides 165–166
Tolman, R.C. 7
Toner 172
Total internal energy per mole 341
Total reflection 254
trans-$(CH)_x$ 158, 160
Transfer characteristics, MOSFET 132
Transformers 312, 313, 318
Transistors 3, 127–136
Transition temperature 90, 92
Translation vectors 46–50
Transmissivity 182–184
Transphasors 263–265
Trans-polyacetylene 158
Transverse effective mass 110
Transverse photodiode 260
Transverse vibrational modes 360
Traveling furnace 142
Traveling-wave laser 252
Trichlorosilane gas 140
Tunnel diode 25, 126–127
Tunnel effect 23–26
Tunneling 25, 95
  ohmic contacts 119
  quantum semiconductor devices 137–138
  Zener diode 122

Ultrahigh recording densities 322
Ultra large-scale integration (ULSI) 139, 150
Ultrasonics 344
Umklapp process 356
Uncertainty principle 4
Undamped vibration 17, 363–364
Undercutting 144
Unipolar transistor 130
Unit cells 44, 47
Universal gas constant 337

Vacancy 77, 85, 93
Vacancy concentration 163
Vacuum tube 139
Valence band 39, 84, 98, 99
Van der Waals forces 157
Vapor deposition 167
Vector product 46, 49
Velocity 7
  of electrons 76
  space 80
Vibration 12, 363
  damped 28–29
  definition 7
  string, vibrating 19, 21
  undamped 17

Vibrational modes 342, 345
Vibrational problems 15
Vibrations of the lattice atoms 237
Vicalloy 2 319
Virgin iron 281, 282, 316
Volatile memory device 150
Voltage–current characteristics 117, 121, 126, 127, 129
Vortex state 92
Vortices 92

Wafer fabrication 140, 142, 144
Wave(s) 6
  damped, differential equation 365
  definition 7, 365
  matter 9, 10
  monochromatic 9–10
  plane, differential equation 365
  standing 22
  undamped, differential equation 365
  velocity of 4
Wave functions 10–11, 19–21
  space dependence 14
Waveguides 254–260
Wavelength 6, 7
  electron 7
  modulation 219
  semiconductor lasers 247–248
Wave number 8, 10, 18, 68
  vector 205, 366
Wave packet 9, 10, 68
Wave vector 18, 39, 44
Wave velocity 9
Wave-particle duality 6–11, 342
Weiss, Pierre 298
Wet chemical etching 143–144
Wiedemann–Franz law 354
Wigner–Seitz cells 44, 45, 50
  for the body-centered cubic (bcc) structure 45, 50
  for the face-centered cubic (fcc) structure 45
Window glass 238
Wire tapes 254
Work 332–333
  function 114
Write once, read many (WORM) 262

Xerography 171–173
X-ray crystallography 46
X-ray diffraction 5
  reciprocal lattice map 40
X-rays 199–200

Yttrium–aluminum–garnet laser 244

Zener breakdown 122
Zener diodes 121–123
Zero point energy 20, 343
Zinc blende structure 112
Zinc, content effect on alloy threshold energy 227
Zinc oxide 113, 166
Zinc selenide 113, 254
Zinc sulfide 113
Zirconia 165
Zone refining 141
Zone schemes 35–40